UNDERGRADUATE TEXTS IN CONTEMPORARY PHYSICS

Springer
New York
Berlin
Heidelberg
Barcelona
Hong Kong
London
Milan
Paris
Singapore
Tokyo

UNDERGRADUATE TEXTS IN CONTEMPORARY PHYSICS

Holbrow, Lloyd, and Amato, Modern Introductory Physics

MODERN INTRODUCTORY PHYSICS

C.H. Holbrow
J.N. Lloyd
J.C. Amato

With 172 Illustrations

C.H. Holbrow
J.N. Lloyd
J.C. Amato
Department of Physics and Astronomy
Colgate University, Hamilton, NY 13346
USA

COVER PHOTOGRAPH: STM scan of molecules: on the upper left, two O_2 molecules; on the lower right a voltage pulse has separated one of the molecules into two atoms. Courtesy of Wilson Ho, Department of Physics, Cornell University.

Drawings by Kevin Hodgson

Library of Congress Cataloging-in-Publication Data
Holbrow, Charles H.
Modern introductory physics / C.H. Holbrow, J.N. Lloyd, and J.C. Amato
p. cm. – (Undergraduate texts in contemporary physics)
Includes bibliographical references and index.
ISBN 0-387-98576-X (alk. paper)
1. Physics. I. Lloyd, J. N. II Amato, J. C. III. Title.
IV. Series
QC21.2.H65 1998 98-24447
530–dc21

Printed on acid-free paper.

Production managed by Lesley Poliner; manufacturing supervised by Thomas King.
Typeset in Janson Text by the Bartlett Press, Inc., Marietta, GA. from files supplied by the author.
Printed and bound by Hamilton Printing Co., Rensselaer, NY.
Printed in the United States of America

9 8 7 6 5 4 3 2 1

ISBN 0-387-98576-X Springer-Verlag New York Berlin Heidelberg SPIN 10659241

... all things are made of atoms—little particles that move around in perpetual motion, attracting each other when they are a little distance apart, but repelling upon being squeezed into one another.

In that one sentence, you will see, there is an *enormous* amount of information about the world, if just a little imagination and thinking are applied.

— Richard P. Feynman

Preface

This book grew out of an ongoing effort to modernize Colgate University's three-term, introductory, calculus-level physics course. The book is for the first term of this course and is intended to help first-year college students make a good transition from high-school physics to university physics.

The book concentrates on the physics that explains why we believe in the existence of atoms and their properties. This story line, which motivates much of our professional research, has helped us limit the material presented to a more humane and more realistic amount than is presented in many beginning university physics courses. The theme of atoms also supports the presentation of more non-Newtonian topics and ideas than is customary in the first term of calculus-level physics. We think it is important and desirable to introduce students sooner than usual to some of the major ideas that shape contemporary physicists' views of the nature and behavior of matter. On the verge of the twenty-first century such a goal seems particularly appropriate.

The quantum nature of atoms and light and the mysteries associated with quantum behavior clearly interest our students. By adding and emphasizing more modern content, we seek not only to present some of the physics that engages contemporary physicists but also to attract students to take more physics. Only a few of our beginning physics students come to us sharply focused on physics or astronomy. Nearly all of them, however, have taken physics in high school and found it interesting. Because we love physics and believe that its study will open students' minds to an extraordinary view of the world and the universe and also prepare them well for an enormous range of roles—citizen, manager, Wall-Street broker, lawyer, physician, engineer, professional scientist, teachers of all kinds—we want them all to choose undergraduate physics as a major. We think the theme and content of this book help us to missionize more effectively by stimulating student interest.

In principle this approach also makes our weekly physics colloquia somewhat accessible to students before the end of their first year.[1]

In parallel with presenting more twentieth-century physics earlier than is usual in beginning physics, this book also emphasizes the exercise and development of skills of quantitative reasoning and analysis. Many of our students come fairly well prepared in both physics and math—an appreciable number have had some calculus—but they are often rusty in basic quantitative skills. Many quite capable students lack facility in working with powers-of-ten notation, performing simple algebraic manipulation, making and understanding scaling arguments, and applying the rudiments of trigonometry. The frustrations that result when such students are exposed to what we would like to think is "normal discourse" in a physics lecture or recitation clearly drive many of them out of physics. Therefore, in this first term of calculus-level physics we use very little calculus but strongly emphasize problems, order-of-magnitude calculations, and descriptions of physics that exercise students in basic quantitative skills.

To reduce the amount of confusing detail in the book, we often omit interesting (to the authors) facts that are not immediately pertinent to the topic under consideration. We also limit the precision with which we treat topics. If we think that a less precise presentation will give the student a better intuitive grasp of the physics, we use that approach. For example, for physical quantities mass, length, time, and charge we stress definitions more directly connected to perceivable experience, and pay little attention to the detailed, technically correct SI definitions. This same emphasis on physical understanding guides us in our use of the history of physics. Many physical concepts and their interrelations require a historical framework if they are to be understood well. Often history illustrates how physics works by showing *how* we come to new knowledge. But if we think that the historical framework will hinder understanding, we take other approaches. This means that although we have tried diligently to avoid saying things that are flat out historically wrong, we do subordinate history to our pedagogical goals.

We believe that it is important for students to see how the ideas of physics are inferred from data and how data are acquired. Clarity and concision

---

[1]These and other aspects of the approach of this book are discussed in more detail in C.H. Holbrow, J.C. Amato, E.J. Galvez, and J.N. Lloyd, "Modernizing introductory physics," Am. J. Phys. **63**, 1078–1090 (1995); J.C. Amato, E.J. Galvez, H. Helm, C.H. Holbrow, D.F. Holcomb, J.N. Lloyd and V.N. Mansfield, "Modern introductory physics at Colgate," pp. 153–157, *Conference on the Introductory Physics Course on the Occasion of the Retirement of Robert Resnick*, edited by Jack Wilson, John Wiley & Sons, Inc., New York, 1997; C.H. Holbrow and J.C. Amato, "Inward bound/outward bound: modern introductory physics at Colgate," in *The Changing Role of Physics Departments in Modern Universities*, pp. 615–622, Proceedings of International Conference on Undergraduate Physics Education, College Park, Maryland, August 1996, edited by E.F. Redish and J.S. Rigden, AIP Conference Proceedings **399**, Woodbury, New York, 1997.

put limits on how much of this messy process beginning students should be exposed to, but we have attempted to introduce them to the realities of experimentation by including diagrams of apparatus and tables of data from actual experiments. Inference from tables and graphs of data is as important a quantitative skill as the others mentioned above.

Asking students to interpret data as physicists have (or might have) published them fits well with having beginning physics students use electronic spreadsheets to analyze data and make graphical displays. Because electronic spreadsheets are relatively easy to learn and are widely used outside of physics, knowledge of them is likely to be useful to our students whether they go on in physics or not. Therefore, we are willing to have our students take a little time from learning physics in order to learn to use a spreadsheet package. Some spreadsheet exercises are included as problems in this book.

The examination of significant experiments and their data is all very well, but nothing substitutes for actual experiences of observation and measurement. The ten or so laboratory experiments that we have developed to go along with this course are very important to its aims. This is particularly so, since we observe that increasingly our students come to us with little experience with actual physical phenomena and objects. We think it is critically important for students themselves to produce beams of electrons and bend them in magnetic fields, to create and measure interference patterns, to observe and measure electrolysis, etc. Therefore, although we believe our book will be useful without an accompanying laboratory, it is our heartfelt recommendation that there be one.

Although our book has been developed for the first of three terms of introductory physics taken by reasonably well prepared and well motivated students, it can be useful in other circumstances. The book is particularly suitable for students whose high-school physics has left them with a desire to know more physics, but not much more. For them a course based on this book can be the one physics course to have when you're having only one. The book can also work with less well prepared students if the material is spread out over two terms. Then the teacher can supplement the coverage of the material of the first several chapters and build a solid foundation for the last half of the book.

The format and techniques in which physics is presented strongly affect student learning. In teaching from this book we have used many innovative pedagogical ideas and techniques of the sort so vigorously presented over the past decade by well-known physics pedagogues such as Arnold Arons, Lillian McDermott, Phyllis Laws, Eric Mazur, David Hestenes, and Alan van Heuvelen. In one form or another they emphasize actively engaging the students and shaping instruction in such a way as to force students to confront and recognize and correct their misconceptions. To apply these ideas we teach the course as two lectures and two small-group recitations each week. In the lectures we use Mazur-style questions; in the recitations we have students work in-class exercises together; we spend considerable effort to make exams and special exercises reach deeper than simple numerical substitution.

Deciding what specific subject matter should go into beginning physics has been a relatively small part of the past decade's vigorous discussions of pedagogical innovation in introductory physics. We hope this book will help to move this important concern further up the agenda of physics teachers. We think the content and subject emphases of introductory physics are a central responsibility of physics teachers and of great importance to the long-term health of the physics community. This book represents our idea of a significant step toward making introductory physics better represent what physics is. Whether or not we have succeeded, we hope this book will stimulate discussion about, encourage experimentation with, and draw more attention to the content of undergraduate introductory physics.

Charles H. Holbrow
James N. Lloyd
Joseph C. Amato
Colgate University
August, 1998

Contents

CHAPTER 1

What's Going On Here?

1.1 WHAT IS PHYSICS?

From earliest times humans have speculated about the nature of matter. The Greeks with their characteristic genius developed a highly systematic set of ideas about matter. They called these ideas "physics," but physics in the modern sense of the word comes into being only in the seventeenth century.

In 1638 Galileo Galilei published *Discourses on Two New Sciences*[1] which summarized a lifetime's work that created the description of motion that we use today. A generation later Isaac Newton made a grand synthesis with his laws of motion and his famous law of universal gravitation.[2]

These two great physicists introduced two exceptionally important ideas that characterize physics still. First, physics is a *mathematical* description of natural phenomena, a description of underlying simple relationships from which the complicated and various behavior of observed matter can be inferred. Second, the predictions or inferences must be checked by measurements and observations. Physicists create quantitative descriptions of the behavior of matter and then examine the consistency and accuracy of these descriptions by philosophical, mathematical, and experimental study.

---

[1]This is a wonderful book available from Dover Publications in a paperback edition. It describes basic features of the science of the strength of materials; and it presents the first mathematical account of that part of physics that we call "mechanics." The mathematics is plane geometry and accessible to anyone with a good high-school education. Here Galileo presents his arguments, both theoretical and experimental, for the law of falling bodies and the resulting possible motions.

[2]These ideas were published in *Principia Mathematica*. This book was written in Latin but is available in English translation. (R.T. Jones, the eminent NASA engineer who played an important role in developing the delta-wing aircraft once said that he learned physics by reading the *Principia*. That is a very strong endorsement.)

Thus, when you say that bodies fall, you are not really doing physics. But when you say all bodies fall with constant acceleration you are propounding a generalization in mathematical form, and you have begun to do some physics. When from that statement you deduce logically that trajectories are parabolas and that the maximum range occurs when a body is launched at 45 degrees, you are doing physics. When you devise arguments and instruments to measure and show that near the surface of the Earth all bodies fall with a constant acceleration $g = 9.8\,\mathrm{m\,s^{-2}}$ and that actual bodies do move almost as you predict, then you are doing more physics. And when you are able to explain quantitatively that observed deviations from your predictions are due to variations in the distance from the surface of the earth and the effects of air resistance, you are doing deep physics. And when you create new concepts in order to construct a quantitative explanation of why falling bodies have constant acceleration in the first place, you are doing physics at a deeper level yet.

Physicists are students of the behavior and structure of matter. This phrase covers a multitude of activities. The 1998 Physics and Astronomy Classification Scheme[3] or PACS-98, as it is also called, lists over 2000 short phrases describing the different things physicists are busy at—from "communication, education, history, and philosophy" through "exotic atoms and molecules (containing mesons, muons and other unusual particles)" to "stellar systems; galactic and extragalactic objects and systems; the Universe." The variety is astonishing.

EXERCISES

1. Look up PACS-98. Look at category 07—Instruments and apparatus—and then at the subcategory .30. List four subject headings pertaining to techniques of producing and measuring vacuum.

2. Look in PACS-98 61.72 and list four subjects having to do with defects in crystals.

3. Look in PACS-98 12.39 and list two kinds of models of strong interactions.

4. Look in PACS-98 98.62 and list four aspects of galaxies that are studied.

---

[3] Use a Web browser and go to http://publish.aps.org/PACS/pacs98.html. A look at the myriad of categories and subcategories and sub-subcategories reveals a wonderland of strange words and jargon. If you like language, you might like to peruse the PACS.

1.2 WHAT IS INTRODUCTORY PHYSICS ABOUT?

You can see that physics can include almost everything. What then is going to be in this book? Well, it is going to be different from the usual introductory physics book intended for people with a serious interest in science. First, let us tell you what the traditional beginning book is like.

Most introductions to physics begin with the mathematical description of motion. They talk about forces, momentum, energy, rotational motion, oscillations. They discuss heat and temperature and the laws of thermodynamics, and they treat electricity and magnetism plus some optics. There is a notoriously numbing quality about this approach. That may be unavoidable, since a goal of the course is to change the structure of your brain, which is full of deeply ingrained misconceptions. The misconceptions have to be straightened out. Also, you need to overcome your resistance to the sharpness and lack of ambiguity that are part of quantitative thought, and you need to be strengthened against blanking out during the long chains of inference by which physicists connect the basic ideas of physics to the observable world. Restructuring your thinking is uncomfortable, and many people are not able to accept the very real "present pain" for the prospective "future pleasure" of greatly enhanced powers of understanding the natural world. Our official recommendation to you is: "Be strong, be brave, be persistent. Hang in there."[4]

1.3 WHAT WE'RE UP TO

This book is based on some different ideas about how to start physics. They are the basis of a significant change in the teaching of the introductory course. Rather than start with seventeenth-century physics and work our way through to the nineteenth century, we are going to emphasize some ideas that have dominated physics in the twentieth century. There are several reasons to do this. Most physicists do twentieth-century physics, and that is quite different from physics of past centuries. It seems to us that introductory physics should introduce you to what we physicists actually do.

Isn't this dangerous? The ideas of physics are cumulative. To talk meaningfully about what is going on deep in an atomic nucleus, you must understand velocity as Galileo used the idea; you need to know about potential energy—an idea developed in the eighteenth century; you need to know about electric

[4]Perhaps Winston Churchill's words say it more firmly: "Never give in. Never give in! Never, never, never, never. Never give in except to convictions of honor and good sense."

FIGURE 1.1 You don't need to know how to build a skyscraper to appreciate the view from it. *Photo courtesy of Mary Holbrow.*

charge, about momentum, about kinetic energy. The usual theory of teaching physics is to introduce these ideas in terms of simple, more directly observable phenomena, and then apply the ideas in increasingly complicated ways. Build the foundation first, then put up the building. By starting with the physics of this century isn't there a danger that we will erect a superstructure with no foundation?

We don't think so. For one thing, you all have a bit of foundation. You know what velocity is, you have heard about acceleration. You have talked about energy and momentum in your high-school physics course. For another, we are not going to be dogmatic about sticking to the twentieth century. If we need to spend some time reviewing or introducing some basic ideas, we will. Furthermore, we are not going to do the hardest parts of modern physics. Our quantum mechanics will not go beyond simple applications of the uncertainty principle. Einstein's special theory of relativity will be treated in a very "nuts and bolts" fashion. You will have to wait for more advanced courses to see the very powerful and elegant mathematical treatments of these two cornerstones of modern physics.

But there is a more important reason why our approach should work. An enormous amount of twentieth-century physics is done with simple ideas and mathematics no more complicated than algebra. Do not think that because ideas are simple, they are trivial. Simple ideas are often used with elegant subtlety to do physics. You can learn enough about waves, particles, energy, momentum, uncertainty, scattering, and mass to make a remarkably comprehensive and consistent picture of the nature of matter without having to know all the underlying connections among the ideas. The more complete elucidation of the connections can wait until later courses.

After all, you would not familiarize yourself with the World Trade Center towers by first studying all their plumbing diagrams and then their wiring diagrams, and then their ductwork, and the arrangement of their girders, and so on until you are familiar with all the parts, and only then assemble them in your mind to create the skyscraper. You must do that if you are building a new building, but if the building is already there, you need to know first where the main doors are, where the express elevators are, and on what floors are the important offices and how some of the suites are connected. You can visit what seems to be of particular interest without knowing the details of the building's construction. Of course, to operate and really appreciate the building you will eventually need to know and understand the details. But not right away.

Physics is a skyscraper of imposing dimensions. This course will show you some of its rooms and some of the furniture in those rooms. You should learn enough so that you can rearrange the furniture in interesting ways as well as get from one room to another. Later courses will go back to the seventeenth century and look at the foundations of physics; then you will go down into the utility rooms of our edifice and see what's there. In this book we will stay upstairs, where the view is better (Fig. 1.1). Once you know how to get to the windows in the skyscraper of physics, you can look out over the entire panorama of nature laid out in the PACS, from subnucleonic quarks and leptons to the ends of the visible universe.

1.4 THIS COURSE TELLS A STORY

The Short Story Of The Atom

Physics helps us to understand the physical world. It extends our perceptions beyond our immediate senses and opens new vistas of comprehension. We think you can understand physics better if the physics you learn tells a story. There are many stories to tell with physics, so we had to choose one. We chose what we think to be the most significant story of the past two centuries. It is the story of the atom and its nucleus. We want you to know both **what** physics teaches us about atoms and their remarkable properties and **why** we believe atoms and nuclei exist and have the properties we think they have.

The story is a good one. It starts in the early nineteenth century with hard, featureless atoms. They become more complicated as more is learned about them. They explain many observations by chemists and many of the observed properties of gases. By the middle of the nineteenth century, the kinetic theory of hard-sphere atoms makes it possible to know that the diameters of atoms are of the order of 10^{-9} m, some nine or ten orders of magnitude smaller than familiar everyday objects. Then their electrical nature is discovered, and by the end of the century atoms are known to be made of positive and negative charges. The negative charges are found to be tiny elementary particles that are named "electrons." Their mass and charge are determined.

At the threshold of the twentieth century, radioactivity reveals new complexity of the atom. The compact core, or "nucleus," of the atom is discovered. It is 10^{-5} the size of the atom and contains 99.97% of its mass. It signals the existence of new elementary particles, the proton and the neutron, and the existence of a previously unknown fundamental force of nature more than 100 times stronger than the familiar electrical forces, and 10^{40} times stronger than gravity. This new force is the agent by which nuclei store the extraordinarily large amounts of energy that can be released by nuclear fission and fusion. The search for political and social controls of these energies remains a major preoccupation in our world as it moves into the twenty-first century.

Special Relativity and Quantum Mechanics

At the heart of our story lie two strange new ideas that revolutionized our view of the physical world. The first idea is that there is a limiting velocity in the universe; nothing can travel in a vacuum faster than the speed of light. This idea and the idea that the laws of physics must not depend on the frame of reference in which they are studied are central to Einstein's special theory of relativity. The consequences of this theory are necessary to understand the behavior of atoms or their components at high energies. This behavior is surprising and unfamiliar to beings whose experience with the physical world is limited to velocities much less than that of light.

EXERCISES

5. What beings might these be?

Stranger still are the ideas of quantum mechanics. The behavior of atoms and their components can only be understood if, unlike the familiar particles of our world—marbles, raindrops, BBs, baseballs, planets, sand grains, bacteria, etc.—they do not have well-defined locations in space, but are spread out in some fashion like water waves or sound waves. In fact, in some sense they must be in more than one place at the same time. To describe atoms and the details of their behavior we must use these peculiar ideas mixed together with a fundamental randomness that physicists schooled in the ideas of Newton have found difficult to accept.

Physics Is Not a Spectator Sport

In this course the "why" of your understanding is extremely important. After all, you already believe in atoms. You don't need convincing. You accept their existence as matters of faith, and you will probably believe most things we tell you about them. Of course, you will need to know many things of and about physics, but it is also important that you learn to make arguments like physicists. We want you to learn what convinces physicists and what does not. In the end, we want you not so much to know the story as to be able to convince yourself and others that it is true. We want you to learn to follow and use quantitative arguments and to be able to describe the posing of questions of physics as experiments.

Of course, physicists, like everyone else, teach and learn as much by authority as by proof. Because there is not time or will, we will often just tell you that something is so in order to pass on to larger issues. Nevertheless, this introductory physics course lays more stress on argument than the traditional course. There are reasons you may not like this. It requires thinking, and thinking is uncomfortable, muddy, difficult, ambiguous, and inefficient. It requires you to participate actively rather than passively. It means that your textbook—this very book—must be different from the traditional text.

Most introductory physics courses greatly emphasize the working of problems. Homework and quizzes and exams are all the working of problems illustrating the topics of the book. Most students respond to this by reading the assigned problems and leafing backwards through the chapter until they find the equations that produce an answer. The text becomes a reference manual for solving problems, and it is not read for any broader comprehension.

We have tried to create a text that has to be read for broader comprehension, a book that does not serve merely as a user's manual for solving assigned problems. We want you to read the book and think about it as you go along. This does not mean that problems are unimportant; they are very important. They are how you test your understanding. Trying to work a problem is the quickest way to show the emptiness of the understanding you thought you gained when you passed your eyes over pages of print without repeated pauses to think. As you read physics, you should be asking yourself questions. To show you how this works we have put questions in among the paragraphs of the text where you should be asking them. In general, you should work exercises as you go along; if they aren't provided, you should make them up yourself.

For starters and to establish our basically kindly nature as authors, we have provided some questions for you. For instance, when you were reading above that a new extremely strong force was discovered, did you ask

EXERCISES

6. How much stronger is the electromagnetic force than the gravitational force?

Or you might have wondered:

7. What does it mean for one force to be stronger than another?

And because we think it might help you with your thinking about that question, we might ask

8. What do we mean when we say, "Lead is heavier than air"? Which weighs more, a pound of air or a pound of lead?

And, of course, you might wonder:

9. If the new force is so strong as we say, why wasn't it discovered much earlier in time?

If you did not already know, you can see that reading physics is slow work. Ten pages an hour is quite fast; five pages an hour is not unreasonable. And for the new and very strange, a page a day or a week is not inconceivable. Reading and working problems go hand in hand.

1.5 WHY THIS STORY?

An Important Idea

We have chosen to make atoms the central theme of introductory physics for three main reasons. First, the idea of the atom is extremely important. Our ideas about atoms color our understanding of all of nature and of all other sciences. One of the greatest physicists of the twentieth century, Richard Feynman, has written[5]

> If, in some cataclysm, all of scientific knowledge were to be destroyed, and only one sentence passed on to the next generations of creatures, what statement would contain the most information in the fewest words? I believe it is the *atomic hypothesis* (or the atomic *fact*, or whatever you wish to call it) that *all things are made of atoms—little particles that move around in perpetual motion, attracting each other when they are a little distance apart, but repelling upon being squeezed into one another.* In that one sentence, you will see, there is an *enormous* amount of information about the world, if just a little imagination and thinking are applied.

Tools for Quantitative Thought

Second, the arguments and evidence we use to infer the existence and properties of atoms are in many ways easier to understand than the arguments of traditional Newtonian physics. Some of the ideas are stranger than Newton's because they are unfamiliar, but, up to a point, the mathematics underlying them is simpler. We can learn a great deal by rough, order-of-magnitude, numerical calculations, and by using proportionality, plane geometry, some trigonometry, and how the sizes of simple functions scale as their variables are changed. These tools of rational argument are basic in all the branches of physics, in all sciences, and in any kind of practical work you may do—from making dinner to running a large corporation.[6] A major aim of this course is to have you become skillful with these simple mathematical tools.

An Introduction to Physics

Third, an introductory physics course built around the theme of atoms will give you a better sense of what physics is and what physicists do than a traditional course would. Most physicists today study atoms or their components and how they interact and behave under different conditions. Many of the deep unanswered questions of physics center on aspects of the behavior of atoms or their parts.

---

[5] Richard P. Feynman, Robert B. Leighton, and Matthew Sands, *The Feynman Lectures on Physics*, pp. 1–2, vol. I (Addison-Wesley, Reading MA, 1963).

[6] Some people argue that these tools are so basic to constructive thought and practical action and physics is such a good place to learn them that every college student should take physics. The same sort of argument could be made for lifting weights.

1.6 JUST DO IT!

This book will teach you the basic physics you need to know in order to understand why we believe in atoms and their properties. This will require learning much traditional physics, but it will be applied in a different context than is usual in beginning physics. We think that the physics you learn this way will make more sense to you, that the larger context will help you perceive that physics is not a disconnected set of formulas used to solve disconnected sets of problems. We also want you to learn what physicists think they know, what they think they don't know, and how they go about learning new physics. The "how" is very important, because as you learn "how" physicists do physics and persuade each other of the truth of what they do, you will be learning how to teach yourself physics. Learning to teach yourself is a goal for the long term. For most people it takes years, but then real mastery becomes possible.

C H A P T E R 2

Some Physics You Need to Know

2.1 A FEW USEFUL IDEAS

Mass, length, and time are concepts basic to all physics. In physics we measure them respectively in units of kilograms, meters, and seconds. Although these units have been very precisely defined by an international committee, there are approximate definitions of these units that are more useful for you to know: A mass of one kilogram is the mass of a liter of water, a little more than a quart; one meter is about the length between your nose and the end of your fingers when your arm is stretched out to the side; one second is about to the time between your heart beats when you are at rest. In what follows we will look at useful ways to understand the magnitudes of units and see how to create new ones to describe and measure velocity, acceleration, momentum, force, and energy. Most of the units used in this book are part of the internationally agreed-upon Système International (SI). There is a good summary of SI units at the end of the book.

All SI units can be scaled by powers of ten by means of standard prefixes such as "micro," "kilo," and "mega" and corresponding standard abbreviations like μ, k, and M. Many of these are introduced during the course of this chapter. Watch for them and make a special effort to learn them.

2.2 MASS

You already have some sense of mass. When you push an object it resists. It takes great effort to get an automobile rolling on a level surface, and most of the car's resistance is due to its mass. When you pull a quart of milk from the refrigerator, you sense its mass of almost one kilogram (10^3 grams.) Using

TABLE 2.1 Masses of Some Familiar Objects

Object	Mass (kg)	Object	Mass (kg)
Golf ball	0.050	Basketball	0.600
Tennis ball	0.057	1 liter of water	1.000
Baseball	0.149	Bowling ball	7.0
Hockey puck	0.160	Jeep 4x4	1500

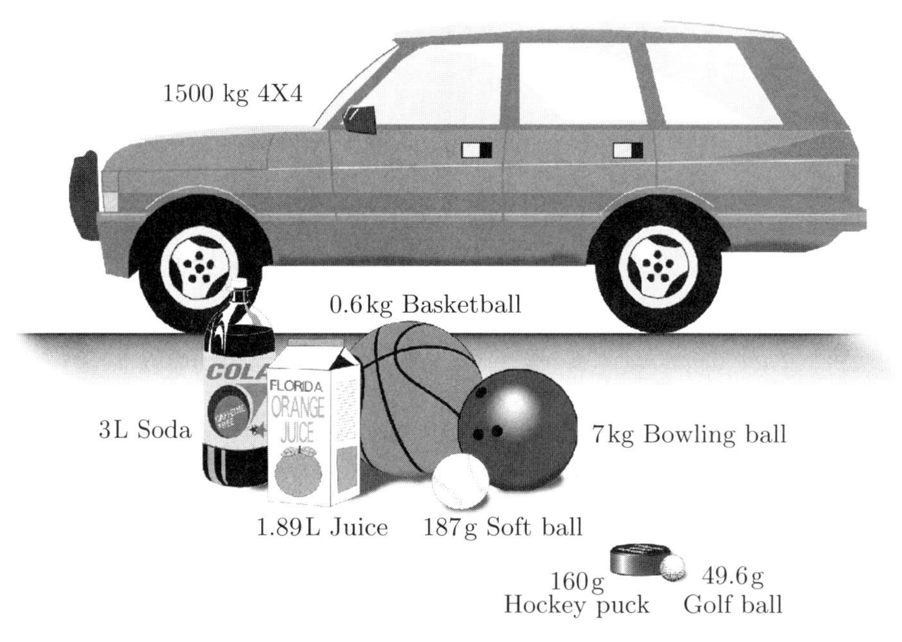

FIGURE 2.1 Some objects that have masses you may have experienced.

the standard abbreviation kg for kilogram, we write or say that a quart of milk has a mass of 0.946 kg. In handling the quart of milk you also resist the weight due to the pull of gravity, but we won't be overly concerned about the difference between weight and mass; we will follow the custom of most of the people of the world who measure the weight of flour, potatoes, and the like in "kilos." Masses of some objects with which you may be familiar are listed in Table 2.1, and some of these objects are shown in Fig. 2.1.

The unit of mass, the kilogram, was originally chosen to be the mass of 1000 cm^3, or one liter, of water at a particular temperature. Since cola and other important beverages are now sold by the liter or the milliliter (1 milliliter = 10^{-3} liter) you can quickly estimate the mass of 2 L of cola exclusive of the mass of the bottle. "L" is the abbreviation for liter, and "mL" is the abbreviation for milliliter.

Because a liter is defined to be 1000 cubic centimeters, one cm^3 of water has a volume of 1 mL, and 1 cm^3 of water has a mass of 1 gram. The metric system was constructed to have all these interrelationships. They are very convenient to know and rich in information. For one thing, you see that every unit volume of water has the same mass. Another way of saying this is that the mass of a unit volume is a constant. Hence the mass of any quantity of water can be determined immediately by finding how many units of volume it contains and multiplying this number by the mass of each unit. We call the mass of a unit volume the "density," which for water is 1 g/cm^3 or, in standard algebraic notation, 1 $g\,cm^{-3}$.

Density

Density is useful for comparing different materials. In everyday language, we refer to things being "heavier" or "lighter" than water. This cannot mean, for example, that any iron object is heavier than any other body of water. You certainly can have a small piece of iron that is lighter than a bucket of water. Yet we are quite clear when we see a stone sink in a lake or a cork floating on water that the first is heavier than water and the second is lighter. The concept of density comes to the rescue here. The important thing is whether the object has more or less mass in a unit volume, i.e., has more or less density than water.

EXERCISES

1. With even such simple ideas as these we can do a lot of science. For example, what is your density? You know that since you can float when you are swimming but only just barely, your density must be a trifle less than that of water. What does that suggest is your principal chemical ingredient? What is your volume in liters and in cubic centimeters? What did you have to assume to answer the third question in this exercise?

EXAMPLES

1. The third question in the above exercise is not really as obvious as it might seem. As a well-trained student, you may well have been led to the answer more by what the instructor seems to want than by thoughtful analysis. But if you were aware that you had to assume that a human body has a uniform density, then you are beginning to do real science. Is that assumption correct? One might imagine continuing a scientific analysis by testing the consequences of the assumption. Get a steak from the store and determine its density, then dry it thoroughly and measure the mass of the dried remains. You would indeed conclude that the meat was mostly water. But now you come across some bones, which are clearly much more dense than water. How, then, can the average density of the body be so close to that of water? How is it that an incorrect assumption led to the right answer? You would have to conclude that there were compensating volumes of much less density, such as the lungs and head cavities. But now an interesting question comes up. Why is the average density so close to that of water? We are getting to some profound evolutionary questions at this point and straying from our main topic. But you perhaps can see how even simple questions can lead to much deeper ones.

2. Suppose you decide to become fabulously wealthy by running the "guess-your-mass concession" at your favorite carnival. If someone 1.8 m tall with a waist size of 0.8 m approaches, what would you estimate his mass to be? Let's model him as a cylinder. His waist size is his circumference $2\pi r$, where r is his radius. The volume V of a cylinder with a height $h = 1.8$ m and a radius $r = 0.8/(2\pi)$ is $V = \pi r^2 h = 0.092\ \text{m}^3$. Notice that we were careful to do this problem in consistent units, so the answer comes out in cubic meters. Since $1\ \text{m} = 10^2\ \text{cm}$, it follows that $1\ \text{m}^3 = 10^6\ \text{cm}^3$. Therefore, the volume V is $92.0 \times 10^3\ \text{cm}^3$, which is 92 L, or about 92 kg.

Now we have been doing some physics here. We have the concepts of volume and mass, and we are using them quantitatively. We have made a mathematical model of our subject. Next, we need to check our results experimentally. Selecting a 1.8 m tall, 0.8 m circumference subject at random, we find that his actual mass is 82 kg.

EXERCISES

2. Suppose our prediction is high by 10%. Is that bad? How might we do better? Why is it off by so much? Why by 10%? Answering these questions will take you into the next round of physics.

2.3 LENGTH

The idea of length is intuitively fairly clear; we need only to make the idea quantitative. About two hundred years ago, after the French Revolution, there was a serious attempt to create a set of units of measure of mass, length, and time that at least in principle could be replicated in some absolute way. The meter was then defined to be one ten-millionth (10^{-7}) of the

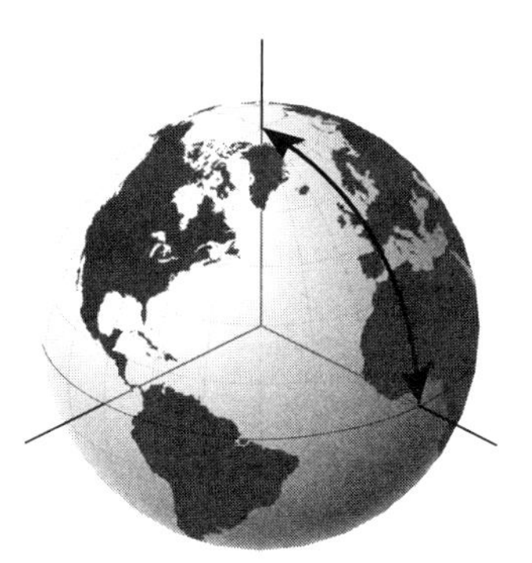

FIGURE 2.2 The meter was originally defined as one ten-millionth (10^{-7}) of a quadrant of Earth's circumference.

circumference of one quadrant of the Earth (see Fig. 2.2), and two marks this distance apart were put on a particular bar of metal that became the international standard. Today, the official definition of the meter used by all scientists depends on the speed of light and how we measure time. For now, we ignore the official definition and use the historical definition: To sufficient accuracy for our purposes the meter is 10^{-7} of one-fourth of Earth's circumference.

Notice that if you know the historical definition of the meter, you know the circumference of the Earth. It is very close to 40 million meters around. That is, 40×10^6 m. As you know, distances on Earth are commonly measured in multiples of thousands of meters, i.e., kilometers, where, as suggested above, the prefix "kilo" stands for 1000 (10^3) wherever it is used in scientific work.

EXAMPLES

3. Knowing Earth's circumference, you can determine its radius or diameter. The circumference of a circle of radius R is $2\pi R$. Therefore, $R_{\mathrm{Earth}} = 40 \times 10^6/(2\pi) = 6.37 \times 10^6$, or 6370 km. The lower case "k" is the standard symbol for "kilo." A kiloanything (ka?) is 1000 anythings.

EXERCISES

3. Show that Earth's volume is 1.08×10^{21} m^3, where m^3 is the usual way of denoting cubic meters.

Notice that the notation m^3 arises from the natural algebraic combination of quantities in the formula for the volume of a sphere: $\frac{4}{3}\pi R^3$. Any formula for any volume must contain three factors of length, so that measures of volume always have three factors of length in their definition. There are many varieties of units of volume: cubic feet, cubic yards, cubic centimeters, cubic kilometers, cubic furlongs. Any time you perform the mathematical operation of cubing a length you obtain a unit of volume. This algebraic property of units is fundamental; if you do not understand it already, here are some more examples of volumes that illustrate it:

Volume formulas always contain three factors of length

Shape	Dimensions	Volume
Cube	edge ℓ	ℓ^3
Box	length ℓ width w height h	$\ell w h$
Cylinder	height h radius r	$\pi r^2 h$
Cone	height h radius r	$\frac{1}{3}\pi r^2 h$
Sphere	radius r	$\frac{4}{3}\pi r^3$

Some Important Masses, Lengths, and Times

Table 2.2 lists some important masses, lengths, and times and gives the SI units in which they are measured. The table illustrates that these quantities are used over ranges from the human scale down to the very small and up to the very large. It makes the point that you should always try to think of concrete examples of physical objects or systems possessing the property you are thinking about.

2.4 THINKING ABOUT NUMBERS

Although it is important to have quantitative values for such things as the volume of the Earth, it is not enough just to have the number. You need to think about it. A number like $10^{21}\,\mathrm{m}^3$ does not spontaneously inform your imagination. Making very large and very small numbers meaningful is a recurring problem in physics. One way to understand them is by comparison.

EXAMPLES

4. Let us try some comparisons for a number that may have some more immediate interest than the volume of the Earth. In a given year the federal budget is on the order of one and a half trillion dollars ($\$1.5 \times 10^{12}$). What

TABLE 2.2 Basic Quantities of Physics

Name of Quantity	SI Unit of Measure	Abbrev-iation	Examples
mass	kilogram	kg	1 liter of water has a mass of 1 kg. The mass of a proton is 1.67×10^{-27} kg. Earth's mass is 5.98×10^{24} kg. Masses of typical American adults range from 50 kg to 90 kg.
length	meter	m	A long stride is about 1 m. A range of typical heights of American adults is from 1.6 m to 1.9 m. Earth is 40×10^{6} m in circumference. An atom's diameter is 0.2×10^{-9} m.
time	second	s	Your heart probably beats a little faster than once a second. Light travels 30 cm in 10^{-9} s. There are 3.15×10^{7} s in a year.

is a trillion dollars? Let us break that down into more manageable units, say 100 million dollars. It takes ten thousand sets of 100 million to make one trillion. One hundred million is still hard to imagine, but its scale is more tangible. For example, a medium-sized liberal arts college has a yearly budget on this order, or quite a nice hospital might be built for $100 million. Even so, fifteen thousand hospitals a year is hard to imagine. But when you discover that 1.5 trillion dollars could build more than 40 new, major, fully equipped hospitals every *day* for a year, you begin to get a sense of what $1.5 trillion means.

EXERCISES

4. In many years the national budget has run a deficit of about $\$2.5 \times 10^{11}$. How high would a stack of $100 bills be if it contained this much money? Notice here, as is often the case in such questions, that you need to make some reasonable estimate of some physical quantity important to your answer—in this case, the thickness of a $100 bill.

5. Write out all the figures in Example 4 numerically, write down the relevant relations, and confirm the statements made.

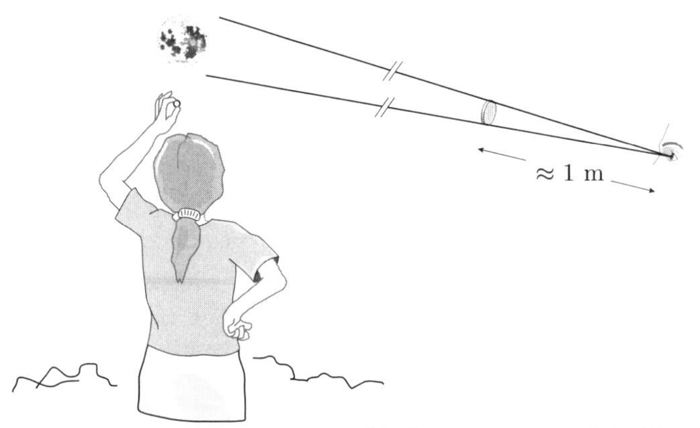

FIGURE 2.3 How a small disk can block out your view of the Moon.

6. What is the largest object for which you have some sense of its size? Estimate its volume. What is Earth's volume in units of your object's volume?

The result of your comparison is probably not very meaningful either. The volume of Earth becomes meaningful only when we compare it with other similar objects: the Moon, the planets, the Sun. The following paragraphs show how to make such comparisons.

EXERCISES

7. The Moon is 3.8×10^8 m from Earth. You can just block out the Moon with a disk 9.28 mm in diameter held 1 m away from your eye. What is the diameter of the Moon? (See Fig. 2.3.) What is its volume?

8. The statement for the previous problem means that the Moon subtends an angle of just about 0.5 degrees when viewed from Earth. What is the exact value of the angle subtended by the Moon? About how many Moons placed side by side would it take to go from horizon to horizon in an arc passing directly overhead?

9. If you like this sort of argument, apply it to the Sun. Curiously enough, the Sun subtends at Earth almost exactly the same angle as the Moon, although the Sun is more distant. It is about 1.5×10^{11} m from Earth. What is its diameter? What is its volume? How does that volume compare to Earth's?

Physicists are always trying to find ways to make numbers meaningful. Here is an example of one approach. A large object often has a large mass, so it is not surprising that a planet with a volume of 1.08×10^{21} m^3 has a mass of 5.94×10^{24} kg, or that a star (our Sun) with a volume of 1.41×10^{27} m^3 has a mass of 1.99×10^{30} kg. But what do these numbers mean? A common trick is to use the large numbers to describe some property that is not itself a large number. For instance, consider how much mass there is in a unit volume, i.e., look at the density. Earth's density is $\rho = 5.52 \times 10^3$ kg m^{-3}. Another trick is to rescale the units. A cubic meter is a pretty large volume; let's look at the density in grams per cubic centimeter, i.e., g cm^{-3}. Then the average density of the Earth is 5.5 g cm^{-3}.

Now that's a number a person can deal with. We already know that water has a density of 1 g cm^{-3}. So Earth is about 5.5 times denser than water. Does that make sense? We could check by measuring the density of some other things. Iron (Fe) has a density of 7.85 g cm^{-3}. Mercury (Hg) has a density of 13.6 g cm^{-3}. More interesting, the granitic rocks of which Earth's crust is made have a density of about 2.8 g cm^{-3}.

EXERCISES

10. But what is going on? We just found that the density of Earth is 5.5 g cm^{-3}. How can Earth be denser than its crust? Make up a reasonable explanation for the discrepancy.

11. Now calculate the density of the Sun and compare it to Earth's. Are you surprised? We hope so. But in any case, do you see how useful it is to play with the numbers from different points of view?

Such rescaling and such comparisons are essential because so little of the universe is set to our scale. The visible universe continues out beyond 10^{26} m; subatomic particles are smaller than 10^{-15} m. We deal casually with the unimaginably large and the inconceivably small. It is hard to know what is important about a mass of 3.35×10^{-27} kg. It becomes more meaningful when you know that it is twice the mass of an atomic nucleus of hydrogen or half that of a helium nucleus. Part of thinking about physics is the search for meaningful comparisons among the numbers we use to describe nature and the interactions of matter.

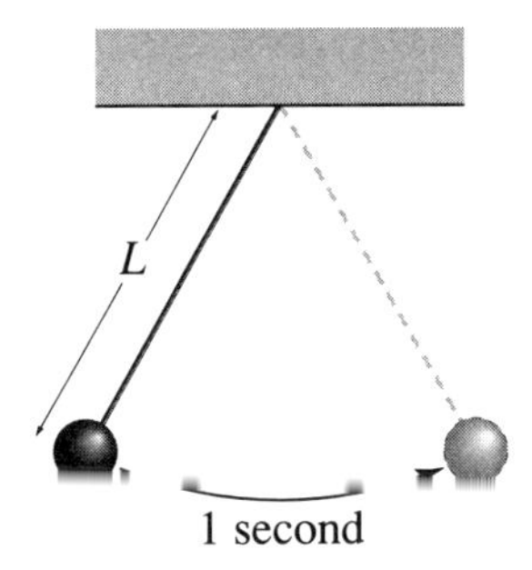

FIGURE 2.4 On the surface of Earth a pendulum of length $L = 0.993 \approx 1$ m will take 1 second to make a small angle of swing from one side to the other.

2.5 TIME

Time is defined for us by repetitive behavior: the swinging of a pendulum, the annual cycle of the seasons, the heartbeat's regular thud. Our unit of time is the second. It is roughly the duration between heartbeats of a person sitting at rest. It is very close to the time that it takes a mass at the end of a string 1 m long to swing from one side to the other near Earth's surface (see Fig. 2.4).

But these guides are not of much use in dealing with atoms and nuclei. We are going to be studying realms of nature where 10^{-9} seconds is considered a very long time. We will become acquainted with things that happen in 10^{-21} seconds. We believe that the Universe has existed for about 10^{18} seconds. Our only hope of developing a sense of what is to be expected at these different time scales is to become familiar with the phenomena associated with them.

EXAMPLES

5. For example, if you are working with the transmission of light in a laboratory between objects a few tens of centimeters apart, then an important time scale is 30 cm divided by the speed of light. Light travels 3×10^8 m s^{-1}, so the time to go 30 cm is 10^{-9} s, i.e., 1 nanosecond (where "nano" is the standard prefix for 10^{-9}—a billionth—of anything), usually written 1 ns. (It is useful to know that light travels just about 1 foot in 1 ns—one of the few times English units produce a convenient number.)

When you study atoms in a small volume so thoroughly evacuated that the atoms very rarely run into each other, collisions with the walls may be important. At room temperature, air molecules have an average speed of about $500\,\mathrm{m\,s^{-1}}$. Thus, in a cylinder 2 cm in diameter the time between collisions with the walls will be roughly $2\,\mathrm{cm}/50000\,\mathrm{cm\,s^{-1}} = 4 \times 10^{-5}$ s, or 40 microseconds as we commonly say. ($1\,\mu\mathrm{s} = 10^{-6}$ s where μ, the lowercase Greek letter mu, is the symbol used to denote "micro," or millionth.) Thus, interactions of atoms with the walls of this cell occur on a scale of millionths of a second. This kind of simple information is often useful. For example, if while studying these atoms you find something that happens in nanoseconds, you know that it has nothing to do with the walls. On the other hand, if the time scale of whatever you are observing is microseconds or longer, you may be seeing some effect of the walls.

2.6 ANGLES AND ANGULAR MEASURE

Although angles are not physical quantities like masses or lengths, a lot of our reasoning about physics involves them. There are several different measures of angles: degrees, fractions of a circle, clock time, and *radians*. Radians will be the measure we use most because they connect most simply and directly to trigonometry.

Let's review some of the vocabulary and ideas associated with angles and their measure.

Vertex and Rays

An angle is the figure formed by the spreading of two rays from a point. That point is called the "vertex" of the angle. In Fig. 2.5 the three points B, A, and C define an angle with A as its vertex and $\overline{AB}$ and $\overline{AC}$ as its rays. It is usual to denote the angle as $\angle BAC$, where the middle letter is the vertex.

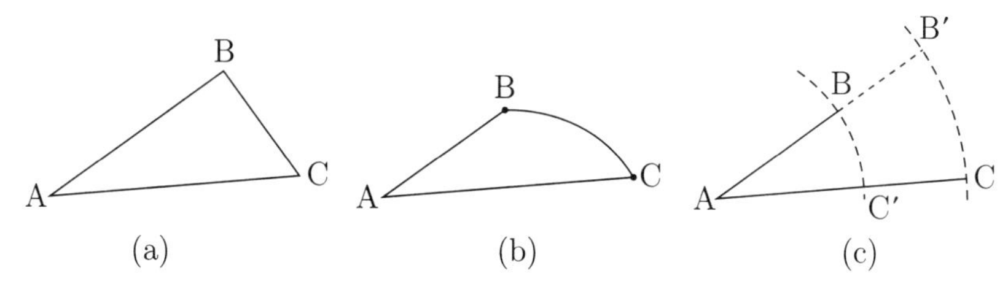

FIGURE 2.5 (a) An angle $\angle BAC$ with its vertex at A. The segment $\overline{BC}$ subtends the angle at A. (b) The arc $\overset{\frown}{BC}$ also subtends this angle at A. (c) Here the circular arcs $\overset{\frown}{B'C}$ and $\overset{\frown}{BC'}$ subtend the angle $\angle BAC$ at A.

What Does "Subtend" Mean?

Imagine that some distance out from the vertex something stretches across between the two rays to form a triangle. For example, imagine two lines diverging from your eye straight to the edges of a white area on the blackboard. The thing that stretches across the diverging lines is said to "subtend" some angle "at the point" from which the rays diverge—here, your eye. So we say that the white mark subtends an angle of some amount at your eye. The phrase tells you two things: what sits at the mouth of the angle and the location of the starting point of the rays that define the angle. In Fig. 2.5 the line $\overline{BC}$ and the arcs $\overset{\frown}{B'C}$ and $\overset{\frown}{BC'}$ each subtend the angle $\angle BAC$ at A.

EXAMPLES

6. The Moon subtends an angle of $\approx 0.5^\circ$ at Earth. The Sun also subtends an angle of $\approx 0.5^\circ$ at Earth (see Fig. 2.3).

EXERCISES

12. What angle does the hypotenuse of a right triangle subtend?

Degrees and, Especially, Radians

There are two principal measures of angles used in physics: degrees and radians. Each of these expresses the angle in terms of segments of a circle. In effect, an angle is measured by specifying what fraction of a circle's circumference is subtended at the vertex by a circular arc. To see how this works consider how you define a "degree."

Degree

Given an angle, construct any circular arc centered on the vertex. The degree is defined as the angle formed by two rays from the vertex that intercept an arc that is 1/360 of the circle's circumference. In other words, we imagine a circle divided into 360 equal arc lengths, and each of these arcs connected by lines to the center of the circle as shown in Fig. 2.6. Expressed in algebraic terms, the angle θ in degrees is

$$\theta = \frac{\overset{\frown}{s}}{2\pi R} 360 \text{ degrees.} \qquad (1)$$

It is common to use the symbol $^\circ$ for degrees. The symbol $\overset{\frown}{s}$ stands for the length of the circular arc subtending θ at the circle's center.

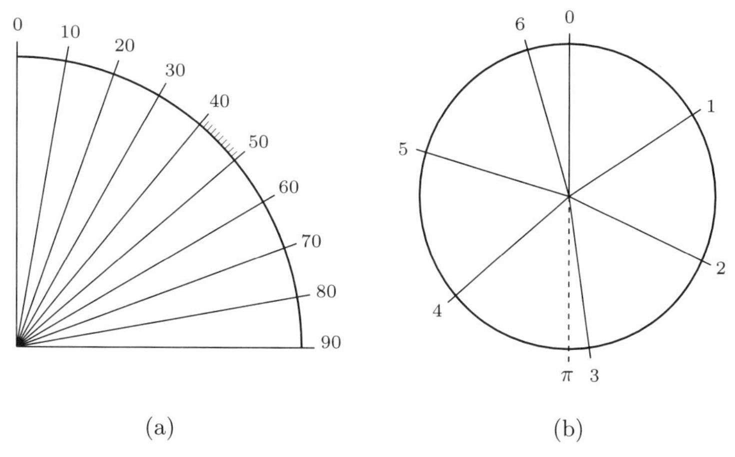

FIGURE 2.6 Angular measure: (a) degrees; (b) radians.

Each degree is in turn divided into 60 equal parts called "minutes," and each minute is divided into 60 equal parts called "seconds." When it is necessary to distinguish between seconds of time and seconds of angular measure, the latter are called "arc seconds." You learned a version of this so-called "sexagesimal" system when you learned to tell time. It is a measure of the strength of cultural inertia that the sexagesimal system survives and is widely used in terrestrial and astronomical measurements of angle, even though the degree and its curious subunits are awkward for many calculations.

EXAMPLES

7. Let's see what angle corresponds to a circular arc $\frown{s} = 2\pi R/8$, one-eighth of the circumference. From Eq. 1,

$$\theta = \frac{2\pi R}{2\pi R}\frac{360}{8} = 45°.$$

EXERCISES

13. Show how Eq. 1 will correctly yield 90° for the angle subtended by a quarter of a circle's circumference.

EXAMPLES

8. For example, given that the Moon is 60 Earth radii distant from Earth, what angle does Earth subtend at the Moon? You can work this problem

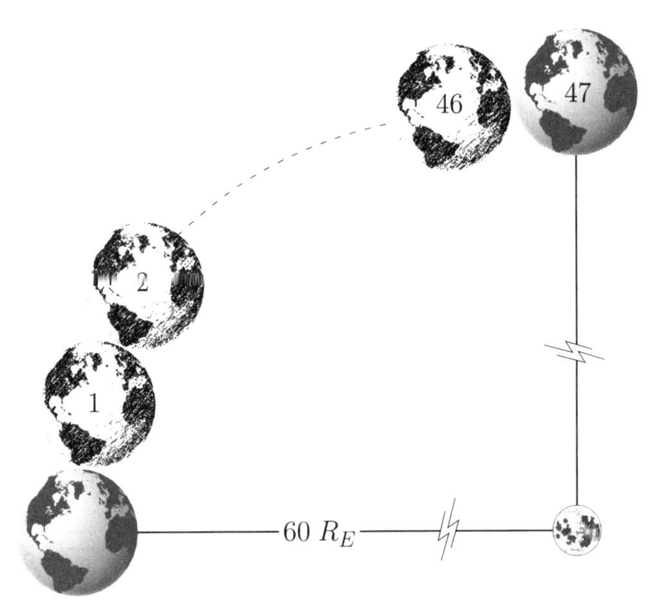

FIGURE 2.7 How many Earths does it take to encircle the Moon?

by imagining a line from Moon to Earth swinging in an arc from Moon horizon to Moon horizon. (Half of this arc is shown in Fig. 2.7). The length of the arc swept out by this line would be half the circumference of a circle of radius $60R_E$, which is $\pi 60R_E$. Standing on the Moon you would see Earth occupy a fraction of that arc equal to $2R_E/(\pi 60R_E) = 1/(30\pi) = 0.00106$. Because you know that from horizon to horizon the arc subtends 180°, you can see that at the Moon, Earth subtends $0.00106 \times 180° = 1.91°$.

The crucial feature of Eq. 1 is that it involves the ratio of two lengths, the circumference and the radius. Thus an angle is a dimensionless number. More important, the angle's measure is independent of what circle you choose, because the *ratio* of the arc length to the radius will not change if the circle is large or small.

Radian

The calculation in Example 8 is basic to the determination of the size of an angle. To get the angle in degrees, the ratio of a circular arc length subtending an angle is first found as a fraction of the circumference of a circle and then multiplied by 360°. As Eq. 1 shows, this means that an answer in degrees is always $360/(2\pi)$ times the ratio of the subtending arc length to the radius of the circle.

By choosing a different definition of the measure of angle, we can make this messy factor of $360/(2\pi)$ disappear. Instead of measuring angle as arc length $\widehat{s}$ over circumference $2\pi R$ times 360°, why not just eliminate the constant factor? All we need to do is to measure angles directly as the ratio of the circular arc length subtending the angle at the center of the circle to the radius of the circle on which the arc length $\widehat{s}$ lies. This gives

$$\theta = \frac{\widehat{s}}{R} \text{ rad},$$

where "rad" is the usual abbreviation for "radians."

Of course, this means that a full circle contains

$$\frac{2\pi R}{R} = 2\pi \text{ radians},$$

and this means that $360^\circ = 2\pi$ rad or that 1 rad $= 57.3^\circ$. This 57.3 is a useful number to remember. Radians may seem strange when you first use them, but you will find that they are more convenient than degrees for measuring angles.

EXAMPLES

9. What angle in radians does the Earth subtend at the Moon? If the Earth–Moon distance is 60 R_E and Earth's diameter is 2 R_E, then the angle subtended at the Moon is $2/60 = 1/30 = 0.0333$ rad.

The Small-Angle Approximation

Radians also facilitate making useful approximations to trigonometric functions. Referred to the large right triangle in Fig. 2.8(a), the trigonometric functions—sine, cosine, and tangent—for θ, are respectively $\sin\theta = y/h$, $\cos\theta = x/h$, and $\tan\theta = y/x$. In Fig. 2.8(b) the very acute triangle shows the important fact that as θ gets small, the lengths of the hypotenuse h and the long leg x of the triangle become almost equal, and the right triangle more and more closely approximates an isosceles triangle. Therefore, as θ gets small, h and x better and better approximate radii of a circle, and the small leg y becomes a better and better approximation to the circular arc length $\widehat{s}$ connecting the two radial legs. This is the basis for approximating the sine or tangent of a small angle by the angle itself in radian measure:

$$\sin\theta = \frac{y}{h} \approx \frac{\widehat{s}}{h} = \theta \approx \frac{y}{x} = \tan\theta.$$

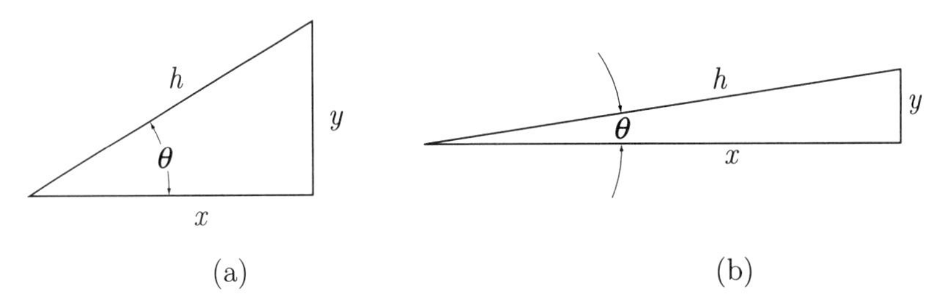

FIGURE 2.8 (a) A right triangle for defining the trigonometric functions. (b) For small angles $h \approx x$.

TABLE 2.3 Small-Angle Approximation

θ in °	θ in radians	$\sin\theta$	$\tan\theta$
1.00	0.0175	0.0175	0.0175
2.00	0.0349	0.0349	0.0349
4.00	0.0698	0.0698	0.0699
8.00	0.140	0.139	0.141
10.0	0.175	0.174	0.176
15.0	0.262	0.259	0.268

EXAMPLES

10. What is the sine of 5.7°? Since 5.7° is about 0.1 rad, and this is fairly small compared to unity, $\sin 5.7° \approx 0.100$.

If you do this with a calculator, you will get $\sin 5.7° = 0.0993$, showing that the approximation agrees with the exact value to better than 1%. The quality of the approximation is apparent from the entries in Table 2.3, which shows that even at angles as large as 15° the sine is only about 1% different and the tangent about 2% different from the radian measure of the angle.

2.7 MOMENTUM AND ENERGY

The next two sections review velocity, acceleration, momentum, force, and energy. The ideas of momentum and energy are especially important and are used repeatedly throughout the rest of this book. They are the basis of two of the most fundamental, universally applicable laws of physics: the conservation of momentum and the conservation of energy.

Momentum

Many of the most important ideas of physics are built up from considerations of mass, length, and time. For example, velocity is the measure of how much length a body travels in a unit time in some particular direction. For bodies in steady motion the idea is simple. If at the end of 1 second we see that a car has moved 24.5 meters and then again at the end of 2 s another 24.5 m, we characterize the car as having a speed of 24.5 $\mathrm{m\,s^{-1}}$. Direction is an important part of velocity, but for now we will worry only about the numerical value, or magnitude, of velocity. This number is called the "speed."

You probably recognize intuitively that when two bodies move at the same speed, the heavier one possesses more of something associated with its motion than the lighter one does. A baseball delivered into the catcher's mitt at 24.5 $\mathrm{m\,s^{-1}}$ is not especially intimidating. In the major leagues such a pitch would be so slow that it very likely would be hit before reaching the catcher. However, a bowling ball delivered at the same speed is quite another story, promising severe bodily damage, and a Jeep Cherokee at the same speed would probably kill you on impact.

Newton thought of moving bodies as possessing different amounts of motion, and he devised a useful measure of this "quantity of motion." It is the product of the mass and the velocity, i.e., mv. Today we use the word "momentum" instead of "quantity of motion," but it means exactly the same thing.

In cases where we are interested in momentum alone and are not calculating it as the product of mass with velocity, it is often convenient to give it a separate symbol, most commonly p,

$$\text{momentum} = \text{mass} \times \text{velocity},$$
$$p = mv.$$

In these terms you see that the baseball[1], with its mass of about 150 g, has a momentum of 3.68 $\mathrm{kg\,m\,s^{-1}}$, while the bowling ball,[2] with a mass of 5.0 kg has a momentum of 122.5 $\mathrm{kg\,m\,s^{-1}}$, and the Jeep,[3] which we take to have a mass of 1500 kg, has a momentum of 36,800 $\mathrm{kg\,m\,s^{-1}}$. The difference between being on the receiving end of 4 $\mathrm{kg\,m\,s^{-1}}$ and 36,800 $\mathrm{kg\,m\,s^{-1}}$ is made evident daily in unpleasant ways.

---

[1]This kind of information can be found in encyclopedias, but often you can find it more directly. We measured the mass of a hardball on a triple-beam balance and got 151.6 g. On September 8, 1987, the "Science Times" section of the *New York Times* mentioned that the regulation American League and National League baseball is 5 to 5.25 ounces; this would be between 142 and 149 g.

[2]Bowling balls range between 4.5 and 7.3 kg.

[3]Curb weight is about 3000 lb, so with driver and a passenger about 1500 kg. [See http://www.jeepunpaved.com/cherokee/comp on the Web.] Incidentally, 24.5 $\mathrm{m\,s^{-1}}$ is about 55 mph.

EXERCISES

14. Estimate the momentum of you and your bicycle together when riding at a typical speed. Compare your answer to the momentums given above for the baseball, the bowling ball, and the automobile.

Force

Newton used his definition of momentum to specify a meaning for another word we use daily: force. Anything that changes the momentum of a given body is a force. The definition includes changes of the direction as well as of the amount of its momentum. The size, or magnitude, of a force depends upon how quickly the momentum changes. In fact, the magnitude of a force is simply the rate of change of momentum, i.e., how much the momentum changes per unit time. Suppose we start with some momentum p_0 at a time we will call t_0. Suppose also that a little later, at time t_1, the momentum has changed to p_1. To find the average rate of change we have to divide the actual change by the number of units of time it took to make the change. In symbols this is

$$F = \frac{p_1 - p_0}{t_1 - t_0}.$$

A more compact notation uses the capital Greek letter delta, Δ, to denote a difference between the final and initial values of a quantity. Thus,

$$F = \frac{\Delta p}{\Delta t}.$$

In terms of mass and velocity, the expression reads

$$F = \frac{\Delta(mv)}{\Delta t},$$

where

$$\Delta(mv) = m_1 v_1 - m_0 v_0$$

and

$$\Delta t = t_1 - t_0.$$

EXAMPLES

11. If the baseball stops in the catcher's glove in 0.01 s, the average rate of change of its momentum is $-3.68/0.01 = -368\ \mathrm{kg\,m\,s^{-2}}$. Similarly, when the bowling ball stops at the end of the lane in perhaps the time of 0.05 s, its rate of change of momentum is $-80/0.05 = -1600\,\mathrm{kg\,m\,s^{-2}}$. These two results mean that a force of $368\,\mathrm{kg\,m\,s^{-2}}$ acted on the baseball and a force of $1600\,\mathrm{kg\,m\,s^{-2}}$ acted on the bowling ball.

You can see that a person might get weary of writing "$\mathrm{kg\,m\,s^{-2}}$" all the time. Moreover, this group of units does not shout "force!" at the reader. To make units more compact and more recognizable, physicists have adopted standard names for certain groups of units. In the SI units the group "$\mathrm{kg\,m\,s^{-2}}$" is called the "newton." We say "the newton is the unit of force when using the meter, the kilogram, and the second as basic units of measurement." The newton is abbreviated "N." Thus to stop the baseball in 0.01 s requires a force of 368 N, while stopping the bowling ball in 0.05 s requires a force of 1600 N. Like velocity and momentum, force has direction as well as magnitude, though for now we will neglect this important aspect of its definition.

Usually, a body's momentum is changed by changing its velocity. When momentum is changed by changing velocity,

$$\Delta p = m\Delta v.$$

We can rewrite the force relation substituting for Δp:

$$F = m\frac{\Delta v}{\Delta t},$$

which is the same as saying mass times the rate of change of velocity. But rate of change of velocity is the definition of acceleration. The average acceleration of a body is defined to be $a = \Delta v/\Delta t$. When the speedometer of your car reads 5 mph more than it did two seconds earlier, you have accelerated 2.5 mph per second. That is an awkward unit, so we much more frequently use units like m/s per s or, equivalently, $\mathrm{ms^{-2}}$ when we are talking about acceleration. Now you see where the famous relation $F = ma$ comes from. A force is required to produce acceleration, that is, to change the velocity in some time interval. $F = ma$ is just a particular way of writing that force is the rate of change of momentum.

EXERCISES

15. Find the average accelerations that occurred while stopping the two objects in Example 11.

To see whether you understand that $F = \Delta p/\Delta t = ma$ means what it says, answer the following questions.

EXERCISES

16. A skydiver without her parachute open falls at a steady 70 m/s. What is the total force acting on her?

17. After the skydiver opens her chute she falls at a steady speed of about 4 m/s. What is now the total force acting on her?

One more word about named units. Once we have defined $\mathrm{kg\,m\,s^{-2}}$ to be a newton, we can express the units of momentum in terms of newtons too. Just from the algebra of the units you see that $368\,\mathrm{kg\,m\,s^{-1}}$ is the same as 368 newton-seconds, or N-s. It is customary to measure momentum in N-s.

EXAMPLES

12. In case you are unfamiliar with the "algebra of units" mentioned above, here is a brief overview. The most important idea when manipulating units is to recognize that when you divide one quantity by another that has the same units you get a pure number, that is, a number without units. Two kilograms is just twice one kilogram, $2\text{ kg} = 2 \times 1\text{ kg}$, so $2\text{ kg}/1\text{ kg} = 2$. This second form is equivalent to dividing out the kg units, top and bottom, as if they were algebraic quantities. Another important manipulation is that you can always multiply a quantity by 1 without changing anything. Thus in the conversion of momentum units:

$$\mathrm{kg\,m\,s^{-1}} \times (\mathrm{s/s}) = (\mathrm{kg\,m\,s^{-2}})\,\mathrm{s} = \text{N-s}.$$

Conservation of Momentum

When we began the discussion of momentum, we used the images of a person trying to stop a baseball and a bowling ball in order to evoke an intuitive sense of the quantity of motion possessed by a moving object. We talked about the forces on the baseball and the bowling ball; we did not talk about the forces on the catcher or on the walls of the bowling alley where the ball stopped. There is, however, an intimate connection between the force exerted by an object A acting on an object B, and the force that B exerts on A. Let's see what it is.

Suppose you were seated on a very slippery surface and tried to stop a bowling ball coming at you. Assuming that you were successful in bringing it to rest relative to you without undue damage, what do you think might happen? You already have some idea that the ball would exert a force on you

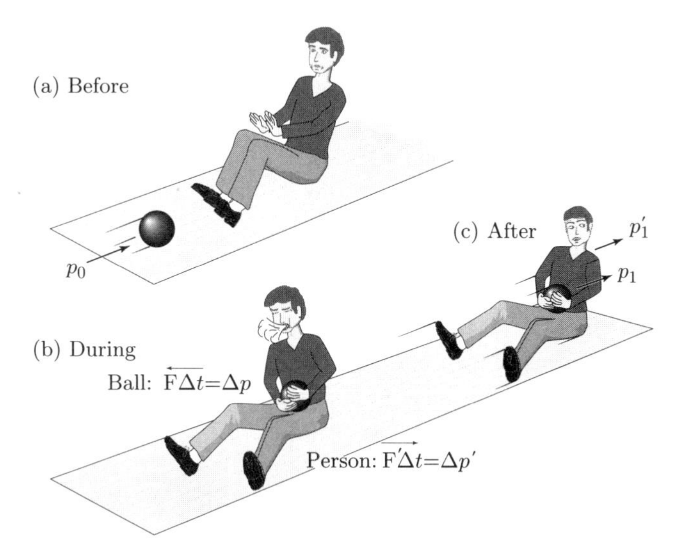

FIGURE 2.9 Collision between a bowling ball and a person on ice. Don't try this at home!

as you try to slow it down. With negligible friction to hold you in place, that force would have to impart some momentum to you. A remarkable thing happens. The momentum imparted to you is exactly equal in magnitude to the amount lost by the bowling ball. In fact, in all such interactions when there is no outside, or *external*, force acting, the *net* change in momentum is zero. This is what we refer to as *conservation* of momentum. The cartoon in Fig. 2.9 depicts the collision interaction. If the initial momentum of the bowling ball is p_0 (the person's initial momentum is zero, since he is at rest), and the final momentums of the ball and person are p_1 and p_1', respectively, the conservation of momentum says that

$$p_0 = p_1 + p_1'.$$

EXAMPLES

13. A 24.5 m/s bowling ball is a bit fearsome, so suppose that a person is on very slippery ice with a ball approaching at 4.0 m/s as suggested by Fig. 2.9. The ball and person slide off together. Take the bowling ball's mass to be 5 kg and the person's mass to be 70 kg. You can calculate the final momentum of each and their common velocity, v, by using conservation of momentum to find the velocity v:

$$\begin{aligned} p_0 &= p_1 + p_1', \\ 4 \times 5 &= 5v + 70v = 75v, \\ v &= 20/75 = 0.27\,\text{m/s}. \end{aligned}$$

Knowing v, the individual momentums are easily calculated.

EXERCISES

18. Calculate the individual momentums.

To analyze collisions in general requires us to understand the vector properties of force and momentum (see the Appendix at the end of this chapter), but in one dimension we can treat vectors simply as algebraic quantities: A force acting to the right is taken as positive, while one acting to the left is negative. The same is true for momentum. To reduce positive momentum, there must be a force applied to the left in order to add some negative momentum. A slight rearrangement of the definition of force allows us to calculate the change in momentum of the ball and the change of momentum of the person:

$$F\Delta t = \Delta p,$$

$$F'\Delta t = \Delta p',$$

where the primed quantities refer to the action on the person.

Since F is negative (to the left), Δp is also negative. Since conservation of momentum means that the sum of the momentum changes is zero, $\Delta p + \Delta p' = 0$, $\Delta p'$ must be positive and of the same magnitude. Further, since the time of application of the force must be the same for the two bodies,

$$F\Delta t + F'\Delta t = 0.$$

This means that $F = -F'$. That is, if you are exerting a force on a body, it will exert a force of the same magnitude and opposite direction back on you. You may already know this fact as Newton's third law of motion.

EXERCISES

19. Suppose the person in Example 13 gave an extra shove that sent the ball backwards at 2 m/s after the collision. Find the person's final momentum.

20. Earlier we calculated the average force required to stop the 5 kg bowling ball moving at 24.5 m/s in 0.05 sec. Suppose the wall were very elastic and the ball rebounded with the same speed, the time of collision being the same. Find the average force on the ball and on the wall for this new situation. Be careful to specify directions, taking the initial momentum to be positive.

TABLE 2.4 Important Quantities for Describing the Physics of Anything

Concept	Compound Units	Name	Abbreviation	Examples
velocity	$m\,s^{-1}$	none	none	A person walks at a rate of about $1.5\,m\,s^{-1}$. A snail goes at a few $mm\,s^{-1}$. The speed limit for U.S. autos on highways in urban areas is $24.6\,m\,s^{-1}$. The speed of light in a vacuum is $3 \times 10^8\,m\,s^{-1}$.
acceleration	$m\ s^{-2}$	none*		A body falling freely near the surface of Earth accelerates at $9.8\,m\,s^{-2}$
force	$kg\,m\,s^{-2}$	newton	N	Earth exerts a force of 98 N on a 10 kg mass near its surface. Earth's atmosphere exerts a force of 1.01×10^5 N on each square meter of Earth's surface.

*An acceleration of $1\ cm\,s^{-2}$ is sometimes called a gal.

21. Sometimes it is useful to average the force over a time that is different from the collision itself. Suppose 20 baseballs are thrown at a wall in 5 seconds and that each rebounds from the wall with the same 24.5 m/s speed that it came in with. Find the average force on the wall during the 5 seconds.

Connecting Concepts to Physical Reality

It is important to consider numerical values illustrating the concepts of physics as you come across them. It will help you understand the physical significance of the concepts if you connect them to specific phenomena. Part of the great power of physics is that it works for very large-scale systems and very small-scale systems. When you are introduced to a new concept, you should try it out at several different scales. Table 2.4 offers some numerical values of real velocities, accelerations, and forces as concrete examples of these concepts for you to consider.

2.8 ENERGY

Although the word "energy" was originally coined by physicists, it is now a word you use in daily speech. You talk about having enough energy to get up and do what has to be done. You hear reminiscences of the "energy crisis" and predictions of the energy shortages to come. People talk about energy needs and energy efficiency and energy conservation.

The idea of energy is fundamental to the story this book tells. Energy is useful for discussing remarkably different phenomena over a huge range of magnitudes—tiny particles, large planets, flowing electric charge, light waves, and colliding atoms or nuclei. Because of this general applicability and because the behavior and interactions of radiation and atomic and subatomic matter are more easily described in terms of energy than in terms of force, we use the idea over and over in this book.

The word actually was made up by physicists to describe a body's ability to do work. That tells you nothing, of course, because "work" needs to be defined. Physicists have a nice precise definition of "work." In the simple case of a constant force pushing parallel to the line of motion of the object on which it acts, and pushing on it over some distance d, the work W done is the amount of force F times the distance d over which the force acts:

$$W = Fd.$$

EXAMPLES

14. Thus, a force of 2 N applied over a distance of 3 m does an amount of work $W = 2 \times 3 = 6$ N-m. A newton-meter has its own name, "joule," so we could as well say that 2 N acting over 3 m does 6 joules of work. The abbreviation for joule is "J," so we would usually write that the work done was 6 J.

All this semantic information has its uses, but it does not answer your central question: What *is* energy? One of the best answers to this question has been given by the renowned American physicist Richard Feynman. He tells a little story that gives some of the flavor of the idea.[4]

Feynman's Energy Analogy

> Imagine a child, perhaps "Dennis the Menace," who has blocks which are absolutely indestructible, and cannot be divided into pieces. Each is the same as the other. Let us suppose that he has 28 blocks. His mother puts him with his 28 blocks into a room at the beginning

[4]Richard Feynman in volume I of *The Feynman Lectures on Physics* by Richard P. Feynman, Robert B. Leighton, Matthew Sands ©1963 by the California Institute of Technology. Published by Addison-Wesley Publishing Co. Inc., 1963. on pp. 4-1 to 4-2.

of the day. At the end of the day, being curious, she counts the blocks very carefully, and discovers a phenomenal law—no matter what he does with the blocks, there are always 28 remaining! This continues for a number of days, until one day there are only 27 blocks, but a little investigating shows that there is one under the rug—she must look everywhere to be sure that the number of blocks has not changed. One day, however, the number appears to change—there are only 26 blocks. Careful investigation indicates that the window was open, and upon looking outside, the other two blocks are found. Another day, careful count indicates that there are 30 blocks! This causes considerable consternation, until it is realized that Bruce came to visit, bringing his blocks with him, and he left a few at Dennis' house. After she has disposed of the extra blocks, she closes the window, does not let Bruce in, and then everything is going along all right, until one time she counts and finds only 25 blocks. However, there is a box in the room, a toy box, and the mother goes to open the toy box, but the boy says "No, do not open my toy box," and screams. Mother is not allowed to open the toy box. Being extremely curious, and somewhat ingenious, she invents a scheme! She knows that a block weighs three ounces, so she weighs the box at a time when she sees 28 blocks, and it weighs 16 ounces. The next time she wishes to check, she weighs the box again, subtracts sixteen ounces and divides by three. She discovers the following:

$$\begin{pmatrix}\text{number of}\\ \text{blocks seen}\end{pmatrix} + \frac{(\text{weight of box}) - 16\ \text{ounces}}{3\ \text{ounces}} = \text{constant}. \tag{2}$$

There then appear to be some new deviations, but careful study indicates that the dirty water in the bathtub is changing its level. The child is throwing blocks into the water, and she cannot see them because it is so dirty, but she can find out how many blocks are in the water by adding another term to her formula. Since the original height of the water was 6 inches and each block raises the water a quarter of an inch, this new formula would be

$$\begin{aligned}\begin{pmatrix}\text{number of}\\ \text{blocks seen}\end{pmatrix} &+ \frac{(\text{weight of box}) - 16\ \text{ounces}}{3\ \text{ounces}}\\ &+ \frac{(\text{height of water}) - 6\ \text{inches}}{1/4\ \text{inch}} = \text{constant}.\end{aligned} \tag{3}$$

In the gradual increase in the complexity of her world, she finds a whole series of terms representing ways of calculating how many blocks are in places where she is not allowed to look. As a result, she

> finds a complex formula, a quantity which *has to be computed*, which always stays the same in her situation.
>
> What is the analogy of this to the conservation energy? The most remarkable aspect that must be abstracted from this picture is that *there are no blocks*. Take away the first terms in Eqs. 2 and 3 and we find ourselves calculating more or less abstract things. The analogy has the following points. First, when we are calculating the energy, sometimes some of it leaves the system and goes away, or sometimes some comes in. In order to verify the conservation of energy, we must be careful that we have not put any in or taken any out. Second, the energy has a large number of *different forms*, and there is a formula for each one.

Don't feel bad if this story leaves you still wondering what energy is. The idea of energy is extremely useful in physics and in all of science, but it is an idea best eased into by becoming acquainted with the various ways it is used. By the time you have read this book you will begin to think of energy as a very real thing. In the meantime, let's look at some applications of the idea to simple systems where it is easy to use energy quantitatively. We will look at a falling body and a mass swinging on the end of a string—a pendulum, and illustrate the important idea of conservation of energy.

Conservation of Energy

Energy comes in many forms. There is heat energy, kinetic energy, gravitational potential energy, electrical potential energy, energies of electric and magnetic fields, nuclear energy. There is energy stored in the compression of a spring, in the compression of gas, in the arrangement of molecules and atoms. Although energy may change from one form to another, the total quantity of energy never changes. We say, therefore, that energy is conserved. It is this property of conservation of energy that makes it so useful. Notice that physicists use the word "conservation" differently from economists and environmentalists, who usually mean "use available forms of energy as efficiently as possible."

Kinetic Energy

One of the most familiar forms of energy is "kinetic energy," the energy a body has by virtue of its motion. It is another kind of "quantity of motion" that was found to be useful at about the same time that Newton began to think in terms of momentum. For reasons we won't go into now, a body of mass m moving with a speed v substantially smaller than the speed of light has a kinetic energy K given by the formula

$$K = \frac{1}{2}mv^2.$$

Knowing this, we can calculate the kinetic energy in the baseball, the bowling ball, and the Jeep that we talked about earlier.

EXERCISES

22. The baseball's kinetic energy is $1/2 \times 0.15 \times (24.5)^2 = 45.0\,\text{J}$. The kinetic energy of the bowling ball is 1500 J. That of the automobile is 0.45 MJ (where "M" is the usual abbreviation for the prefix "mega," which stands for 10^6). Verify that these are correct numbers. Notice that since each mass has the same speed, their kinetic energies vary only by the ratios of their masses, i.e., $K_{\text{bowling ball}} = 5.0/0.15 \times K_{\text{baseball}}$. (Useful insights like this make calculating easier.)

23. A pitcher warms up by throwing a baseball to the catcher at 45 mph; this means it has a kinetic energy of about 30 J. During the game he throws a fastball at 90 mph. What is its kinetic energy then?

Gravitational Potential Energy

When you lift a mass m to some height—call that height h, you add energy to the mass. This energy is called "gravitational potential" energy, or for brevity just "potential" energy. The amount of gravitational potential energy U you add by lifting the mass m a height h is

$$U = mgh,$$

where g is the acceleration due to gravity near Earth's surface, $9.80\,\text{ms}^{-2}$.

Why is it called "potential"? Perhaps because it has the potential for becoming kinetic energy. If you lift the mass a height h and then release it, experience shows you that its speed increases as it falls. This means that its kinetic energy increases. Of course, as m falls, the height changes from h to some smaller height y, so the gravitational potential energy of m diminishes all the while its kinetic energy increases. What you are seeing here is the conversion of one form of energy—gravitational potential energy—into another—kinetic energy—as the mass falls. What makes this way of looking at the fall so useful is that the sum of the two forms of energy remains constant throughout the fall. This constancy is an example of the conservation

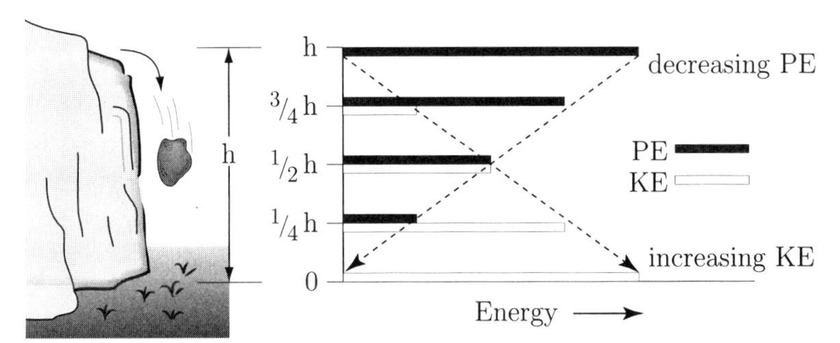

FIGURE 2.10 A rock falls off a cliff of height h. Its potential energy is converted into kinetic energy, but the total energy remains constant.

of energy:

$$mgh = mgy + \frac{1}{2}mv^2.$$

Figure 2.10 graphically illustrates this remarkable property of energy. As a stone falls off a cliff of height h its PE decreases steadily (dark bars) while its KE increases steadily (light bars). At any instant the sum of the dark and light bars is always the same.

EXERCISES

24. What is the value of the sum of the two bars when $y = h/2$?

25. Suppose you lift a baseball ($m = 150\,\mathrm{g}$) 1 m above a table. By how much do you increase its gravitational potential energy?

26. Suppose you drop the baseball. What will be its kinetic energy when it is 0.5 m above the tabletop? By how much will its gravitational potential energy have changed at that point?

27. Suppose there is a hole in the table and the ball falls through it. What will be the ball's gravitational potential energy when it is 20 cm *below* the table? What will be its kinetic energy at this point? What will be the sum of its kinetic and potential energy?

What happens after the ball hits the table and stops? Clearly, its kinetic energy becomes 0 J. Also, the ball has reached the point from which we chose to measure potential energy, and so its potential energy is 0 J. What has become of its 1.47 J? It has gone into heating up the point of impact, into

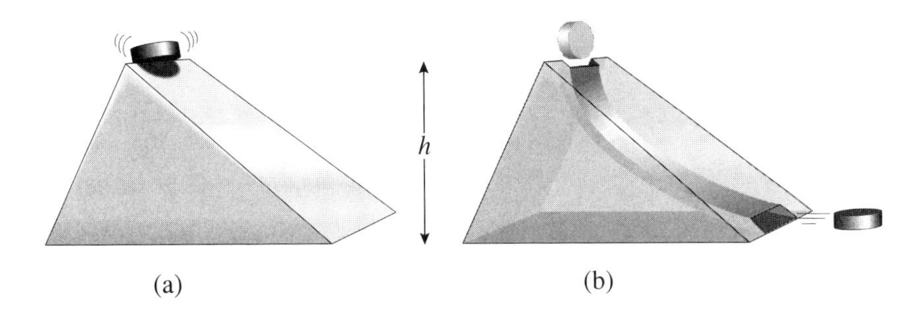

FIGURE 2.11 (a) A frictionless hockey puck teeters indecisively. Regardless of which way it slides, it will have the same kinetic energy when it reaches the bottom. (b) Now the puck may slide down through a tunnel drilled in the block, but its kinetic energy is still the same at the bottom.

the compression of the spot on which it is resting, and into acoustic energy—the sound of its impact. A fascinating aspect of energy is that so far in the history of physics there has always been an answer to the question: What has become of the initial energy? And very often the answer casts revealing light on the nature or behavior of matter.

It is an interesting and useful fact that gravitational potential energy depends only on *vertical* distance, and sideways movements of a body do not change its gravitational potential energy. This means that no matter how a body falls from one height to another, the change in gravitational potential energy will be the same. Then as long as the only other form of energy can be kinetic, the change in kinetic energy will also be the same. Look at Fig. 2.11, where a flat object teeters at the top of two different inclines. It can slide frictionlessly down the left side or the right side, but either way it slides, it will have the same kinetic energy at the bottom.

EXAMPLES

15. Suppose that in Fig. 2.11 the puck has a mass of 160 g and that $h = 20$ cm. What will be its kinetic energy when it reaches the bottom of the left-hand incline? The bottom of the right-hand incline?

Relative to the bottom of the inclines, the puck has a gravitational potential energy of $mgh = 0.16 \times 9.8 \times 0.2 = 0.314$ J. As the puck slides frictionlessly down either incline, this amount of gravitational potential energy is converted to kinetic energy. The kinetic energy of the puck is the same at the bottom of either side; it is 0.314 J.

EXERCISES

28. Suppose a chute was drilled through the block, curving off to the side and arriving at the bottom right-hand corner as shown in Fig. 2.11(b). If the puck fell down the chute, what would be its kinetic energy when it arrived at the bottom? How fast would it be moving?

29. Notice that Table 2.4 has no entries for energy or momentum. Make up appropriate entries for these two quantities.

Energy Costs Money

There is one aspect of energy that should convince you that it is something real. Whatever energy is, if you want some you usually have to pay for it. You buy electrical energy to run your household appliances; you buy oil to heat your house; you buy gasoline to run your automobile; you buy food to run your body.

A kilowatt-hour of electricity is the same thing as 3.6 MJ. In the U.S., the average cost of a kW-h is about $0.07. This is not very expensive, which is one of the reasons many Americans' lives are often so pleasant. A representative price for all forms of energy is $10 for 10^9 joules. Scientists use the prefix "giga" (pronounced jeega) to represent the factor 10^9 (an American billion), so 10^9 joules is called a gigajoule. Giga is abbreviated G, so you could write that energy costs roughly $\$10\,\mathrm{GJ}^{-1}$.

Actual prices vary a great deal depending on special features of the energy: is it easy to handle? Is it very concentrated? Is the energy accessible easily? Can you get out a lot of energy quickly? In this book our concerns with energy will usually be quite remote from practical considerations of cost and availability. For us energy will be a guide to studying atoms and their structure.

2.9 PENDULUMS AND ENERGY

A pendulum is a concentrated mass hanging by a tether from some pivot point. Its motion is familiar if you have ever swung on a swing or looked inside a grandfather clock. Now you can understand that what you observe when you watch a mass swing back and forth at the end of a string is the cyclical conversion of gravitational potential energy into kinetic energy. The pendulum's motion is begun by pulling its bob to one side; this has the effect of lifting it some vertical distance h, as shown in Fig. 2.12(a), and it acquires gravitational potential energy. When released, the pendulum swings back to

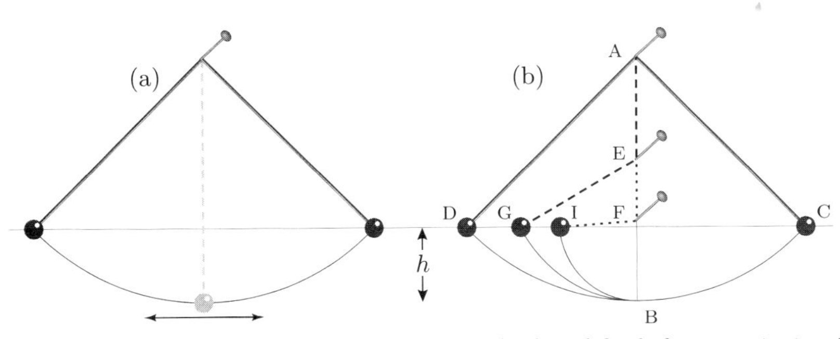

FIGURE 2.12 (a) A pendulum converts energy back and forth from gravitational potential energy to kinetic energy. (b) For any given amount of total energy, changing the pivot point will not change the height to which the pendulum can rise.

its lowest position, where it is moving its fastest because all its gravitational potential energy has been converted into kinetic energy. As the bob rises to the other side, it slows down because of the conversion of kinetic energy into gravitational potential energy. It reaches the highest point of its swing when all its kinetic energy has been converted to potential energy; then it moves back toward the lowest point, beginning another cycle of conversion.

Figure 2.12(b) illustrates an argument made by Galileo that you can explain on the basis of the conservation of energy. He writes:

> Imagine this page to represent a vertical wall, with a nail driven into it; and from the nail let there be suspended a lead bullet of one or two ounces by means of a fine vertical thread, AB, say from four to six feet long; on the wall draw a horizontal line DC, at right angles to the vertical thread AB, which hangs about two finger-breadths in front of the wall. Now bring the thread AB with the attached ball into the position AC and set it free; first, it will be observed to descend along the arc CBD, to pass the point B, and to travel along the arc BD, till it almost reaches the horizontal CD, a slight shortage being caused by the resistance of the air and the string; from this we may rightly infer that the ball in its descent through the arc CB acquired a momentum [he means kinetic energy; the difference between momentum and kinetic energy was not clear until two hundred years after Galileo] on reaching B, which was just sufficient to carry it through a similar arc BD to the same height. Having repeated this experiment many times, let us now drive a nail into the wall close to the perpendicular AB, say at E or F, so that it projects out some five or six finger-breadths in order that the thread, again carrying the bullet through the arc CB, may strike upon the nail E when the bullet reaches B, and thus compel it to traverse the arc BG, described about E as center. From this we can see what can be done by the same momentum [kinetic energy] which previously starting at the same point B carried the same body through

the arc BD to the horizontal CD. Now, gentlemen, you will observe with pleasure that the ball swings to the point G in the horizontal, and you would see the same thing happen if the obstacle were placed at some lower point, say at F, about which the ball would describe the arc BI, the rise of the ball always terminating exactly on the line CD. But when the nail is placed so low that the remainder of the thread below it will not reach to the height CD (which would happen if the nail were placed nearer B than to the intersection of AB with the horizontal CD) then the thread leaps over the nail and twists itself about it.[5]

EXERCISES

30. Using the conservation of energy, give your own explanation of the demonstration described here by Galileo.

2.10 FORCES AS VARIATIONS IN POTENTIAL ENERGY

Figure 2.11 illustrates an important feature of potential energy. Wherever there is a spatial variation of potential energy, there is a force. Notice that as the puck slides down the right slope, its potential energy changes more gradually than it does when it slides down the left slope. The steeper the spatial change of potential energy, the greater is the force.

EXAMPLES

16. Suppose the angle θ of the incline in Fig. 2.11 is 30°. Then as the puck slides along the right slope a distance $\Delta s = 1$ cm, it drops a vertical distance of $\Delta z = 0.5$ cm because $\Delta z/\Delta s = \sin\theta$ and $\sin 30° = 0.5$. The force F produced by this change in potential energy is just

$$F = -\frac{\Delta U}{\Delta s} = -\sin\theta\frac{\Delta U}{\Delta z}, \qquad (4)$$

which means that $F = -mg\sin\theta$.

[5]In Galileo Galilei *Dialogues Concerning Two New Sciences*, Northwestern University, 1939 (Dover, New York), pp. 170–171.

You may have known this already, but the point here is not an amazing discovery that the force down an incline is proportional to the sine of the angle of the incline. The point is that the force is equal to the spatial change of the potential energy.

In less clumsy language, the force is equal to the negative derivative of the potential energy with respect to its spatial coordinates:

$$F = -\frac{dU}{ds} = -\sin\theta\frac{dU}{dz}.$$

The force is the negative of the slope of the potential energy in space.

Changes in potential energy in space may be hard to visualize at first, but you will see that this way of looking at things is very useful. You can think of space as having "hills" and "valleys" of potential energy. Masses tend to move from the places in space where gravitational potential energy is high to those where it is low. Near the surface of Earth this idea is not immediately obvious because the shape of potential energy is so simple: It is just a steady downward slope toward the surface.

At the end of the sixteenth century the Flemish mathematician Stevinus realized that the force on a body on an incline had to be smaller if the angle of incline was smaller. He showed that it was exactly proportional to the sine of the angle of the incline, $\sin\theta$. To prove this he imagined a chain of balls laid over a double incline as shown in Fig. 2.13. The triangle shown here is a 3-4-5 right triangle for convenience. If the forces on a ball were the same on each side of the incline, the three balls would be subject to a total force that was less than that acting on the four balls, and the chain would start rotating by itself and accelerate indefinitely. Although he had no idea about energy or its conservation, Stevinus realized that such a perpetual motion machine was impossible. The total forces on the two sides must be equal. This can be

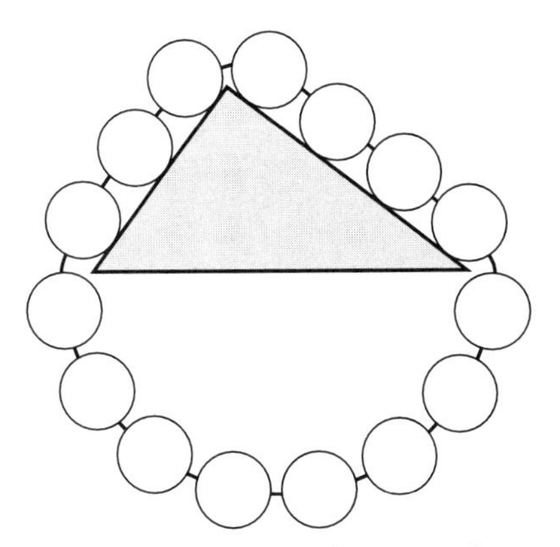

FIGURE 2.13 Stevinus's diagram shows that the force must be greater along the steeper slope than along the shallower one.

true only if the force acting on each of the 4 balls is 3/4 the force acting on each of the 3 balls.

EXERCISES

31. Show that this is so if Eq. 4 is correct.

Stevinus had Fig. 2.13 carved on his gravestone. He was justly proud of what we today can recognize as the use of the conservation of energy to do physics.

2.11 SUMMARY

In this review of basic concepts and units of physics many interesting subtleties have been omitted, and the vector nature of many of the quantities has been ignored. Nevertheless, you should have gained some familiarity with mass, length, time, velocity, momentum, and energy, their units, the prefixes that give the powers of ten that become parts of the units, and the way they are used.

With our basic ideas of length and time we describe motion in terms of velocity, the direction and rate at which a body covers distance, and in terms of acceleration, the rate at which velocity changes.

The amount of motion in a body is the product of its mass and its velocity, mv, and is called momentum. A body's momentum is changed only by a force; a force is anything that causes momentum to change. Force is measured as the rate of change of momentum $\Delta(mv)/\Delta t$. In a closed system momentum is conserved.

A moving body also has kinetic energy which at velocities of familiar objects is given by $\frac{1}{2}mv^2$. Energy comes in many forms. In a closed system energy may change from one form to another, but the total amount of energy does not change. In a closed system energy is conserved.

These concepts and ideas are important, but so are the techniques for thinking about them. Practice assigning numbers to them and developing a sense of their physical scale and significance. You will need those techniques to make the best use of the rest of this book.

Another exhortation on SI prefixes

SI prefixes—micro, mega, kilo, nano, giga—have been introduced and used several places in this chapter. You *must* know them, their abbreviations, and their numerical values well enough so that you can convert among them quickly and accurately. There is a list of all the SI prefixes at the end of the book. Maybe you can learn them by osmosis, but if you can't, then just memorize them. Do whatever it takes, but learn them.

2.12 APPENDIX: VECTORS

Some of the quantities discussed above have a special property that we will use occasionally. When the direction is of importance in the full description of the quantity, the quantity is a "vector." Examples of vectors we will use are change of position of an object, its velocity, and its momentum. Examples of quantities that are *not* vectors are mass, energy, and time.

Representing vectors

Because change of position, or "displacement," is easily visualized, it is useful for our purposes to take the basic idea of a line segment that has a direction attached to it as a prototype for vector properties. We will not try to make a highly formal definition that covers all possible physical and mathematical situations.

Figure 2.14 depicts a typical displacement vector, $\vec{R}$, as a line segment. We use the notation of a letter with an arrow over it to indicate a vector quantity, one with both magnitude and direction. The plain letter, such as R, just represents the magnitude, or length, of the vector. In this case it is the shortest distance and object could move while traveling from the origin to the point (X, Y).

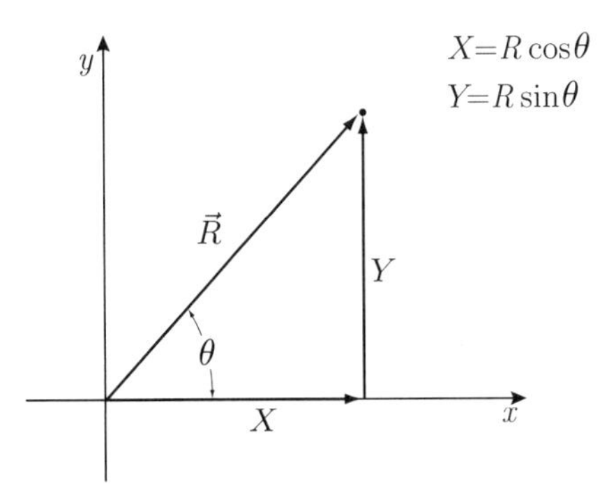

FIGURE 2.14 Components (X and Y) of the vector $\vec{R}$.

Components

The diagram in Fig. 2.14, suggests another way of looking at the displacement. Arriving at the point (X, Y) could have been accomplished by a displacement in the x direction by an amount X and then in the y direction by an amount Y. If we make these two quantities themselves vectors, the vector $\vec{R}$ is entirely equivalent to adding the two displacements in the coordinate directions, $\vec{X}$ and $\vec{Y}$. The magnitudes of these two displacements are particularly handy quantities for dealing with vectors. They are called the "Cartesian components" of the vector. Figure 2.14 exhibits the trigonometry and geometry used to go back and forth between a vector and its components using the Pythagorean theorem and basic definitions of the sine and cosine:

$$X = R\cos\theta,$$
$$Y = R\sin\theta,$$
$$X^2 + Y^2 = R^2.$$

EXERCISES

32. Prove that if $X = R\cos\theta$ and $Y = R\sin\theta$ then $R^2 = X^2 + Y^2$.

We see that the magnitude of a vector is just the square root of the sum of the squares of its components.

What if there were a displacement in the third dimension, z? The generalization is easy:

$$R^2 = X^2 + Y^2 + Z^2.$$

Adding Vectors

Adding vectors can get somewhat more complicated in the general case. We have already seen that reconstituting a vector from its components is equivalent to adding two vectors at right angles. One other case is important for us to deal with at this time. Adding vectors that are all along one particular coordinate axis is a simple matter of addition and subtraction. Imagine walking ten paces west. If you reverse and come east for three paces, your second displacement undoes some of the first, so it is in effect negative. The systematic way of handling this is to take displacements in the positive coordinate direction to be positive and displacements in the negative coordinate direction to be negative. Then a simple algebraic sum of displacements gives the net displacement as long as we stick to the one dimension. Other vector quantities have the same kind of additive properties as does displacement, so we can follow exactly the same rules for adding them up, even if a geometric picture seems inappropriate.

PROBLEMS

1. What is the order of magnitude of your height? What meaning are you using for "order of magnitude"?

2. What is the circumference of Earth? How do you know this?

3. Estimate the volume of your body.

4. Without looking at a ruler or other measuring device, draw a line 1 cm long.

5. A large speck of dust has a mass $m = 0.00000412$ g (grams).

a. Rewrite the mass in grams using scientific notation.

b. Express m in (i) kg; (ii) mg (iii) μg

6. Solid aluminum has density 2.7 g cm^{-3}. What is this density in units of kg m^{-3}? Use scientific notation and show your calculation.

7. a. A rectangular box has dimensions $L = 2$ cm, $W = 5$ cm, and $H = 10$ cm. How many of these boxes will fit into a cube of volume 1 m^3?

b. If each box has a mass $m = 0.01$ kg, what is the density (mass per unit volume) of the assembly of boxes?

8. The speed of light is 3×10^{10} cm s^{-1}; the circumference of Earth is 40 Mm. How long would it take light to circle the Earth?

9. The electron volt is a unit of energy, abbreviated eV. What is the ratio of the energies of a 20 GeV electron and a 3 keV electron?

10. A mass of 2 kg travels at 4 m s^{-1} towards a wall. It hits the wall and bounces directly back, now traveling at 2 m s^{-1} away from it. By how much does the momentum of the mass change in this collision?

11. Suppose an 80 kg mass falls off a table 30 inches high and moves toward the floor. Does its momentum remain constant? Why?

12. What will be the kinetic energy of the mass in the preceding question just before it hits the floor? Explain how you know.

13. An object of mass 2 kg, initially at rest, is dropped from a tower in order to determine its height. The object's velocity is 0.02 μm/ns just before it hits the ground.

a. Find the object's velocity in m s^{-1}.

b. What is its kinetic energy just before impact?

c. What is the height of the tower?

14. If a 20 N force is applied to a 10 kg mass for 10 ms, by how much does the momentum of the mass change?

15. In icy weather, cars A (mass 1000 kg) and B (mass 2000 kg) collide head-on while traveling in opposite directions in the same lane of a highway.

a. A and B each had speed 15 m s^{-1} just before the collision, and A was heading due east. Sketch vectors (arrows) representing the momentums of the cars before the collision, and calculate their total momentum.

b. The cars stick together immediately after the collision. What is their speed immediately after the collision, and in what direction are they moving?

c. The average force on a body is defined as the change in momentum divided by the time it takes for that change to occur. If the duration of the impact is 0.125 s, find the average force exerted on car A by car B.

16. Suppose you have a circle 1.2 m in diameter. Imagine that lines are drawn from the center of the circle to two points on the circumference 0.05 m apart. What is the (small) angle between the two lines? Give your answer both in degrees and in radians.

17. For the circle in Problem 16, what would be the angle if the length of the arc between the two points was 1.2 m? Find your answer both in degrees and in radians.

18. Complete the following table of standard prefixes for units

prefix	abbreviation	value
nano		
	M	
		10^{-2}
milli		
	p	
		10^{9}
kilo		
	μ	
	T	

19. How many μg are there in 2 kg?

20. List all the SI prefixes used in this chapter, giving their names, abbreviations, and numerical values. Use them with units to describe various physical phenomena.

CHAPTER 3

The Chemist's Atoms

3.1 INTRODUCTION

The idea of an atom has a long history. Around twenty-five hundred years ago, Greek philosophers argued that matter must be built up of small, hard, identical pieces. Because these pieces were thought to be irreducible they were called "atoms." The word "atom" is derived from the Greek for "uncuttable." The pieces, or atoms, come in only a few kinds, said the Greek philosopher Thales, and the complexity we observe in nature arises from the variety of ways in which these building blocks go together and come apart.

This simplified picture of the Greek concept of atom contains the essentials of the modern idea. Why then did it take until the early 1800s for chemists and physicists to produce convincing physical evidence for the existence of atoms? The answer, in part, is that to obtain and interpret their evidence they needed both the modern concept of the chemical element and the development of quantitative techniques of measurement. These became precise enough to yield useful information at the end of the eighteenth century. Only then was the stage set for obtaining and understanding the physical evidence for atoms that is the subject of this chapter.

3.2 CHEMICAL ELEMENTS

A major step toward an atomic theory of matter was the recognition of the existence of the "chemical elements." Centuries of study of the qualitative properties of matter had made it possible to recognize and distinguish among a large variety of substances. Chemists, notably among them Lavoisier, recognized that while many chemical substances could be broken down into others, certain chemicals, like carbon, sulfur, and oxygen, could not be broken down by heat or grinding or other known chemical processes. These irreducible chemicals were viewed as "elements," basic species of which all other chemical substances were compounded. This is the origin of the idea of chemical "compounds" built up from chemical elements.

Although the idea of the chemical element was established by the end of the eighteenth century, notice that the idea does not necessarily imply that there are such things as atoms. All that has been established is that there seem to be a number of distinct types of chemicals from which others can be constructed. What then would be good evidence that elements are composed of atoms?

3.3 ATOMS AND INTEGERS

An essential feature of atomicity is countability. If discrete building blocks of matter exist, they should be countable like chairs, students, or money. Countability is closely associated with the integers. The number of chairs in your classroom is probably an integer. So is the number of students in your classroom, or the number of coins in your piggy bank. Experience suggests an important general idea: Countable sets are integer multiples of some basic, individual unit: one chair, an individual student, a coin. These units are indivisible at least in the sense that a set of these objects can only be subdivided so far and no farther without changing the nature of the elements of the set: to kindling, to a scene in a horror movie, to bits of metal.

Atomicity is closely related to integer countability. The observation of integer relationships in the formation of chemical compounds was early, important evidence for the existence of atoms as discrete building blocks of these compounds. By the beginning of the nineteenth century, techniques for weighing small masses had become precise enough to permit accurate quantitative measurements of the amounts of different chemicals before and after combination. By their ingenious analysis of such data, Proust, Dalton, Gay-Lussac, Avogadro, and other scientists at the beginning of the nineteenth century revealed that *elements combine in integer ratios that convincingly imply the existence of atoms.*

Proust's Evidence: The Law of Constant Proportions

The chemist Joseph-Louis Proust observed that when chemicals combine to make a particular substance, the reacting elements always make up the same percentage of the weight of the new substance. For example, stannous oxide always is made up of 88.1% tin (Sn) and 11.9% oxygen (O); and stannic oxide always is made up of 78.7% Sn and 21.3% O. This behavior suggests that a fixed amount of tin can combine only with one or another fixed amount of oxygen. This kind of constancy was verified by many careful measurements on many different substances. Table 3.1 shows the constant proportions of the weights measured by Sir Humphrey Davy for three different compounds of nitrogen (N) and oxygen. Because Proust was the first to notice that all chemicals are made up of fixed percentages of their constituents, this behavior is known as Proust's law of constant proportions.

Dalton's Evidence: The Law of Multiple Proportions

It was by looking at the above data in a different way that John Dalton obtained the first convincing scientific evidence that substances are built up out of small, individual blocks of matter. Dalton's work was first published in 1808.[1]

Instead of dealing in percentages, Dalton first asked what weights of one element would combine with a fixed amount of another. Thus, he found that 13.5 g of oxygen would combine with 100 g of tin to make stannous oxide, while 27 g of oxygen would combine with 100 g of tin to make stannic oxide. (Nowadays, we would write that he was observing the formation of SnO and SnO_2.) Then he examined the ratio of the two different amounts of oxygen. Quite obviously, the ratio is one of simple integers, 1:2. Examining many cases, he found that always when different amounts of one element combine with another to make different chemical compounds, the amounts are in simple integer ratios to each other.

Dalton realized that this law of multiple proportions, as he called it, was evidence that chemicals combined in integer multiples of some basic unit, or *atom*. His data could be explained if an integer number—one or two—of atoms of one element combined with an integer number—one, two, three, or more—of atoms of another element. More important, Dalton's way of

TABLE 3.1 Davy's Percentage of Mass of Nitrogen and Oxygen in Oxides of Nitrogen

Compound Gas (Modern Names)	% Mass of Nitrogen	% Mass of Oxygen
Nitrous oxide	63.30	36.70
Nitric oxide	44.05	55.95
Nitrogen dioxide	29.50	70.50

[1] See *From Atomos to Atom* by Andrew G. Van Melsen, Harper & Brothers, New York, 1960. This book quotes extensively from a later edition of Dalton's work *A New System of Chemical Philosophy*, London 1842.

comparing the masses of chemicals before and after combination was a quantitative method that could be used by other scientists and applied to many chemical reactions. On the basis of such analysis he proposed the following ideas.

Dalton's ideas about atoms:

- All elements consist of minute discrete particles called atoms
- Atoms of a given element are alike and have the same mass
- Atoms of different elements differ, each element having unique atoms
- Chemical changes involve the union or separation of *undivided atoms* in fixed simple numerical ratios. Atoms are not created or destroyed when chemical change occurs.

EXAMPLES

1. A striking example of the law of multiple proportions can be obtained from Davy's data in Table 3.1 on the combinations of nitrogen and oxygen if you look at them from Dalton's point of view. To do this, let's calculate how much oxygen reacts with a fixed amount of nitrogen, say 100 g. The first line of the table states that nitrous oxide is 63.3% nitrogen and 36.7% is oxygen. To scale the amount of nitrogen to 100 g, just multiply by 100/63.3; use this same scale factor to get the amount of oxygen

$$\frac{100}{63.3} \times 36.7 = 58\,\mathrm{g}.$$

A similar procedure applied to the next two lines in the table implies that 127 g and 239 g of O combine with 100 g of N to form nitric oxide and nitrogen dioxide, respectively. It is the near-integer ratios of these different amounts of oxygen reacting with the fixed amount of nitrogen that supports the idea of atoms:

$$58 :: 127 :: 239 \approx 1 :: 2 :: 4.$$

Although Davy's data deviate somewhat from simple integer ratios, later more careful measurements confirm Dalton's law of multiple proportions very well. More complete studies show that 100 g of nitrogen will combine with 57 g or 113 g or 171 g or 229 g or 286 g of oxygen. These are respectively the compounds of nitrous oxide, nitric oxide, nitrous anhydride, nitrogen dioxide, and nitric anhydride.

EXERCISES

1. What are the "simple proportions" of these quantities? What would you guess to be the chemical formulas for these compounds? Hint: nitric oxide is NO.

2. Given that the density of nitrogen at one atmosphere of pressure and 0°C is 1.2506 kg m^{-3} and that of oxygen at the same temperature and pressure is 1.429 kg m^{-3}, calculate the relative volumes of the two gases that combine to form the compounds named above. If you do this problem correctly, you will discover what Gay-Lussac discovered. These results are shown in Table 3.2.

Gay-Lussac's Evidence: The Law of Combining Volumes

The appearance of simple integer relationships is even clearer in an interesting version of Dalton's law that was discovered by the chemist and physicist Joseph Louis Gay-Lussac (1778–1850). He studied the chemical combination of gases and observed that at the same temperature and pressure their *volumes* combined in ratios of small integers. For example, he found that 100 cm^3 of oxygen combined with 198.6 cm^3 of hydrogen. Within the uncertainty of his measurements, this is a ratio of 1:2. Using contemporary values of densities, he also converted Davy's data on the various compounds of nitrogen and oxygen (Table 3.1) to volumes. Table 3.2 shows the striking result, which is known as the "law of combining volumes."

Clearly the volumes of these gases combine in simple ratios of small integers like 2:1 and 1:1. Furthermore, for gases, the simplicity of the ratio shows up in a single reaction. There is no need to examine different compounds of the same elements, as is required in order to exhibit Dalton's law of multiple proportions. The implication is plausible that the gases react by combining simple building blocks in whole-number, i.e., integer, amounts.

TABLE 3.2 Combining Volumes of Nitrogen and Oxygen

Compound Gas	Volume of Nitrogen	Volume of Oxygen	Final volume (approximate)
Nitrous oxide	100	49.5	100
Nitric oxide	100	108.9	200
Nitrogen dioxide	100	204.7	200

When Gay-Lussac performed experiments himself, he observed that not only did the gases combine in simple proportions, but the final volume of the reacted gas was simply related to the combining volumes. For example, when two volumes of hydrogen combined with one volume of oxygen, the resulting water vapor occupied two volumes. It is simple to interpret this result as two atoms of hydrogen combining with one atom of oxygen to make a water molecule. But if that were so, the end product should occupy only *one* volume of water. The appearance of *two* volumes of water was, therefore, puzzling.

A similar puzzle occurs with the combination of nitrogen and oxygen. If you combine 100 cm^3 of nitrogen gas with 100 cm^3 of oxygen you get 200 cm^3 of nitric oxide gas, as suggested in Table 3.2 after rounding off the oxygen volume. Apparently, the combination of one elemental unit of nitrogen gas with one elemental unit of oxygen produces two such units, or *molecules*, of nitric oxide gas. This is strange because one *atom* of nitrogen combining with one *atom* of oxygen should produce just one molecule of nitric oxide.[2]

How can this be? In 1811 the Italian chemist Amadeo Avogadro suggested the solution to the puzzle.[3] He proposed that the volumes of hydrogen, nitrogen, and oxygen contained molecules made up of two identical atoms: H_2, N_2, O_2. His idea helped us to understand that molecules are structures made up of atoms.

EXERCISES

3. Gay-Lussac's work made it possible to determine chemical formulas. To see how, use Avogadro's idea and Gay-Lussac's data (Table 3.2) to find

[2]The oxides of nitrogen are confusing. Here is a list of their modern names and chemical formulas. Don't they make a nice application of Dalton's law of multiple proportions?

Nitrous oxide	N_2O
Nitric oxide	NO
Dinitrogen trioxide	N_2O_3
Nitrogen dioxide	NO_2
Dinitrogen tetroxide	N_2O_4
Dinitrogen pentoxide	N_2O_5

[3]A. Avogadro, "Essay on a Manner of Determining the Relative Masses of the Elementary Molecules of Bodies and the Proportions in Which They Enter into These Compounds" *Journal de Physique*, 1811, as excerpted in *Readings in the Literature of Science*, ed. W.C. Dampier and M. Dampier, Harper Torchbook, Harper & Brothers, New York, 1959. You might be interested to know that Avogadro's idea that hydrogen, nitrogen, and oxygen were molecules consisting of two atoms of the same kind was not accepted for several decades. His contemporaries, among them Dalton and Gay-Lussac, considered the idea too strange to be likely.

the chemical formula for nitrogen dioxide, which he called "nitric acid." (You see that it is quite different from the HNO_3 that we today call nitric acid.)

4. What are the chemical formulas for the other two compounds in Table 3.2?

Avogadro's Principle

Avogadro and the Swedish chemist Jons Berzelius realized that if volumes of gases (at the same temperature and pressure) always combine in simple ratios, then equal volumes of gas contain equal numbers of molecules. To see why this is so, consider our earlier example of two volumes of hydrogen combining with one volume of oxygen to form two volumes of water. You can read the chemical equation

$$2H_2 + O_2 \rightarrow 2H_2O$$

as though it is a statement about the combination of volumes, but you can also read it as a statement that two molecules of hydrogen combine with one molecule of oxygen to form two molecules of water. *There is a one-to-one correspondence between volumes and numbers of molecules.* There must be twice as many molecules in two volumes as in one. Because this statement does not depend on the kind of molecule, we, like Avogadro, conclude that equal volumes of gases (at the same pressure and temperature) contain equal numbers of molecules independent of their kind. This is a very useful feature of gases, and you should remember it.

EXERCISES

5. Make an argument like the one above, only do it for combinations of nitrogen and oxygen instead of hydrogen and oxygen.

The full and precise statement of Avogadro's principle is this:

> Equal volumes of all gases, under the same conditions of temperature and pressure, contain the same number of molecules.

Obviously, the inverse is also true: At a given pressure and temperature equal numbers of molecules occupy equal volumes.

3.4 ATOMIC WEIGHTS

The law of combining volumes and Avogadro's principle enable us to determine the relative masses of atoms. They make it possible to establish a table of atomic weights that will be useful to us. Let's see how this works out.

From the fact that equal volumes hold the same number of gas molecules, it follows that the ratio of the masses of these volumes must equal the ratio of the masses of the individual molecules. Imagine a 1-liter jar containing oxygen at atmospheric pressure and from which you can pump out the gas. Weigh the jar full; then weigh it empty. You will find that it contains 1.429 g of oxygen. If you replace the oxygen gas with xenon gas and repeat the weighings, you will find that the jar contains 5.86 g of xenon. Because each volume contains the same number of molecules, it follows that the ratio of the mass of a xenon molecule to the mass of an oxygen molecule is $5.86/1.429 = 4.10$.

It takes additional experimentation to learn the atomic composition of a molecule. We have already seen how Avogadro concluded that gaseous oxygen is diatomic, O_2. Some gases are more complicated, e.g., carbon dioxide is CO_2, and some are simpler, e.g., when xenon was discovered at the end of the nineteenth century, it was quickly apparent that it is monoatomic Xe.

EXAMPLES

2. Once we know the atomic composition of the gas molecules, it is easy to see that an atom of xenon must have a mass that is $5.86/(1.429/2) = 8.20$ times the mass of an oxygen atom.

A complete set of relative masses of atoms can be established by weighing equal volumes of different gases and by other techniques. For convenience, physicists and chemists have set up a standard scale of relative masses. Because relative masses are experimentally determined using various chemical combinations of atoms, it is helpful to choose for a standard mass some kind of atom that chemically combines with many others and is convenient to use. Carbon is such an atom, and by international agreement, the scale of relative masses of the chemical elements has been set by assigning to carbon a chemical atomic weight of $m_C = 12.011$ u.

The unit mass of this scale is called the "atomic mass unit" and is abbreviated u. By assigning an atomic mass of about 12 u to carbon, the lightest element, hydrogen, has an atomic mass of about 1 u; the masses of most of the other elements also come out to be nearly integers. Why the standard chemical atomic weight is 12.011 u and not some nice integer like 12.00 will

TABLE 3.3 Some Chemical Atomic Weights and Gas Densities

Element	Symbol	Chemical Atomic Weight[a]	Gas	Density at STP (kg/m^3)
hydrogen	H	1.00794	Air	1.293
helium	He	4.002602	O_2	1.429
lithium	Li	6.941	N_2	1.251
beryllium	Be	9.01218	Cl_2	3.21
boron	B	10.811	H_2	0.0899
carbon	C	12.0111	CO_2	1.965
nitrogen	N	14.0067		.
oxygen	O	15.9994		.
fluorine	F	18.9984		.
neon	Ne	20.179		.
chlorine	Cl	35.453		.
argon	Ar	39.948		.
krypton	Kr	83.80		.
xenon	Xe	131.30		.
radon	Rn	220.		.

[a] These data are taken from *The Chart of the Nuclides*, 13th ed., revised to July 1983, by F. William Walker, Dudley G. Miller, and Frank Feiner, distributed by General Electric Company, San Jose, CA.

be explained later when we discuss isotopes and the physical scale of atomic masses. Masses of atoms measured in atomic mass units are often called "atomic weights." Masses of molecules in these units are called "molecular weights."

Table 3.3 lists some elements and the masses of their atoms relative to carbon's chemical atomic weight of 12.011 u. You should know the nearest integer values of the atomic weights of H, C, N, and O.

Example 2 shows you how to use Avogadro's principle to find the ratio of the masses of two different molecules, Xe and O_2. Exercises 3.3 and 3.4 ask you to find the chemical composition of a molecule using the law of combining volumes. Together these two pieces of information can be used to determine an unknown atomic mass.

For example, to find the atomic mass of oxygen you could use the results of another of Gay-Lussac's experiments. Two volumes of CO were observed to combine with one volume of O_2 to form two volumes of another carbon–oxygen compound. The results imply that this compound is CO_2.

EXERCISES

6. Show that the experimental results imply the product of the reaction to be CO_2.

From Avogadro's principle and measured densities it is easy to find the ratio of the molecular weights of CO_2 and O_2. The densities of these two gases are such that at room temperature and pressure a cubic meter of O_2 has a mass of 1.429 kg and a cubic meter of CO_2 has a mass of 1.965 kg. The ratio of these two numbers is the ratio of the molecular weights.

EXAMPLES

3. Here is a direct way to find the atomic mass of oxygen from these data. Using the atomic idea greatly simplifies the analysis.

Avogadro's principle tells you that equal volumes contain equal numbers of molecules. In our example one liter of O_2 has a mass of 1.429 g, and one liter of CO_2 has a mass of 1.965 g. Imagine separating the CO_2 molecules into C atoms and O_2 molecules. Then the liter would contain 1.429 g of O_2 and $1.965 - 1.429 = 0.536$ g of C. The molecular weight of O_2 must then be

$$\frac{1.429}{0.536} \times 12.011 = 32.0\,\mathrm{u},$$

and since we know that the oxygen molecule contains 2 atoms, it follows that the atomic weight of oxygen is 16.0.

EXERCISES

7. Using the data given earlier for the combining masses of gases of nitrogen and oxygen, find the atomic mass of nitrogen.

You can use atomic masses to determine the number of atoms in a sample of one element relative to the number in a sample of a different element. The simplest case occurs when the samples have masses in the same ratio as their atomic masses. Such samples contain the same number of atoms. Thus 65.65 g of xenon has the same number of atoms as 110 g of radon or 8 g of oxygen.

EXERCISES

8. What masses of CO_2 and O_2 would contain the same number of molecules? This question and the next are for practice in distinguishing between molecules and atoms.

9. What masses of CO_2 and O_2 would contain the same number of atoms?

3.5 THE MOLE

One gram of hydrogen contains the same number of atoms as 12 grams of carbon or 16 grams of oxygen. In general, an atomic weight in grams of any element contains the same number of atoms as an atomic weight in grams of any other element. This useful property is a direct consequence of the idea of atomic weights: Because they are defined as the relative weights of atoms, the masses of any two elements that have the same ratio as their atomic weights must contain the same number of atoms. More specifically, if a lump of a chemical element has a mass M in grams equal to its atomic weight, the mass M is called its "gram atomic weight." In a similar way a mass of molecules equal in grams to the molecule's molecular weight is called its "gram molecular weight."

How many atoms are there in a gram atomic weight of an element? How many molecules are there in a gram molecular weight of any particular molecule? We know the number pretty well now, but finding it was an important question during the nineteenth century. It was only precisely answered in the early years of the twentieth century. The answer is 6.022×10^{23}.

This number is quite important. Knowing it enables us to find the mass and the size of an individual atom or molecule. It is so important that it has two common names. First, it is known as "Avogadro's constant." We have already referred to it and write it as N_A. Second, it is the number that defines a quantity called a "mole," abbreviated "mol."

"Mole" is the name for any collection of 6.022×10^{23} objects. We say that 12.011 g of carbon contains 1 mole of carbon atoms. Forty-four grams of CO_2 contains 1 mole of CO_2 molecules. It also contains 1 mole of C atoms, 1 mole of O_2 molecules, 2 moles of O atoms, and 3 moles of atoms of all kinds. The word mole is used the way you use the word "dozen" or "score."

EXAMPLES

4. One way to remember the mass of Earth is that it is 10 moles of kilograms.

EXERCISES

10. Rewrite the opening line of the Gettysburg Address to use "moles" instead of "score."

Fairly often we need to refer to the number of moles of something. In these notes we use the symbol n_M to represent the number of moles. Try not to confuse n_M with the symbol we use to tell how many of something there are in a unit volume. We call the number of something per unit volume "the number density" and write it as n. For example, we noted above that equal volumes of any gases at the same temperature and pressure contain equal numbers of molecules. At 0°C and atmospheric pressure the number density turns out to be $n = 2.7 \times 10^{19}\ \mathrm{cm}^{-3}$.

EXERCISES

11. How many molecules of gas are there in 11.2 liters under the above conditions? ($1\ \mathrm{L}=10^3\ \mathrm{cm}^3$)

12. What is the number of moles n_M of gas molecules in 11.2 liters of gas under the above conditions?

13. How many liters of gas does it take to hold a mole of molecules at the above temperature and pressure?

14. If you have $n_M = 3.2$ moles of gas molecules in 4 liters of volume, what is the number density of this gas?

15. If the molecular weight of the above gas is 28 u, what is the mass density of the gas in the previous problem?

Once you know Avogadro's constant it is easy to find the mass of an atom. From the definition of N_A we know that 18.998 g of fluorine contains 6.02×10^{23} atoms. Therefore, one fluorine atom has a mass of $18.998/(6.02 \times 10^{23}) = 3.16 \times 10^{-23}$ g.

EXERCISES

16. What is the mass of a hydrogen atom? A Be atom?

17. From the density of liquid water, estimate the diameter of a water molecule.

It took nearly a hundred years to find ways to measure N_A to three significant figures. It was important to find its value because it relates the scale of atomic sizes to the macroscopic scale of our everyday world. The next few chapters will have as one important theme the gradual development of a precise value for N_A.

3.6 THE CHEMIST'S ATOM

Summary

In the space of a few years around 1810 atoms were established as an important concept in chemistry and physics. These atoms have some interesting properties. First, there are different kinds, one for each "chemical element." "Chemical element" is defined operationally: If the mass and chemical behavior of a given substance are not changed by heat, electricity, or other chemical reactions, it is said to be a chemical element. By this criterion chemists identified nearly sixty chemical elements before 1850. We now know of 112 elements, 22 of which do not occur naturally on Earth but can be made in the laboratory or in nuclear explosions.

Second, the atoms connect to one another. The atoms seem to have "hooks." For instance, a hydrogen atom has one hook, so two atoms of H can be hooked together to make H_2. Oxygen has two hooks, so we can hook a hydrogen atom on to each and make H_2O, or we can hook an H atom and an O atom onto one O, and another H onto the second O, and make hydrogen peroxide, H_2O_2. The somewhat whimsical diagram in Fig. 3.1 gives the idea. Chemists call the number of hooks the "valence" of the atom.

Questions

Much of chemistry in the first two-thirds of the nineteenth century was the sorting and describing of chemical compounds and reactions. The idea of atoms was used by many chemists to produce useful ideas and good science, but it raised as many questions as it answered.

What are these "hooks" that connect one atom to another? Why do some elements have several different valences? For example, nitrogen seems to have valences of 1, 2, 3, 4, or 5 depending upon circumstances. What makes one element different from another? Hydrogen is very reactive; helium is

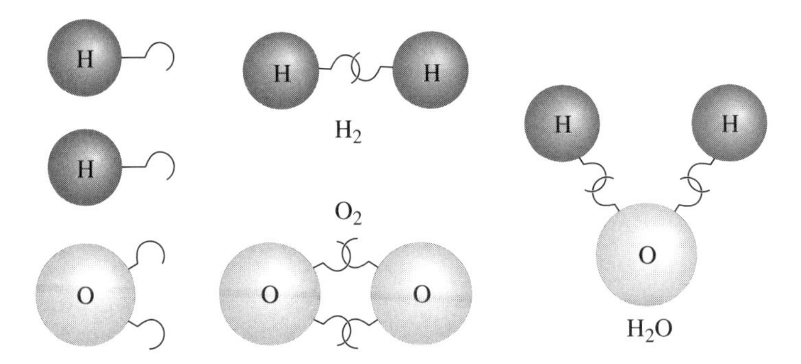

FIGURE 3.1 Ball-and-hook model of simple chemical bonds.

not. Yet both are gases and very light in mass. What makes some elements surprisingly similar to others? Lithium, sodium, potassium, cesium, rubidium, francium all show similar chemical behavior.

These questions are related to more general, deeper questions: Do atoms have internal parts, i.e, do they have structure? If so, what are the insides of atoms like? How do the parts connect? What forces hold them together? How does the behavior of the parts explain the similarities and differences of atoms?

Answers

It took physicists over one hundred years to find satisfactory answers to these difficult questions. The answers required surprising elaboration of our concept of the atom itself and radical changes in our formulations of the principles that govern the behavior of matter. The central material of this book will be the physics needed to begin answering these questions. But before looking inside the atom, let's examine the very simple model of the atom that physicists used in the nineteenth century to answer two basic questions: How big is an atom, and what is its mass? The answers are closely related to the value of Avogadro's number, i.e., the number of atoms in a mole.

These questions were hard to address directly because until very recently the extremely small size of atoms made it impossible to observe them individually. The first answers were obtained by studying large numbers of identical atoms or simple molecules in gases at low pressures, where they are separated by relatively large distances and therefore unaffected by each other. If the physical behavior of a gas depends only on the properties of the individual atoms or molecules of which it is composed, then by studying gases we can hope to learn something about the individual atoms and molecules themselves. The next two chapters trace the connection between atoms and the nature of gases.

PROBLEMS

1. The table below on the left shows hypothetical data obtained from a chemical reaction

$$X + Y \rightarrow A, B, C$$

in which three different compounds A, B, and C are formed from the reaction of elements X and Y.

Proust's law of constant proportions

	X	Y
A	57%	43%
B	73%	27%
C	47%	53%

Dalton's law of multiple proportions

	X	Y
A	100 g	
B	100 g	
C	100 g	

a. To exhibit Dalton's law of multiple proportions find the mass of Y that interacts with 100 g of X and fill in the blanks in the table. Explain how these results illustrate Dalton's law.

b. From your results write down possible expressions for the chemical compounds A, B, and C. For example, X_2Y_5.

2. Explain Gay-Lussac's law of combining volumes. Why does this law support the existence of atoms?

3. Why when 100 cm^3 of nitrogen gas is combined with 100 cm^3 of oxygen gas do you get 200 cm^3 of the compound gas at the same pressure and temperature?

4. How did Avogadro explain that when 200 cm^3 of hydrogen at STP is combined with 100 cm^3 of oxygen at STP they form 200 cm^3 of water vapor at the same temperature and pressure?

5. Dalton noticed that 100 g of tin combined with either exactly 13.5 g or exactly 27 g of oxygen. He said that this result was evidence for the existence of atoms.

a. Why? Explain his reasons.

b. Take the atomic weight of oxygen to be 16. From Dalton's data calculate the atomic weight of tin. State what you assume in order to get an answer.

6. Observer A notes that water always consists of 11.1% hydrogen and 88.9% oxygen by weight, while hydrogen peroxide always consists of 5.88% hydrogen and 94.1% oxygen by weight. Observer B notes that in the formation of water 800 g of oxygen always combines with 100 g of hydrogen, while in the formation of hydrogen peroxide 1600 g of oxygen combines with 100 g of hydrogen.

a. Whose observations illustrate the law of

i. combining volumes?

ii. constant proportions?

iii. multiple proportions?

b. Show how to deduce what B sees from A's observations.

c. Explain in what way B's observations suggest that atoms exist.

7. Stearic acid is described by the chemical formula $C_{17}H_{35}COOH$.

a. Find the mass of 1 mole of stearic acid in grams.

b. Use the fact that there are 2.1×10^{21} molecules in a gram of stearic acid and your answer to (a) to find a value of Avogadro's constant (N_A).

c. If the density of stearic acid is 8.52×10^{2} kg m^{-3}, how many atoms are in 1 cm^3?

8. Obeying Proust's law of constant proportions, stannous oxide is always 88.1% tin and 11.9% oxygen by weight; stannic oxide is 78.7% tin and 21.3% oxygen. Show that these numbers imply that in forming stannous oxide 100 g of tin combines with 13.5 g of oxygen.

9. Certain volumes of nitrogen gas and hydrogen gas react to form 4 L of ammonia gas, according to the reaction shown below.

$$N_2 + 3H_2 \rightarrow 2NH_3.$$

The atomic masses of nitrogen *atoms* and hydrogen *atoms* are 14.0 u and 1.0 u, respectively. The mass of NH_3 formed in this reaction is 3.4 g.

a. How many moles of NH_3 were formed?

b. How many hydrogen *atoms* took part in the reaction?

c. Determine the initial volumes of N_2 and H_2.

10. What is a "mole" as the term is used in physics and chemistry? Give both qualitative and quantitative answers.

11. You wish to determine the identity of element X in the gaseous chemical reaction

$$X_2 + 3\,H_2 \rightarrow 2XH_3.$$

The densities of H_2 and XH_3 are measured to be 8.4×10^{-5} g cm^{-3} and 7.14×10^{-4} g cm^{-3}, respectively.

a. Find the mass of H_2 in a 30 L volume of gas.

b. If you react enough X_2 to combine with 30 L of H_2, what volume of XH_3 is generated?

c. From the information given, determine the atomic weight of element X and identify the element.

12. The chemical reaction for making nitrogen dioxide is given by

$$2\,O_2 + N_2 \rightarrow 2\,NO_2.$$

Suppose we combine nitrogen and oxygen to make 23 g of NO_2.

a. How many moles of NO_2 does that correspond to?

b. How many moles of N_2 are required?

c. How many molecules of O_2 are required? Express your answer in terms of N_A.

d. If the reaction produces 10 L of NO_2, what are the volumes of O_2 and N_2? Explain your reasoning.

C H A P T E R

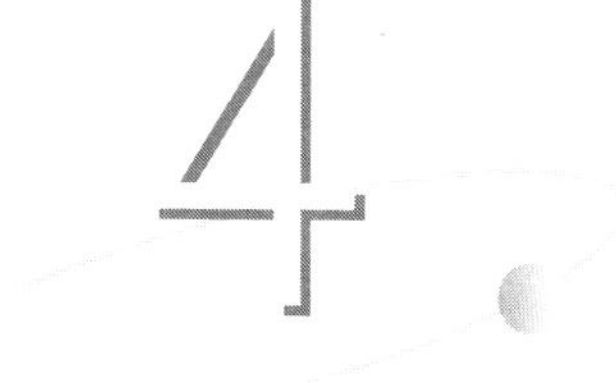

Gas Laws

4.1 INTRODUCTION

The last chapter reviewed the chemical evidence for atoms that was discovered at the beginning of the nineteenth century. In the decades that followed physicists asked, What are the basic properties of atoms? How are atoms alike? How are they different? What properties must they have to produce the observable properties of matter?

Until recently it has not been possible to trap or otherwise follow the progress of individual atoms, and we could study them only in large collections. Gases of simple atoms or molecules are especially suitable for such studies because they behave in unusually simple ways. Gases flow with only a little resistance. Unlike solids and liquids they can be compressed and expanded easily by applying or relaxing pressures, and small changes in temperature will produce large changes in volume or pressure. The temperature, pressure, and volume of gases turn out to be related by a simple mathematical function. It was a major triumph of physics in the mid-nineteenth century to understand these properties and how they relate to each other in terms of a very simple model of the atom. Before discussing that model and its physics, let's review both pressure and the gas laws.

4.2 PRESSURE

The Idea of Pressure

Imagine a tiny, empty hole in a liquid. Your intuition correctly tells you that on Earth the weight of the surrounding fluid will cause it to squeeze in and

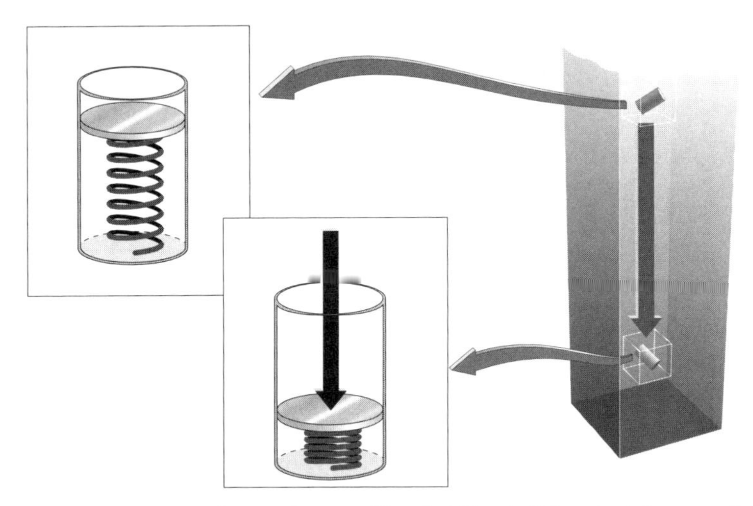

FIGURE 4.1 A tiny evacuated can submerged in a fluid has its lid pushed down until the spring under it compresses enough to exert a force just equal to that exerted by the fluid outside the can. The deeper the can the more the spring is compressed.

make the hole disappear. This squeeze, which will be from all directions, is what we call pressure. It is what you feel in your ears when you swim under water.

Now imagine that instead of the hole, there is in the fluid a tiny, completely empty can with a lid that slides in and out of the can like the piston in the cylinder of an automobile engine. If there is really nothing in the can, the weight of the column of liquid and atmosphere above the can will push the lid to the bottom of the can. But suppose, as in Fig. 4.1 there is a tiny spring between the lid and the bottom of the can. Then as the weight of the liquid and atmosphere pushes the lid into the can, the spring compresses and pushes back. The lid will slide in only to the point where the force of the spring equals the weight of the liquid and atmosphere.

If you move the can deeper into the water, the spring compresses more. If you raise it to a shallower depth, the weight of water on the lid diminishes and the lid moves away from the bottom of the can. If you hold the can at any given depth and turn it around in different directions, the compression of the spring does not change because, as noted above, the squeeze of the fluid is the same in all directions.

If you use a smaller can with a lid of smaller area but the same size spring, then at any given depth the spring under the smaller lid will be compressed less than the spring under the larger lid. This is because the force squeezing the lid into the can is equal to the weight of the column of matter above the can, and the lid of the bigger can with its bigger area has a larger and heavier column of fluid above it than the smaller lid has.

Definition of Pressure

As always in physics, we want a way to measure the squeeze, to assign a number to it. We could use a device like the little can: The amount the spring is compressed is a direct measure of how much squeeze there is. But that is not so good, because, as we just saw, the observed result depends on how big the can is. A better measure is the ratio of the force to the area of the can cover. This will work because even though the force on the can and the amount of compression of the spring goes to zero as the area of the lid goes to zero, the ratio of force to area approaches a definite value. This is both conceptually and practically useful, and therefore, we define pressure P to be the ratio of the force F to the small area A on which it acts perpendicularly,

$$P = \frac{F}{A}.$$

Because of this definition, we use "force per unit area" as a measure of pressure.

Discovery of Vacuum and the Atmosphere

Around the year 1640 Evangelista Torricelli produced the first vacuum, i.e., a volume from which matter—including air—has been excluded, and showed the existence of the pressure of the atmosphere. His technique was very simple. He poured mercury into a glass tube closed at one end, put his thumb over the open end, inverted the full tube, put the open end into a bowl of mercury and removed his thumb. Without flowing entirely out of the tube, the mercury fell away from the tube's closed end and left an empty space, a vacuum. As shown in Fig. 4.2, a column of mercury about 760 mm high remained standing in the tube above the level of the mercury in the bowl.

Torricelli realized that the empty space above the mercury column was a vacuum; he also realized that the weight of the column of mercury was being balanced by a force produced by the pressure of the atmosphere arising from the weight of the atmosphere pushing on the surface of the mercury in the bowl. This pressure, transmitted through the mercury, produces a force over the cross section of the tube at the base of the column just equal to the weight of the mercury in the tube. Torricelli realized that he was observing the balance between the weight of a column of mercury and the weight of a column of air. This meant, as he put it, that we all live at the bottom of an ocean of air.

It is remarkable that the height h of the column for which this balance occurs does not depend on the column's cross-sectional area A. This is so because pressure is force per unit area. Suppose the cross-sectional area of the tube was $10\,\mathrm{cm}^2$. The weight of mercury supported by the force of the

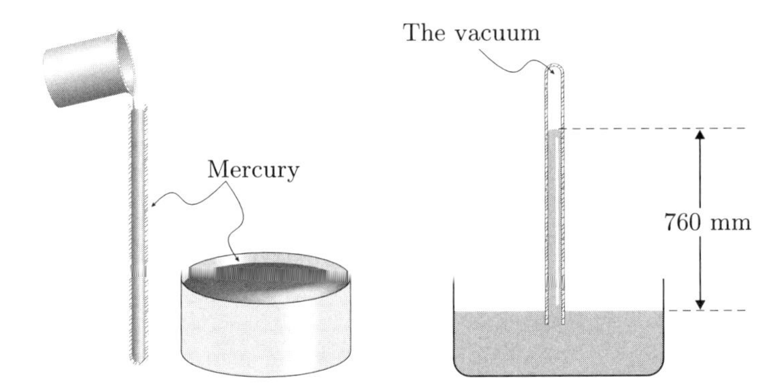

FIGURE 4.2 Apparatus for producing a Torricellian vacuum.

air would then be the mass of the mercury in the tube times the acceleration of gravity, g. The mass of the mercury is just its volume (that is, its height of 0.76 m times its cross-sectional area of $10^{-3}\,\mathrm{m}^2$) times its density, $\rho = 13.6 \times 10^3\,\mathrm{kg\,m}^{-3}$. The force that balances this volume must be

$$0.76 \times 10^{-3} \times 13.6 \times 10^3 \times 9.80 = 101\,\mathrm{N}.$$

If we made the cross section of the tube 10 times bigger, i.e., $100\,\mathrm{cm}^2$, the force necessary to balance the mercury would need to be 10 times larger. But because the area is bigger, the force exerted across that surface by the pressure of the atmosphere would also be bigger by that same factor of 10, and the same height of column would be balanced.

Let's express this using algebra. Call the height of the column h, the cross-sectional area A, the density of mercury ρ, the acceleration of gravity g, and the pressure of the air P. The weight of the mercury is its volume hA times its density ρ times g. Now, the force exerted by the air is $F = PA$, but since F balances the weight of the mercury, it must be true that $F = hA\rho g = PA$. From this follows the important fact that

$$P = \rho g h \tag{1}$$

because A divides out from both sides. This result means that regardless of the size of the mouth of the tube, atmospheric pressure will support the same height h of mercury. This height h is a direct measure of the magnitude of atmospheric pressure, and Eq. 1 is often called "the barometer equation."

From h we can calculate directly what the atmospheric pressure is:

$$P = 13.6 \times 10^3 \times 9.8 \times .76 = 101 \times 10^3\,\mathrm{N\,m}^{-2},$$

where you see that the units are force divided by area.

EXERCISES

1. Check for yourself that the units in the above calculation are, indeed, force divided by area.

2. Suppose on a day when the atmospheric pressure is $101\,\mathrm{kN\,m^{-2}}$ you inflate a balloon to a diameter of 20 cm. Assuming that the elastic force of the balloon rubber is small enough to neglect, what will be the size of the force pushing out on $1\,\mathrm{cm}^2$ of the balloon? What will be the size of the force pushing in on $1\,\mathrm{cm}^2$?

Several different kinds of units are used to measure pressure.

$\mathbf{N\,m^{-2}}$ The most straightforward is $\mathrm{N\,m^{-2}}$.

pascal The combination of units $\mathrm{N\,m^{-2}}$ is given its own special name, pascal, abbreviated Pa. Thus, we might say that atmospheric pressure is about 101 kPa, where as usual we use the prefix "k" to mean "kilo" or 10^3.

mm Hg Often U.S. physicists measure pressure in terms of the height of the column of mercury that it would balance. Often the units are given as millimeters of mercury, or mm Hg. Thus we say that atmospheric pressure is 760 mm Hg. If you need to convert this to more conventional units, you must multiply by the density of mercury and the acceleration of gravity g.

Torr The mm Hg is also given its own name, the torricelli, or Torr. This unit is frequently used for measuring low pressures inside vacuums. It is rather easy to obtain a vacuum of 10^{-3} Torr; a good vacuum of 10^{-6} Torr usually requires more than one kind of pump; an ultra-high vacuum of 10^{-10} to 10^{-12} Torr is routine but expensive and tedious to achieve, especially in large volumes.

in Hg U.S. weather reporters usually give barometric pressure in inches of mercury. The conversion from mm Hg to in Hg is obvious: Divide by 25.4. Then 760 mm Hg becomes 29.9 in Hg, a number you often hear from your TV weatherman.

bar There is another unit of pressure developed by meteorologists called the "bar." It is *almost* equal to the pressure of one atmosphere: 1 bar = 100 kPa exactly. All the sub- and supermultiples are used: mbar, kbar, Mbar, etc.

atm Finally, pressures are often measured in "atmospheres." By convention, a "standard atmosphere" is defined to be 760 mm Hg.

psi In American engineering practice pressures are measured in pounds per square inch, or psi, or p.s.i. Atmospheric pressure in these units is about 14.7 psi. In physics we sometimes write this as 14.7 lb/in^2.

gauge pressure Quite often, high pressures in tanks are measured as the *difference* between the internal tank pressure and the external atmospheric pressure. These are called "gauge" pressures. Sometimes the pressure-measuring device will tell you that the pressure being measured is gauge pressure by giving the units as psig or p.s.i.g., but frequently it won't.

Gas Pressure

Gases in closed containers exert pressure on the container's walls. A simple experiment shows that this is so. Put an air-filled balloon inside a glass jar, pump air from the jar, and the balloon will expand; let air back into the jar, and the balloon contracts. The balloon is like the little can talked about above, and the gas in it acts like the spring, except that now the "spring" pushes in all directions against the rubber surface of the balloon with some force on a unit area, say $101 \times 10^3\ \mathrm{N\,m^{-2}}$. As long as the force on a unit area due to the air pushing on the outside of the balloon plus the force due to the stretching of the balloon's rubber is equal to the force on a unit area due to the air inside, the balloon stays the same size. When air is pumped out of the jar, the external pressure on the balloon decreases, and the balloon expands.

What causes the gas to push out like a spring? As you shall soon see, the pressure of a contained gas arises from the momentum and energy of its moving atoms and molecules. But in order to connect the behavior of contained gases to the behavior of atoms, we first need a quantitative description of how the pressure of a gas depends on its temperature and the volume of its container.

4.3 BOYLE'S LAW—THE SPRINGINESS OF GASES

A modified form of Torricelli's barometer was used by the English physicist Sir Robert Boyle to perform what he called "Two New Experiments Touching the Measure of the Force of the Spring of Air Compressed and Dilated."[1]

[1]Reprinted in *A Treasury of World Science*, ed. Dagobert Runes, Philosophical Library, New York, 1962.

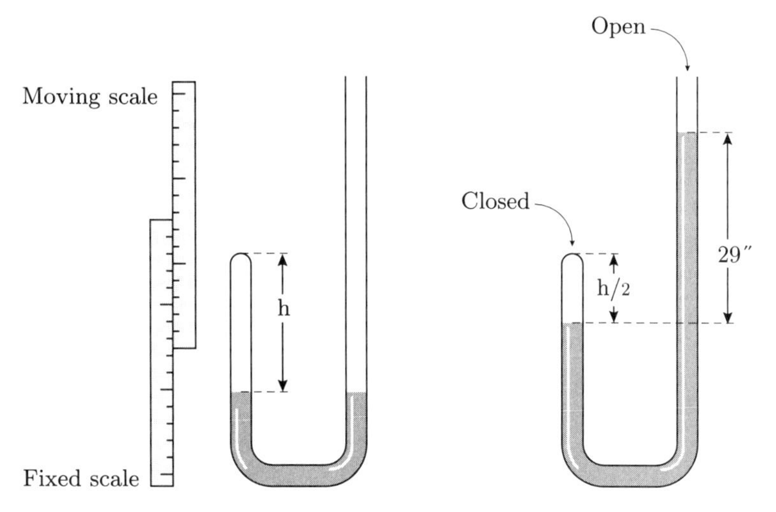

FIGURE 4.3 Apparatus for establishing Boyle's law.

Boyle's Experiment

Boyle made a J-shaped tube sealed at one end and open at the other. He pasted a scale along both arms of the tube (only one scale is shown in Fig. 4.3). He poured mercury into the open end and trapped a small volume of air under the closed end. He measured the volume of the trapped air by measuring the height of the column of air between the mercury and the end of the sealed tube, and he measured the pressure of the trapped air by measuring the height of the column of mercury in the open end. He varied the pressure by pouring in more mercury to compress the trapped gas. He also did a separate experiment in which he pumped out some air from the open end of the tube and saw the trapped volume of air expand. In his own words:

> ... we took care, by frequently inclining the tube, so that the air might freely pass from one leg into the other by the sides of the mercury (we took, I say, care) that the air at last included in the shorter cylinder should be of the same laxity with the rest of the air about it. This done we began to pour quicksilver into the longer leg of the siphon, which by its weight pressing up that in the shorter leg, did by degrees streighten the included air: and continuing this pouring in of quicksilver till the air in the shorter leg was by condensation reduced to take up by half the space it possessed ... before; we cast our eyes upon the longer leg of the glass, on which was likewise pasted a list of paper carefully divided into inches and parts, and we observed, not without delight and satisfaction, that the quicksilver in that longer part of the tube was 29 inches higher than the other .... For this being considered, it will appear to agree rarely well with the

> hypothesis, that as according to it the air in that degree of density and correspondent measure of resistance, to which the weight of the incumbent atmosphere had brought it, was able to counterbalance and resist the pressure of a mercurial cylinder of about 29 inches, as we are taught by the Torricellian experiment; so here the same air being brought to a degree of density about twice as great as that it had before, obtains a spring twice as strong as formerly. As may appear by its being able to sustain or resist a cylinder of 29 inches in the longer tube, together with the weight of the atmospherical cylinder, that leaned upon those 29 inches of mercury; and, as we just now inferred from the Torricellian experiment, was equivalent to them.

This experiment is illustrated by Fig. 4.3. The data that Boyle published are given in Table 4.1. These data support the conclusion that the pressure P and volume V were inversely proportional, i.e., $P \propto 1/V$, or as we write it in the form we today call "Boyle's law,"

$$PV = \text{constant}$$

at a constant temperature.

Notice that in the note for column E in Table 4.1 Boyle states his law clearly. How well do his data agree with his hypothesis? It appears that Boyle assumed that his first data point V_0, P_0 was exactly correct and then used his subsequent measured values of V_m to calculate values of P_c. These are given in column E of the table for comparison with the measured values P_m in column D.

EXERCISES

3. Check Boyle's work by calculating the value of P_c at three or four rows spread throughout Table 4.1. (Notice that Boyle did not use decimals. When did decimals come into use?)

Another technique for seeing whether your data agree with your theory is to put the theory in a form such that a plot of your data comes out a straight line. Your eye is quite a good judge of the straightness of a line, and so of the quality of agreement between theory and experiment.

EXERCISES

4. A good test of Boyle's law is to plot the pressure P against the reciprocal of the volume $1/V$. (Why might this be a useful thing to do?) Do this for

TABLE 4.1 A Table of the Condensation of the Air

Volume V_m	Excess Pressure	Atmospheric Pressure P_0	Total Pressure P_m	$P_c = P_0 V_0/V_m$ P_c
A	B	C	D	E
(1/4 inches)	(inches)	(inches)	(inches)	(inches)
48	00	29 2/16	29 2/16	29 2/16
46	01 7/16	"	30 9/16	30 6/16
44	02 13/16	"	31 15/16	31 12/16
42	04 6/16	"	33 8/16	33 1/7
40	06 3/16	"	35 5/16	35
38	07 14/16	"	37	36 15/19
36	10 2/16	"	39 5/16	38 7/8
34	12 8/16	"	41 10/16	41 2/17
32	15 1/16	"	44 3/16	43 11/16
30	17 15/16	"	47 1/16	46 3/5
28	21 3/16	"	50 5/16	50
26	25 3/16	"	54 5/16	53 10/13
24	29 11/16	"	58 13/16	58 2/8
23	32 3/16	"	61 5/16	60 18/23
22	34 15/16	"	64 1/16	63 6/11
21	37 15/16	"	67 1/16	66 4/7
20	41 9/16	"	70 11/16	70
19	45	"	74 2/16	73 11/19

Column A: The number of equal spaces in the shorter leg, that contained the same parcel of air diversely extended [and hence proportional to the volume].

Column B: The height of the mercurial cylinder in the longer leg that compressed the air into those dimensions.

Column C: The height of the mercurial cylinder that counterbalanced the pressure of the atmosphere. (This was measured to be 29 1/8 inches.)

Column D: The aggregate of the two last columns, B and C, exhibiting the pressure sustained by the included air.

Column E: What the pressure should be according to the hypothesis, that supposes the pressures and expansions to be in reciprocal proportion.

TABLE 4.1 (continued)

Volume V_m	Excess Pressure	Atmospheric Pressure P_0	Total Pressure P_m	$P_c = P_0 V_0/V_m$ P_c
A	B	C	D	E
(1/4 inches)	(inches)	(inches)	(inches)	(inches)
18	48 12/16	"	77 14/16	77 2/3
17	53 11/16	"	82 12/16	82 4/17
16	58 12/16	"	87 14/16	87 3/8
15	63 15/16	"	93 1/16	93 1/5
14	71 5/16	"	100 7/16	99 6/7
13	78 11/16	"	107 13/16	107 7/13
12	88 7/16	"	117 9/16	116 4/8

the data in Table 4.1. Use reasonable scales and a nice layout of your graph. Graphs are quantitative tools, not just pretty pictures, but they need to be easy to read.

5. Boyle, writing in 1660, presented his data in a format that today we call a "spreadsheet." Today we use computer programs to enter and manipulate data in the form of a spreadsheet. Popular programs for this purpose are Borland's "Quattro," Lotus's "Lotus 1-2-3," and Microsoft's "Excel." Enter Boyle's data into one of these spreadsheet programs, and then perform his calculations and make a graph to show how the data conform to Boyle's Law. Hand in a printout of your spreadsheet and a printout of your graph.

Now you see why we believe that for many gases under conditions that are not too drastic, PV = constant. It is important to understand that this law is an idealization based on measurement. Because measurements are never exact, we can never be sure that the law will continue to hold when we make more and more precise measurements or when we change the conditions under which the measurements are made. The incomplete verification of the law is evident just from examining the data. There are a couple of disagreements between theory and experiment that are close to 1%. However, Boyle's experimental uncertainties prevented him from attributing such discrepancies to inaccuracies in the law. We would say that Boyle's law is valid within his experimental uncertainty.

EXERCISES

6. What is the largest percent variation between theory and experiment in the data of Table 4.1?

4.4 GASES AND TEMPERATURE

Thermal Expansion

At constant pressure any given volume of a gas expands as it is heated. Gay-Lussac measured the temperature dependence of that expansion and discovered two important properties.

First, if you compare the volume V_t at some temperature t with the volume V_0 at $t = 0\,°C$ you find that it obeys a simple linear relationship:

$$V_t = V_0(1 + k_V\, t) \tag{2}$$

where k_V is a constant.

Second, to within his experimental accuracy the constant k_V was measured to be the same for all gases regardless of their chemical composition. Its value turns out to be just about 1/273. This means, according to the equation, that if you increase the temperature of any gas from 0°C to 273°C while keeping the pressure constant, the volume will double in size. This result suggests that the underlying mechanism of thermal expansion of gases is simple, that in some basic way gases are quite similar to one another.

It is useful to look at this equation in terms of *changes* in volume and temperature. For a given temperature change $\Delta t = t - 0$ what will be the change in volume $\Delta V = V_t - V_0$? To answer the question either differentiate Eq. 2 with respect to temperature t, or subtract V_0 from both sides of the equation and use the above definitions of ΔV and Δt to get

$$\frac{\Delta V}{V_0} = k_V\, \Delta t. \tag{3}$$

Equation 3 shows that the fractional change in the volume relative to the volume at zero degrees, $\Delta V/V_0$, is directly proportional to the temperature change Δt, and Gay-Lussac's experiments showed that for a number of different gases the constant of proportionality is $k_V \approx 1/273$.

EXERCISES

7. Show that if its pressure is kept constant, a given volume of gas at 0°C will expand by $\sim$ 0.37% when its temperature is increased by 1°C.

The proportionality constant

$$k_V = \frac{\Delta V}{V_0 \Delta t} \tag{4}$$

is called the "volume coefficient of thermal expansion." For gases it is nearly independent of temperature and chemical composition over a large range of temperatures and volume changes. For other substances this is not so, but k_V is still useful for describing how much the volume of a substance will change when it is heated a small amount; you just need to measure and use different values of k_V at different temperatures. Equation 4 can also be read as a prescription for measuring k_V: Measure the volume and temperature of a chunk of stuff; heat it a little and measure its new volume and temperature; calculate k_V using Eq. 4. Values of k_V obtained in this way for aluminum and water at room temperature (20°C) are shown in Table 4.2. They are quite different from each other and from the value 1/273 obtained for gases, but they make the point that the idea of thermal expansion is applicable to any physical system, any state of matter.

A Useful Way to Think About Physics

You want to avoid thinking of physics as formulas. Ideas in physics (and formulas) connect to the real world by experiment. Whenever you can, try to think how to make such connections. For example, you should try to imagine how you might measure k_V for a gas. Such a mental exercise will help you better to understand the physical quantities and ideas involved.

TABLE 4.2 Coefficients of Thermal Expansion at Temperature t and Pressure 101 kPa

Substance[2a]	k_V (10^{-6}°C^{-1})	t (°C)
Aluminum	69	20
Water	208	20
Nitrogen	3670	0
Helium	3659	0
Chlorine	3883	0
Carbon dioxide	3728	0
	3725	0
Gay-Lussac's gases	$3663 \approx 10^6/273$	0

[a] The values of the volume thermal expansivity of the gases are taken from J. R. Partington, *An Advanced Treatise on Physical Chemistry*, Longmans, Green, and Co., London, 1949. p. 547.

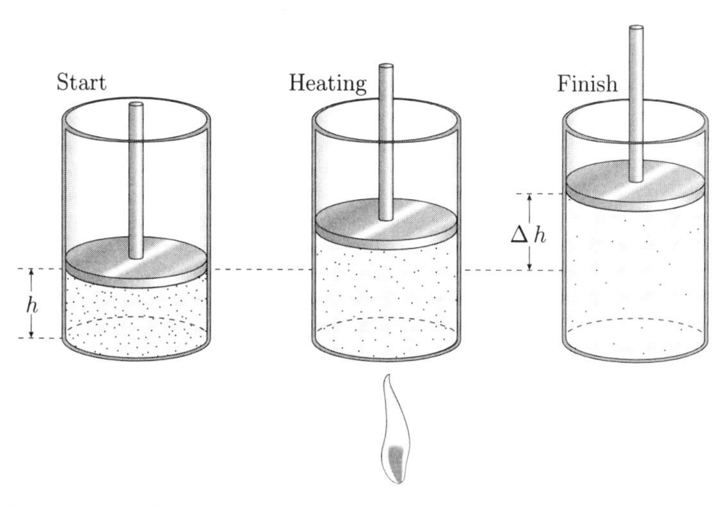

FIGURE 4.4 Device for measuring the thermal expansion of a gas.

You might measure the thermal expansivity of a gas in the following way. Make a cylinder of height h and cross-sectional area A. Confine the gas by a piston of some mass m. Then if you add energy to the cylinder by heating it, the gas will expand by raising the piston to a height $h + \Delta h$.

Notice that the pressure of the expanded gas does not change. Before expansion, the pressure of the gas is enough to exert an upward force equal to the downward weight of the piston and the downward force from atmospheric pressure. After expansion the gas pressure in the cylinder exerts the same force on the piston because it balances the same weights of piston and atmosphere as before. Therefore, *in this apparatus* pressure does not change upon heating (or cooling). Because expansion can occur only vertically, the change in volume is proportional to the change in height, and we can write from Eq. 4

$$\Delta V = \Delta h A = k_V \Delta t V_0,$$

or

$$\Delta h = V_0 k_V \Delta t / A.$$

▼ EXAMPLES

1. Suppose we have a piston with an area of $100\,\text{cm}^2$ sitting on a volume of 10 L of nitrogen gas at $t = 20°\text{C}$. If the gas is heated and warmed to $t = 30°\text{C}$, by how much does its volume change?

This is a bit tricky, because V_0 is the volume at 0°C, and the given volume of 10 L is the volume at 20°C. In order to use Eq. 2 you must find V_0 first. From $10 = V_0(1 + 20 \times 0.00366)$, it follows that $V_0 = 10/1.0732$.

Then because $\Delta t = 30 - 20 = 10°C$, Eq. 3 gives

$$\Delta V = \frac{0.00366 \times 10 \times 10}{1.0732} = 0.341 \text{ L}.$$

EXERCISES

8. Using the values from Example 4.1, find how far the piston would rise when the gas is warmed from 20 to 30 degrees.

9. Suppose the gas in the cylinder were chlorine. What percentage difference would there be in the expansion of chlorine compared to nitrogen?

10. If you understand how Boyle made his measurements, you can make up a description of how one might measure the thermal expansion coefficient of a gas at constant pressure. Do it.

Imagining an Ideal Gas

Table 4.2 shows some interesting features.

- First, as you would anticipate, gases are more thermally expandable than liquids, which in turn are more expandable than solids.
- Second, the values for the gases are all quite close to being the same.
- Third, the values for the gases are not exactly the same. Two values for the coefficient of thermal expansion of CO_2 are given just to suggest the kind of experimental variation that occurs. That variation is only 0.08% compared to the variation between, say, nitrogen and helium which is 0.2%. This suggests that the difference in k_V for He and N_2 is real. This result might lead you to ask, Why do the two different gases have different values of k_V?
- Fourth, none of these values are exactly the $1/273 = 3663 \times 10^{-6}$ that Gay-Lussac got, although they are close.

EXERCISES

11. What is the maximum percentage variation from Gay-Lussac's value among the coefficients of thermal expansion for the gases in Table 4.2?

The modern measurements in Table 4.2 show that although the values of k_V are similar for many gases, they are also measurably different. Both the similarities and the differences are important, but here we will learn more by emphasizing the similarities. To that end we want to do experiments that suggest what properties of gases explain their similarities. By understanding what properties make one gas like another, we can imagine a gas that has only those properties and none of the ones that produce differences. Then we can devise a model gas that is a simplified idealization of a real gas, i.e., an "ideal gas."

For example, if you measure k_V for different gases at pressures other than 760 mm Hg (101 kPa) with enough precision, you will get values different from each other and from those in Table 4.2. But, as shown in Fig. 4.5, an interesting result emerges when you graph the value of k_V versus the gas pressure at which the measurement was made. As the pressure goes down, all the values of k_V tend toward the same limit. And wonderfully enough, the limit turns out to be

$$k_V = 3661 \times 10^{-6} = 1/273.15^\circ\text{C}^{-1}.$$

Apparently, Gay-Lussac's physical law, Eq. 2, is a limiting case of the behavior of real gases. Because in all these experiments pressure was reduced by reducing the density while holding the temperature constant, at lower pressures the gas atoms or molecules were farther apart from each other than at higher pressures. Therefore, the tendency of all gases to behave alike as density is reduced suggests that as the average distance between the atoms or molecules is increased, it becomes less important what particular kind of molecule or atom is involved. Perhaps, then, an ideal gas is one in which the atoms or molecules all have the same effect on each other when they are close together and no effect at all when they are apart.

The idea of limiting behavior as in the case of the coefficient of thermal expansion can also be applied to the expansion or compression due to pressure changes described by Boyle's law. For a given number of gas molecules, we then find that as the pressure is decreased and the volume expands, the product of P and V approaches a constant value that is the same for all gases. This is an important generalization of Boyle's work, which after all was performed only for air. For one *mole* of gas, i.e., N_A of atoms, the limiting product at $t = 0^\circ$C is $P_0V_0 = 2271\,\text{kg}\,\text{m}^2\,\text{s}^{-2}$. It is more than coincidental that these units are joules: $P_0V_0 = 2271\,\text{J}$.

EXERCISES

12. Show that the units of PV are joules.

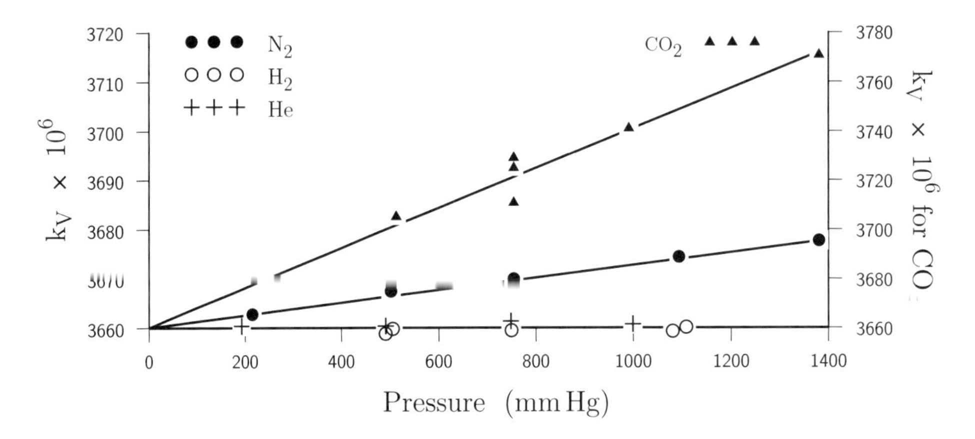

FIGURE 4.5 Coefficient of thermal expansion vs. pressure for several gases.

4.5 THE IDEAL GAS LAW

Boyle's law and Gay-Lussac's law are quite good representations of the behavior of certain gases at temperatures from 0°C to some hundreds of degrees C, and at pressures from a few atmospheres downwards. In general, the higher the temperature and the lower the pressure, the better the two laws apply. It is useful to combine the two laws into a single mathematical statement called "the ideal gas law."

The Kelvin Temperature Scale

Boyle's law relates pressure and volume at constant temperature:

$$PV = \text{constant};$$

Gay-Lussac's law relates volume to temperature at constant pressure:

$$V_t = V_0\left(1 + \frac{t}{273.15}\right).$$

Combining the two gives a particularly simple result if we invent and use a temperature scale different from the Celsius scale. To see how this works, consider the following three-step argument.

First, choose an arbitrary pressure and volume, P and V. Now imagine that at constant temperature t you squeeze or expand the gas to some volume V' such that the pressure becomes P_0, the value it would have if the temperature were 0°C. Then by Boyle's law,

$$PV = P_0V'.$$

Second, using the best value for k_V, we know from Gay-Lussac's Law that $V' = V_0(1 + 0.003661t)$. We can substitute this into the above equation and get

$$PV = P_0V_0\left(1 + \frac{t}{273.15}\right),$$

where as before, P_0 and V_0 are the pressure and volume of a fixed amount of gas at $t = 0°\mathrm{C}$.

Third, now notice how this relationship suggests a different, convenient, temperature scale. If you redefine temperature to be measured on a scale such that $T = t + 273.15$, the gas law we have just derived can be written

$$PV = \frac{P_0V_0(273.15 + t)}{273.15} = \frac{P_0V_0}{273.15}T.$$

This temperature scale is measured in units of kelvins, abbreviated K. The scale is named in honor of Lord Kelvin, a British physicist who made major contributions to our understanding of this scale. A temperature in kelvins is found simply by adding 273.15 to a temperature in degrees Celsius. A kelvin is the same size as a degree Celsius, but the zero point of the Kelvin scale is shifted 273.15 degrees below the Celsius zero point: Water freezes at 273.15 K and boils at 373.15 K (at 1 atm of pressure).

For one mole of gas the constant is $P_0V_0/273.15 = 2271/273.15 = 8.314\,\mathrm{J\,K^{-1}}$. This quantity is called the "universal gas constant" and is usually represented by the symbol R. The quantity R has units of $\mathrm{J\,K^{-1}}$ to be consistent with our newly devised temperature scale and the fact that the units for PV are joules. For n_M moles of gas the constant is just $n_M \times 8.314$. (We have already used the letter n for number of atoms per unit volume, so to avoid confusion, we distinguish the number of moles by adding the subscript M.)

The Ideal Gas Law

In terms of R and n_M we can write our combined gas law as

$$PV = n_M RT, \tag{5}$$

where P is the pressure in pascals, V is the volume in $\mathrm{m^3}$, n_M is the number of moles, R is $8.314\,\mathrm{J\,K^{-1}}$, and T is the temperature in kelvins.

You have already seen that neither Boyle's law nor Gay-Lussac's law is exactly true for all gases, but as gas density becomes low, the laws become increasingly accurate. Obviously, the combined gas law that we just derived suffers the same limitations. However, it is useful to imagine that there is some gas that behaves exactly as described by the simple law of Eq. 5. Such a gas is called an "ideal gas," and the simple law of Eq. 5 is called "the ideal gas law."

Some real gases behave very much like an ideal gas. A look at Fig. 4.5 shows that He and H_2 gases obey the law very well. A bulb filled with one of these gases makes an excellent thermometer for measuring temperatures on the kelvin scale. Nevertheless, you know that real gases cannot obey $PV = n_M RT$ over all values of T and P. If the gas gets cold enough, it liquefies and even freezes solid. This means that V changes abruptly at certain temperatures and does not vary linearly with T, so the ideal gas law cannot hold for real gases at low temperatures and non-zero pressure.

Although it is an idealization, the gas law allows us to calculate many features of changes in pressure, volume, and temperature with more than adequate precision.

EXAMPLES

2. Two moles of gas (perhaps determined by weighing) are to be confined in a volume of 10 liters at room temperature. What will be the pressure? This is not a trivial situation, because there might be real concerns about the ability of the container to withstand the resulting pressure. Since we know the number of moles, the temperature and the volume, only one unknown is left in the ideal gas law equation. As the units in the equation are joules and kelvins, we need to get our data into proper units: $10\,\text{L} = 0.01\,\text{m}^3$, and 20°C equals 293 K. Thus

$$0.01P = 2 \times 8.314 \times 293 = 4870.$$

From this result it follows that $P = 4.87 \times 10^5\,\text{N}\,\text{m}^{-2}$, or about 4.8 times atmospheric pressure. We probably need not be too concerned about the strength of the container unless it is made of ordinary glass.

EXERCISES

13. What is the volume of the container necessary to hold the two moles of gas in the above example at atmospheric pressure?

14. Assume that the piston of Example 4.1 has a mass of 10 kg and that the outside air pressure is the standard atmospheric pressure of 760 mm Hg. How many moles of N_2 does the cylinder contain?

15. Suppose instead of nitrogen we put hydrogen in the cylinder, keeping the same volume, temperature, and pressure. How many moles of H_2 would the cylinder contain?

Some Useful Details

It is important to know that the temperature of 0°C and the pressure of 760 mm Hg are called "standard" temperature and pressure. Furthermore, they are referred to so often that it is customary to use the abbreviation STP.

EXAMPLES

3. Thus if we have 100 L of He at STP, we know that the volume contains

$$n_M = \frac{101.3 \times 10^3 \times 0.1}{273.16 \times 8.314}$$

moles of He, i.e.,

$$n_M = 4.46 \quad \text{moles of He.}$$

EXERCISES

16. How many atoms of He are there in 100 L at STP?

It is also very useful to know the volume occupied by one mole of an ideal gas at STP, i.e., when $T = 273.15$ K and $P = 101.3$ kPa. Obviously, that will be

$$V_{\text{mole}} = RT/P$$
$$V_{\text{mole}} = 22.4 \times 10^{-3}\ \text{m}^3 = 22.4\ \text{L}$$

which is a useful number to **remember** because it is true for all gases that resemble an ideal gas.

With this number you can calculate the density of most of the familiar gases at STP.

EXAMPLES

4. To find the density of carbon monoxide, CO, at STP use the fact that one mole of CO will have a mass of 12 + 16 = 28 g and will occupy a volume of 22.4 L. The density of CO at STP must therefore be

$$\rho = 28/22.4 = 1.25\ \text{g L}^{-1}.$$

EXERCISES

17. Calculate the density of nitrogen and oxygen gas at STP and compare your results to the data given in Chapter 3.

18. Calculate the density of air at STP. (Take air to be four parts nitrogen and one part oxygen. Find an average molecular weight.)

19. Using the ideas of the last two chapters and the fact that a body will float if it displaces a mass of fluid exactly equal to its own mass, explain why helium-filled balloons float in air and baseballs do not.

4.6 WHAT UNDERLIES SUCH A SIMPLE LAW?

We have found that to useful accuracy over a range of temperatures and pressures many gases obey the ideal gas law:

$$PV = n_M RT.$$

But why? What is there about gases that makes them behave so similarly and so simply? The temptation is strong to model the gas from simple components. That is our cue to try to understand this law in terms of atoms. The next chapter shows how a simple atomic model of gases explains a wide variety of their physical properties.

PROBLEMS

1. Have your instructor tell you the location of a nearby barometer. Go there and find out what the pressure of the atmosphere is today. Make a careful drawing of the pointer and scale on the barometer showing the pressure as you read it.

Alternatively, you can use the World Wide Web.
Go to http://covis.atmos.uiuc.edu/java/weather0.5/Weather.shtml, and once you have reached this site put the cursor on the approximate location of your school. Choose "station report." It will give you the atmospheric pressure in millibars. If you select isobars for the map, you will also get a rough idea of what the pressure is in your area.

Now, to run yourself through all these terms and units, make a table with two rows and eight columns. In the first row label each column with the name of a unit of pressure. Using the appropriate units, put the value of the pressure that you measured into each column of the second row.

Why is the number you measured different from the value of the "standard atmosphere"?

2. Suppose we have a piston with an area of 100 cm^2 sitting on a volume of 10 L of nitrogen gas at $t = 20°C$. If the gas is heated and warmed to $t = 30°C$, by how much does the volume change? Give your answer accurate to $\pm 1\%$.

3. How is the ideal gas law derived?

4. State the ideal gas law, identify the variables in the law, and use it to find the pressure of 1 mole of gas contained in a volume of 11.2 L at 0°C.

5. State the basic properties assumed for atoms in the atomic model of an ideal gas.

6. A 45-liter container at room temperature is known to have in it a total of $N = 15 \times 10^{23}$ diatomic molecules of oxygen gas.

a. What is the number density n of this gas? Give your answer in units of cm^{-3}

b. What is the number of moles, n_M, of this gas?

c. What is the pressure of this gas?

7. An experimenter measures pressure of gas as its volume is changed and gets the following data:

Pressure (Pa)	Volume (m^3)
1000	1.5
1500	1.0
2000	0.75
2500	–

a. Do these data obey Boyle's law? Explain how you know.

b. What value of volume would you expect to measure for the missing entry in the above table?

8. Suppose the above data were taken at a temperature of 27°C and then later the gas was cooled to −73°C. At what volume would the cooled gas have a pressure of 1000 Pa?

9. Consider the following table of measurements of pressure vs. volume for O_2 ($M_{O_2} = 32$ u) in a closed container at 300 K.

Pressure (kPa)	Volume (cm^3)
90.9	24.89
73.6	30.73
59.5	37.93

a. Is it reasonable to conclude that the gas obeys Boyle's law? How did you come to this conclusion? Discuss clearly.

b. How many molecules are contained in this volume? How do you know?

c. If the oxygen molecules are replaced by an equal number of helium gas molecules ($M_{He} = 4\,u$) at the same temperature, what would be the gas pressure when the volume $V = 30.73\ cm^3$? (Hint: See the above table.)

10. Boyle's measurement of the air pressure outside his apparatus (Fig. 4.6) was $29\frac{2}{16}$ inches of mercury (Hg). [The density of Hg is $\rho_{Hg} = 13.6\,g\,cm^{-3}$.]

a. What was the air pressure in mm Hg?

b. What was the air pressure in Torr?

c. What was the air pressure in Pa?

11.a. Imagine that you are holding a balloon with a volume of 10 liters (L) filled with He gas ($M_{He} = 4\,u$). The temperature is a pleasant 27°C. Assuming that the He gas is at atmospheric pressure (P_0), what is its mass?

b. An object floats (moving neither up nor down) in air when its "buoyancy force" equals its weight. Archimedes first discovered that the buoyancy force on an object equals the weight of the volume of air displaced by the object. For example, a $10\,m^3$ block will experience a buoyancy force upwards equal to the weight of $10\,m^3$ of air. Assuming that air is 20% O_2 ($M_{O_2} = 32\,u$) and 80% N_2 ($M_{N_2} = 28\,u$), what is the buoyancy force on the balloon?

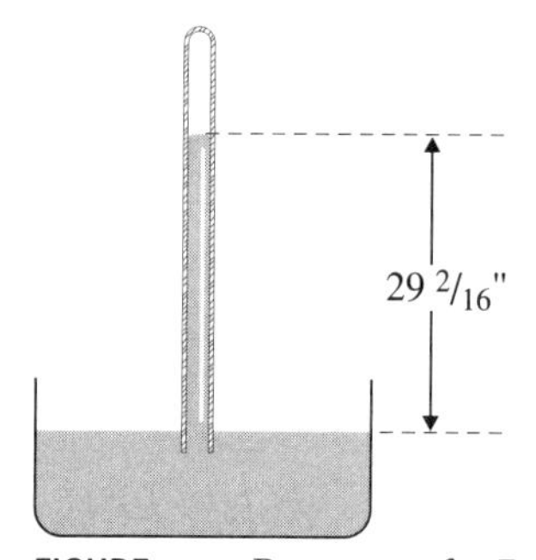

FIGURE 4.6 Barometer for Problem 10.

c. What volume of helium should the balloon contain in order to float carrying its own weight of 20 kg plus the weight of a 70 kg person?

12. Certain volumes of nitrogen gas and hydrogen gas react to form ammonia gas, according to the reaction shown below:

$$N_2 + 3H_2 \rightarrow 2NH_3.$$

The atomic masses of nitrogen *atoms* and hydrogen *atoms* are 14.0 u and 1.0 u, respectively. The mass of NH_3 formed in this reaction is 3.4 g.

a. If the pressure and temperature of each gas (N_2 and H_2 before reaction, and NH_3 after reaction) are equal to 83.1 kPa and 300 K, find the final volume of NH_3.

b. What would be the final volume if you arranged to double the final pressure?

13.a. What is the mass of a column of air 1 cm^2 in cross section rising from Earth's surface?

b. How many molecules are there in that column?

c. How many moles of molecules?

C H A P T E R

Hard-Sphere Atoms

5.1 INTRODUCTION

Why do gases obey the simple ideal gas law so well? Even gases made up of more complex entities than just individual atoms behave the same way. This fact suggests that the internal structure of the entities has little, if anything, to do with the gas law. The observed similarities of the properties of all gases must arise from similar properties of the atoms or molecules themselves. What then is the least we need to assume about the particles of gas that will explain the gas law?

To answer this question we try to imagine what atoms and molecules are like, and we make up a *model* to represent the atom or molecule. For gases we will use a model so simple that it does not distinguish between atoms and molecules, and therefore in this chapter we use the word "atom" to mean "atoms and molecules." We call our description a "model" to remind ourselves that it is surely incomplete, that we are abstracting only a few important features of atoms to explain a limited set of properties.

Assuming that a gas is made of atoms, how can we connect the ideal gas law $PV = n_M RT$ to their behavior? The volume V is a geometric property fixed by the choice of container; the temperature T is somewhat mysterious; the pressure P seems the best place to look for connections between the gas law and atoms. Indeed, we know that pressure is related to forces, i.e., to changes in momentum, so this might be a good place to look for connections between the gas law and the Newtonian mechanics of little particles.

You saw in Chapter 4 that while the volume of solid materials responds weakly to changes in temperature or external pressure, a gas responds much more dramatically. The solid acts as though its parts were in close contact with each other; the gas acts as though its atoms had plenty of space between them. A gas can expand indefinitely, but at any volume it will continue to resist compression. Something must be happening to keep its atoms apart. We are led to guess that because atoms don't pile up like sand grains on the floor of the container, they are in rapid, incessant motion. Other similar considerations lead to the following assumptions about gas atoms (and gas molecules):

- An atom or molecule is a tiny, hard sphere, too small to be observed by eye or microscope.
- An atom has mass.
- Every atom is in constant random motion.
- Atoms exert forces on each other only when they collide.
- When atoms collide, both momentum and mechanical energy are conserved.

It is difficult to get direct evidence for atoms. Attempts to subdivide matter in straightforward ways do not show discreteness or granularity. It seems that even very small chunks of matter must contain vast numbers of atoms. (Modern imaging techniques now provide direct observation of atoms, but there was certainly nothing so convincing in the nineteenth century.)

5.2 GAS PRESSURE FROM ATOMS

With this picture of the gas as jostling myriads of tiny atoms, it is reasonable to imagine that pressure arises from the force on a wall that comes from numerous collisions of the gas atoms with the walls. The average of the forces of their collisions with the walls gives rise to the observable pressure.

To see how this can occur, consider a cubical box full of gas at some pressure P. Let's say that there are N atoms, each of mass m, contained in the box (see Fig. 5.1(a)), which has a length L in the x direction and walls with area A at each end. To simplify the discussion, assume that the walls are smooth, so that there are no changes in momentum of the particles in directions at right angles to x. While Fig. 5.1(b) shows one atom colliding with the right-hand wall, it is all the atoms hitting this wall that exert the average force on the wall and produce the observed pressure. The average force exerted by the atoms on the wall is the sum of their changes in momentum divided by the time interval over which the collisions occur.

Although there are many atoms, and there are many different values of velocity they might have, there are several ways to calculate the pressure. We follow the simplest one here, but the appendix to this chapter shows a

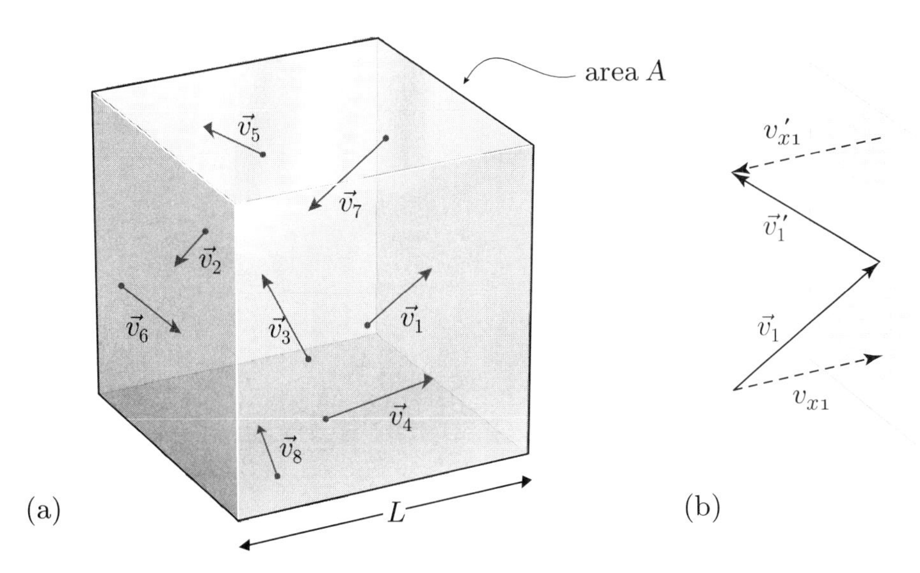

FIGURE 5.1 (a) Gas atoms move in random directions. (b) An atom's change in momentum upon colliding with a wall.

somewhat more rigorous development based on ideas that will be used later in the book. In the volume shown in Fig. 5.1(a), consider an atom with a velocity component in the positive x direction v_{x1}. When it rebounds from the right wall, it must have its x component of velocity reversed, because by assumption, there is no change in the other components and energy is conserved. Therefore, each time one of these molecules hits the wall, its momentum changes from mv_{x1} to $-mv_{x1}$. The change in momentum of a molecule, that is, the final momentum, p', minus the initial momentum, p, is thus

$$p' - p = \Delta p = 2mv_{x1}.$$

These statements are illustrated in Fig. 5.1(b).

We do not know how to describe the instantaneous force on the wall, but in the long run this atom will exert an average force given by the changes in its momentum divided by the time between collisions with the wall,

$$\langle F_1 \rangle = \frac{\Delta p}{\Delta t_1} = \frac{2mv_{x1}}{\Delta t_1},$$

where the notation $\langle ... \rangle$ means "average value."

The time between collisions with the wall must be the distance traveled, $2L$, divided by the velocity, v_{x1}, so

$$\langle F_1 \rangle = \frac{mv_{x1}^2}{L}.$$

The total force on the wall would then be the sum of these little average forces from all the atoms in the container:

$$F = \frac{mv_{x1}^2}{L} + \frac{mv_{x2}^2}{L} + \cdots,$$

where $\cdots$ represents the contributions from the rest of the atoms. This expression can be written more compactly using the notation for a summation

$$F = \frac{m}{L}\sum_{i=1}^{N} v_{xi}^2, \tag{1}$$

where v_{xi} is the x velocity of an arbitrary atom labeled i, and the symbol $\sum_{i=1}^{N}$ stands for the sum over all atoms with i ranging from 1 to N, the total number of atoms in the box.

One thing may (should!) be bothering you. We have assumed that the atoms travel freely back and forth between the walls. However, when the density gets large enough, there must be collisions between atoms away from the walls. What then happens to the x momentum of each atom? Clearly, the values will change around. Conservation of momentum comes to the rescue. Whatever is lost by one atom will be gained by another, so the total incoming momentum at the wall in any time interval is the same, regardless of the number of intervening collisions between atoms. (The treatment in the appendix avoids this problem.)

Equation 1 is still not much help. After all, for 1 mole of gas there will be 6×10^{23} terms in the summation. We get around this problem by defining an average square speed:

$$\langle v_x^2 \rangle = \sum_{i=1}^{N} \frac{v_{xi}^2}{N}.$$

Then the sum in Eq. 1 is just N times the average square speed. The force can then be written

$$F = \frac{Nm}{L}\langle v_x^2 \rangle. \tag{2}$$

To find the pressure P, divide both sides of Eq. 2 by the area A of the end wall of the box, and note that the product LA is just the volume of the container, so we obtain

$$P = \frac{m}{LA}N\langle v_x^2 \rangle = \frac{N}{V}m\langle v_x^2 \rangle. \tag{3}$$

The quantity N/V is the number density of the gas, which, as described in Chapter 3, tells you how many atoms or molecules there are in a unit volume.

N/V is often given its own separate symbol n; n is very useful when we don't want to talk about some particular volume.

EXAMPLES

1. What is the number density of a gas at STP? Remember that at STP a mole of ideal gas occupies 22.4 L, or 22,400 mL. Thus

$$n = 6 \times 10^{23}/22,400 = 2.7 \times 10^{19}\,\text{cm}^{-3}.$$

EXERCISES

1. What is the number density for a gas in an ultra-high-vacuum chamber at a pressure of 10^{-12} Torr? Take the temperature to be 0°C.

2. It is useful to *remember* the number density of 1 Torr of ideal gas at 0°C. What is it?

It is more meaningful, as we shall see, to use the magnitude of the total velocity and not just the x-component. As with any vector magnitude in three dimensions,

$$v^2 = v_x^2 + v_y^2 + v_z^2,$$

which when averaged becomes

$$\langle v^2 \rangle = \langle v_x^2 \rangle + \langle v_y^2 \rangle + \langle v_z^2 \rangle.$$

Nature cannot tell the difference between the x direction and the y or z directions, so it must be true that

$$\langle v_x^2 \rangle = \langle v_y^2 \rangle = \langle v_z^2 \rangle,$$

from which it follows that

$$\langle v^2 \rangle = 3\langle v_x^2 \rangle.$$

The quantity $\langle v^2 \rangle$ is called the "mean square velocity." We can rewrite Eq. 3 in terms of the mean square velocity and obtain

$$P = \frac{Nm}{V}\frac{\langle v^2 \rangle}{3},$$

which we can rearrange to get

$$P\,V = N\,m\frac{\langle v^2 \rangle}{3}. \tag{4}$$

EXERCISES

3. It is sometimes useful to express Eq. 4 in terms of the mass density ρ of the gas. Show that Eq. 4 implies

$$P = \frac{1}{3}\rho\langle v^2\rangle. \tag{5}$$

4. Show that PV equals $\frac{2}{3}$ of the average kinetic energy of the gas.

You now can begin to see the point of all the calculations above. First, pressure and volume have been related to a single number, the mean square velocity of the atoms. Equation 4 begins to look familiar as well. There is PV all by itself looking like the left half of the ideal gas law. Moreover, PV equals 2/3 of the average kinetic energy of the gas. We now see why all gases at the same temperature might have the same value of PV. Remember from Chapter 3 that experiment showed that for a mole of any gas at STP, $PV = 2271\,\mathrm{J}$. Apparently, the number depends only on the average kinetic energy of the gas and not on any other properties, e.g., not on how many hooks the atoms have, not on how big the atoms are, not on how complicated the molecules are.

Notice also that we get a measure of the velocity of an atom without any knowledge of the size or number of atoms. If we multiply both sides of Eq. 5 by the volume of one mole of gas, V_{mol}, and use the fact that $\rho V_{\mathrm{mol}} = M$, the molecular weight, i.e., the mass of 1 mole in 1 kg of the gas, we get

$$PV = \frac{1}{3}M\langle v^2\rangle,$$

where P, V, and M are all measurable quantities. For example, for nitrogen at STP, $PV = 2271\,\mathrm{J}$ and $M = 0.028\,\mathrm{kg}$, so

$$\langle v^2\rangle = 3 \times 2271/0.028 = 243 \times 10^3\ \mathrm{m^2\ s^{-2}},$$

which implies

$$\sqrt{\langle v^2\rangle} = 493\,\mathrm{m\,s^{-1}}. \tag{6}$$

This number, called the "root mean square" (rms) velocity, gives a measure of the speed of N_2 molecules at 0°C. Atoms and molecules travel quickly.

5.3 TEMPERATURE AND THE ENERGIES OF ATOMS

Let's look at some other possibilities suggested by the model. We know that an ideal gas obeys $PV = n_M RT$. Thus, if our atomic model of an ideal gas is correct, it follows from Eq. 5 that

$$RT = \frac{1}{3}\frac{Nm}{n_M}\langle v^2\rangle = \frac{1}{3}M\langle v^2\rangle.$$

This equation relates temperature T to the average kinetic energy of the atoms. We can make the connection to kinetic energy more explicit by writing the above equation as

$$RT = \frac{2}{3}\frac{M\langle v^2\rangle}{2}, \tag{7}$$

which is the same as

$$n_M RT = \frac{2}{3}\,\text{K.E.}$$

where K.E. is the kinetic energy of the entire mass of gas in the given volume.

This result suggests that we can interpret temperature to be a measure of the energy of random motion of the atoms. You probably already believe this because you have often been told that it is so. What we have tried to do here is show you the kinds of arguments (above and in the appendix) that helped to convince physicists that temperature is closely related to the kinetic energy of moving atoms.

How large is the kinetic energy of each atom? To answer you need to know Avogadro's number N_A, which historically was not yet known at this point in our story. Let's jump ahead a bit in our story and answer the question anyway.

Energies of Atoms: Boltzmann's Constant

Equation 7 gives the amount of energy in 1 mole of gas. Physicists usually express the result in terms of the average energy of a single atom or molecule. To do this, realize that the total number of atoms or molecules N is the same as the number of moles n_M times the number of entities in a mole, N_A, so that $N = n_M N_A$. Then using the fact that the mass M of 1 mole is the number of entities in a mole multiplied by the mass m of a single entity, $M = mN_A$, Eq. 7 can be rewritten

$$\frac{1}{2}m\langle v^2\rangle = \frac{3}{2}\frac{R}{N_A}T. \tag{8}$$

The ratio R/N_A appears over and over again in atomic physics, and so it is given its own symbol, "k_B," and name. It is called "Boltzmann's constant" and has a value

$$k_B = \frac{8.314}{6.02 \times 10^{23}} = 1.38 \times 10^{-23}\,\mathrm{J\,K^{-1}}.$$

The usual form of Eq. 8 is thus

$$\frac{1}{2}m\langle v^2\rangle = \frac{3}{2}k_B T. \tag{9}$$

This result is the average kinetic energy of a single atom.

EXAMPLES

2. For a numerical example let's find the average kinetic energy of a hydrogen molecule at room temperature. Notice that we do not have to know the mass or the velocity or any details; all we need to know is the temperature and Boltzmann's constant, which we can find from measurements of the gas constant R and Avogadro's constant N_A. The average kinetic energy of an H_2 molecule at room temperature is then

$$1.5 \times 1.38 \times 10^{-23} \times 293 = 6.1 \times 10^{-21}\,\mathrm{J},$$

a rather small number.

Because atomic energies are important in atomic physics and we refer to them frequently, it is convenient to have a unit of energy scaled to the size of atomic happenings. This unit is the "electron volt," abbreviated eV. It has what may seem to you a strange definition:

$$1\,\mathrm{eV} = 1.60 \times 10^{-19}\,\mathrm{J}.$$

However, a few chapters from now you will see that it is a very reasonable unit. *It is so reasonable that from now on you should always use the eV as the unit of energy when talking about phenomena at the atomic scale.* All the usual multiples are used: meV, keV, MeV, GeV, and even TeV for tera electron volts where T and tera stand for a U.S. trillion, i.e., 10^{12}.

EXERCISES

5. Show that the average kinetic energy of the hydrogen molecule in the previous example is 0.038 eV.

6. What is the average kinetic energy of an O_2 molecule at room temperature? What about N_2? He?

The number obtained in Exercise 5.5 is important. It tells you the magnitude of thermal energy associated with room temperature (usually taken to be 20°C (293 K), but sometimes 300 K). Most physicists *remember* that at room temperature

$$k_B T = 0.025\,\text{eV} = \frac{1}{40}\,\text{eV}, \tag{10}$$

or

$$\frac{3}{2}k_B T = 0.038\,\text{eV} \approx 1/25\,\text{eV}.$$

EXERCISES

7. The temperature near the surface of the sun is thought to be about 10,000 K. What would be the average kinetic energy of a hydrogen atom there? A He atom?

Obviously, if we know the mass of an atom and its average kinetic energy, we can determine some kind of average speed. Knowledge of temperature T and mass m give $\langle v^2 \rangle$. The square root of $\langle v^2 \rangle$, i.e., the "root mean square" velocity, is often referred to as r.m.s velocity, or v_{rms}. Thus

$$v_{\text{rms}} = \sqrt{\frac{3k_B T}{m}}. \tag{11}$$

EXAMPLES

3. The mass of a hydrogen atom is 1.67×10^{-27} kg. What is the rms velocity of hydrogen at room temperature? At room temperature we know that hydrogen comes in molecules, each with a mass of 3.34×10^{-27} kg. Therefore,

$$v_{\text{rms}} = \sqrt{\frac{3 \times 0.025 \times 1.60 \times 10^{-19}}{3.34 \times 10^{-27}}},$$

$$v_{\text{rms}} = 1906\ \text{m s}^{-1}.$$

EXERCISES

8. Do we need to know the mass of an individual molecule in order to find its rms velocity? Explain using Eq. 7.

9. Calculate the rms velocity of oxygen at room temperature. What is it for nitrogen? Helium?

10. Compare the speed of sound in air at room temperature to the rms velocity of nitrogen.

11. An object moving with a velocity of 11 km s^{-1} will, if not deflected, leave the Earth and never return. At what temperature would hydrogen atoms have $v_{\rm rms}$ equal to this escape velocity?

5.4 SUMMARY THUS FAR

The results of our atomic model of an ideal gas are very satisfactory. In terms of the model, we understand why all gases obey the same ideal gas law. The model says that the pressure of the gas depends only on its amount of kinetic energy per unit volume. The model also suggests a reasonable interpretation of temperature. Measured on the kelvin scale, temperature is directly proportional to the average kinetic energy per unit volume. Thus, at a given temperature all gases have the same kinetic energy per unit volume and so have the same pressure.

The consistency of the model and its simplicity make it very appealing. With knowledge of the molecular weights of atoms or molecules we can deduce a measure of their velocities. We can understand why gas will stream into an evacuated vessel when a hole is opened in it. (Why?)

We can understand some other phenomena as well. By a more careful description of interatomic collisions, we can understand how gases conduct heat (thermal conductivity), how two different gases will mix together without being stirred (diffusion), and why gases resist flow and dissipate energy when they rub across a surface (viscosity). The viscosity of gases is an interesting property that provided the first experimental basis for determining molecular sizes and Avogadro's number, so we will look at this subject in some detail.

5.5 VISCOSITY

Gas atoms and molecules are in ceaseless motion, repeatedly bumping into the walls of the container and into each other. For an ideal gas of atoms rattling around in a closed container, the collisions of atoms with each other do not matter because the conservation of momentum evens out their effects. However, in nonequilibrium situations such as different temperatures at each end of the container, or the gas streaming over a surface, the frequency of collisons between the atoms has important effects.

To explain these effects we must extend our model slightly by introducing the finite size of the atom. That collisions occur at all implies that atoms are not point particles. Point particles with no physical extent occupy no space, and collisions between them would be vanishingly improbable. Therefore, ascribe to each atom or molecule a radius r_m. With this extension of the model we can develop a simple description of the likelihood of collisions.

Mean Free Path

The frequency of collisions is determined by the speed of the atoms and the distance they travel between collisions. Although this distance will vary greatly from collision to collision for a given atom, there will be some average distance that atoms travel before colliding. We call that average distance between collisions the "mean free path," denoted by ℓ.

To get an idea of how the mean free path relates to the size of an atom and its speed, consider Fig. 5.2. Figure 5.2(a) shows a molecule of radius r_m moving past other similar molecules. Obviously, two molecules cannot get closer than a distance $2r_m$ to each other without colliding.

It is convenient to describe this situation by the equivalent situation shown in Fig. 5.2(b), where one atom of radius $2r_m$ moves among point atoms. In effect we put all the physical size in one atom by treating it as a sphere of cross-sectional area $\pi(2r_m)^2$. Now imagine a disk of this area moving through a volume containing point atoms with a density of n points per unit volume. While moving a distance d, the disk will sweep out a volume of $\pi(2r_m)^2 d$. This volume contains $n\pi(2r_m)^2 d$ molecules, where n is the number of molecules per unit volume. If the molecules were all at rest and equally spaced, then the moving disk would suffer n collisions every time it traveled a distance d. Let ℓ be the distance for just 1 collision. Then

$$\ell = \frac{1}{\pi(2r_m)^2 n}.$$

This cannot be quite right because the point molecules are not all at rest. They are rushing toward and away from the moving molecule of radius $2r_m$. With considerable more study we could learn how to correct for the relative motion, but let's just write down the answer, which is almost the same as the

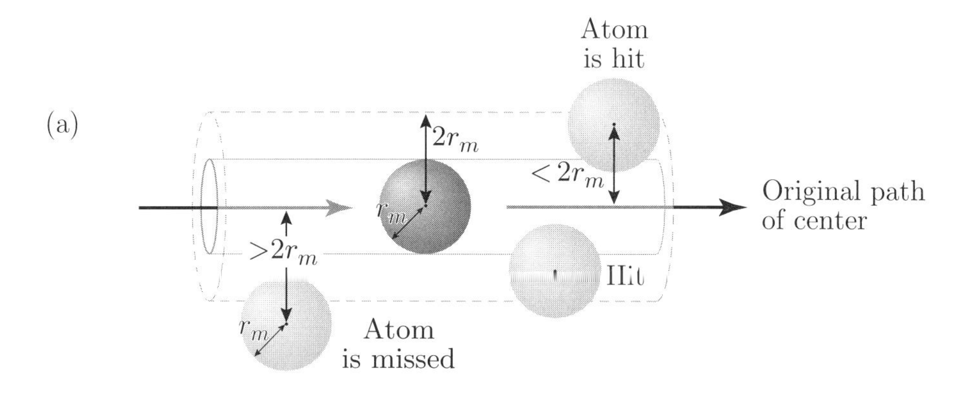

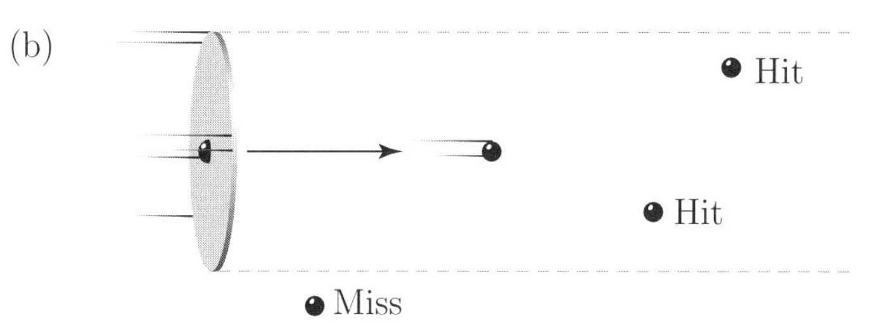

FIGURE 5.2 (a) Atoms will collide if their center-to-center distance is $< 2r_m$. (b) Equivalent picture of an atom's cross section of radius $2r_m$ colliding with point-sized atoms.

above equation:

$$\ell = \frac{1}{\sqrt{2}\pi(2r_m)^2 n}. \tag{12}$$

EXAMPLES

4. An N_2 molecule has a radius of $\approx 1.9 \times 10^{-10}$ m. What is the mean free path of an N_2 molecule in a volume of nitrogen gas at STP? You need to know the number density n. This is found from the fact that at STP,

$$n = \frac{N_A}{V_m} = \frac{6.02 \times 10^{23}}{22.4 \times 10^{-3}} = 2.69 \times 10^{25}\ \text{m}^{-3} = 2.69 \times 10^{19}\ \text{cm}^{-3}.$$

Then

$$\ell = \frac{1}{\sqrt{2}\pi \times (3.8 \times 10^{-10})^2 \times 2.69 \times 10^{25}} = 5.80 \times 10^{-8}\ \text{m} = 58\,\text{nm}.$$

This may seem like a small distance, but notice that it is ≈ 300 molecular radii.

EXERCISES

12. What is the value of ℓ for these molecules at room temperature?

13. On average, how many collisions will an N_2 molecule have in 1 second at STP?

Resistance to Flow

Imagine a flat plate moving with a velocity u through a layer of gas above a flat surface. Right at the upper plate, molecules will adhere to its surface and flow with it at a velocity u. At the bottom layer of gas, at the lower plate, the molecules on average will have no net flow velocity. Because there has to be a smooth transition in flow velocity from u to 0, we expect that below the top layer, the molecules have a flow velocity a little slower than u. Further down they will be flowing more slowly yet, and so on, until on the bottom surface they will not be flowing at all. Figure 5.3 illustrates the situation.

The upper plate is found to have a retarding force on it, a drag. Experiments show that this force depends on the kind of gas and its temperature. Because the force goes to zero when good vacuums are produced, it must be the gas that imposes the drag on the moving surface. We say that this drag is produced by the "viscosity" of the gas.

Viscosity is a measurable quantity. In principle you might pull one plate past another and measure the force of the drag. This is difficult in practice. It is much easier to rotate one cylinder inside a larger one and measure the force

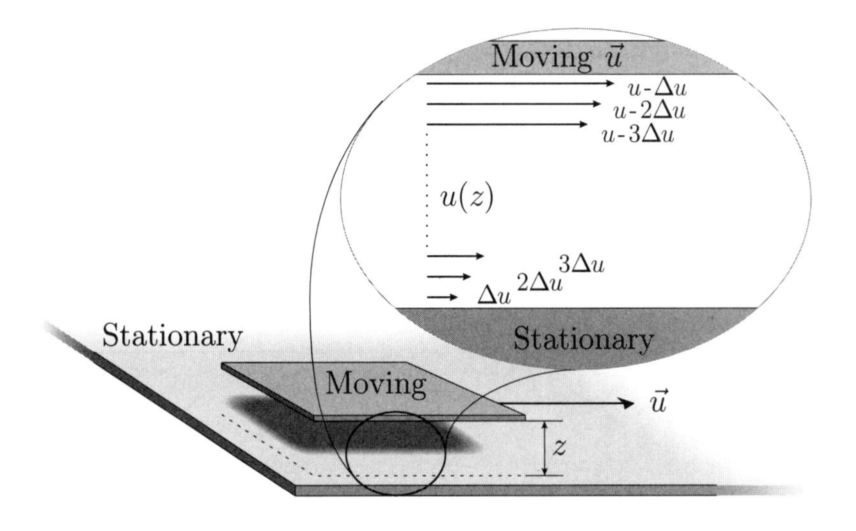

FIGURE 5.3 Change in velocity of gas flow near a moving surface.

resisting the rotation. If the gap z between the cylinders is small compared to their radii, the geometry is equivalent to the flat plate diagrammed in Fig. 5.3.

EXERCISES

14. Sketch for yourself a likely experimental setup of the two-cylinder arrangement for measuring viscosity.

In such measurements the force is found to be proportional to the area A of the cylinder and also proportional to the difference in velocity, u, divided by the width of the gap, z, between the two cylinders. We write

$$F = \eta A \frac{u}{z}, \tag{13}$$

where the coefficient of proportionality—conventionally represented by the lowercase Greek letter eta, η—is a measure of the strength of the viscous resistance and is called the "coefficient of viscosity." Frequently, the coefficient is called just "the viscosity" of the gas or liquid in question.

When at 20°C a cylinder of area $50\,\text{cm}^2$ inside a similar cylinder and separated from it by 2 mm of air is rotated such that its circumferential velocity is $10\,\text{m}\,\text{s}^{-1}$, it feels a drag of $450\,\mu$N. If water at 20°C is poured between the two cylinders, the drag force becomes 25 mN. If you use SAE 10 motor oil at 30°C, the drag becomes 5 N. All of which seems to show that, as you might expect, gooey things have more viscosity than runny things.

EXERCISES

15. Derive the units of the coefficient of viscosity.

In SI units viscosity is measured in pascal-seconds, i.e., Pa s. Although the Pa s is the official SI unit, many tables give viscosities in terms of a historical unit called the "poise." The conversion is 1 Pa s = 10 poise.

From the experimental results given above we can determine the coefficient of viscosity.

EXAMPLES

5. For SAE 10 motor oil we find that the viscosity is

$$\eta = \frac{F\,z}{A\,u} = \frac{5}{0.005} \times \frac{0.002}{10} = 200\,\text{mPa s}.$$

TABLE 5.1 Viscosity in (μPa s) of Some Common Gases at 20°C

Gas	Viscosity	Gas	Viscosity
N_2	17.57	H_2	8.87
O_2	20.18	He	19.61
CO_2	14.66	Ne	31.38
Cl_2	13.27	Ar	22.29

EXERCISES

16. Use Eq. 13 to find the viscosity of air at 20°C. Do the same for water.

Some values of the viscosity of various gases are given in Table 5.1.

5.6 AN ATOMIC MODEL OF VISCOSITY

We can use our atomic model to explain the viscosity of gases and derive an expression for η. Although in beginning physics we cannot derive the most accurate relationship between η and atomic parameters, we can do quite well, and, as was the case historically, the result yields a quantitative estimate of the size of atoms based on experimental measurements.

The viscous force arises from a transfer of momentum from the faster-flowing gas to the slower-flowing gas. Figure 5.3 suggests how this might occur. Across the gap, for every change Δz there is a change Δu in the flow velocity. To understand what is happening we must distinguish clearly between the velocity u with which the gas flows and the much larger and randomly directed velocities of the gas molecules, which we call v. We know what magnitudes of v might be expected, because we have seen that for N_2, $v_{\mathrm{rms}} = 500\,\mathrm{m\,s^{-1}}$. Even though v_{rms} is a special way of calculating an average, it gives a characteristic velocity for gas molecules.

Thus the molecules are moving randomly with the great speed v to which a small amount of velocity u has been added to give the overall flow. This means that a molecule moving with a velocity v away from the surface of the upper plate in Fig. 5.3 goes from a region of flow velocity u to one of $u - \Delta u$; similarly, molecules moving toward the upper surface with velocity v move from a region where the flow velocity is $u - \Delta u$ to one where it is u. The result is that these slower-moving molecules get speeded up by collisions with faster-moving gas molecules; momentum is added to the lower-momentum gas. This transfer of momentum must come from the moving surface, which is just a way of saying that a force is being exerted on the surface, i.e., there is viscous drag.

We can find the viscous force on the surface by finding how much momentum moves away from a unit area of the plate in a unit of time. Although this quantity has the units of force per unit area, it is not pressure. Pressure would be a force perpendicular to the surface, but this force per unit area is tangent to the surface of the plate. It is the viscous force dragging on a unit area of the plate.

The rate at which momentum leaves the surface is controlled by the difference between the rate at which molecules with flow velocity u leave the surface and the rate at which those with flow velocity $u - \Delta u$ come to it. On average a number of molecules proportional to $n\langle v\rangle$ will cross a unit area of surface each second.[1] The average velocity $\langle v\rangle$ is not the same as the rms velocity, but the difference is small enough so that for our purposes we can use the two interchangeably in most circumstances.

Because we repeatedly use this argument for estimating the number of molecules crossing a surface, let's go through it again.

Look at Fig. 5.4. Imagine a cylinder of 1 cm^2 cross section. In time Δt all the atoms traveling with a velocity $\langle v\rangle$ will travel a distance $\langle v\rangle\Delta t$. The volume of a cylinder this long ending at a portion of the surface having an area A will thus be $\langle v\rangle A\Delta t$. You can imagine this volume of gas extending down and passing through the surface A in the time interval Δt. On average, half the atoms will be traveling upward and half downward. Therefore, of the atoms with flow velocity u, the number that pass downward through the surface will be $(n/2)\langle v\rangle A\Delta t$, where n is the number per unit volume as before. An equal number of molecules with a flow velocity of $u - \Delta u$ will

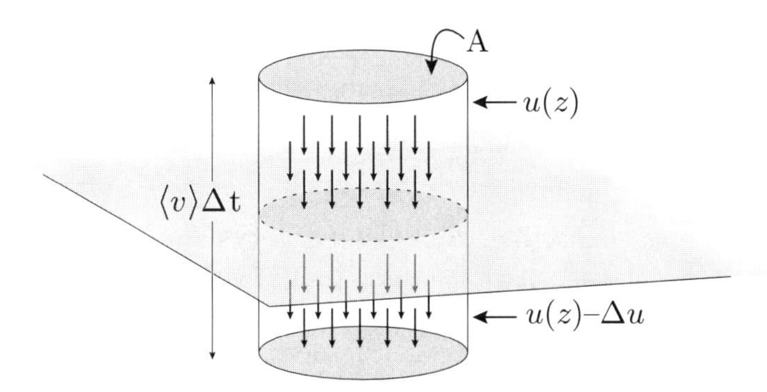

FIGURE 5.4 A net flow of momentum occurs from the faster moving layer to the slower moving layer.

[1]The average velocity $\langle v\rangle$ used here is slightly different from v_{rms}; $v_{\text{rms}} = 1.085\langle v\rangle$. $\langle v\rangle$ is calculated by a different method and represents the simple average of the *magnitudes* of the velocities.

travel upward from the more slowly flowing gas. Thus in the time interval Δt the amount of momentum carried across the area A will be

$$\begin{aligned}\Delta p &= \left(\frac{n}{2}\langle v\rangle A\Delta t m u\right) - \left(\frac{n}{2}\langle v\rangle A\Delta t m(u-\Delta u)\right)\\ &= n\langle v\rangle A\Delta t m\Delta u, \end{aligned} \tag{14}$$

where Δu is the change in the velocity of flow of the gas over the length of the cylinder, and m is the mass of a molecule of the flowing gas.

We need one other piece of information. We need to decide how big a time interval to take, because otherwise Δu would be arbitrary. It seems reasonable to choose Δt to be the time interval between collisions, i.e., the time it takes to travel one mean free path. The idea is that after this time interval one batch of atoms will have mixed completely with the next layer and changed its flow velocity from u to $u - \Delta u$. It also seems reasonable to assume (and experiment confirms) that the flow velocity drops linearly from u to 0 across the distance z between the moving and stationary plates. Then the change in flow velocity across one mean free path ℓ is the fraction ℓ/z of the total u,

$$\Delta u = \ell\frac{u}{z}.$$

We can rewrite Eq. 14 as

$$\frac{F}{A} = n\, m\, \langle v\rangle\, \ell\frac{u}{z}.$$

Comparing this result with Eq. 13, we see that it implies that

$$\eta = n\, m\langle v\rangle\ell.$$

Notice that $nm = \rho$, the mass density of the gas.

The above derivation gives the right form of the result, but it is too simplified to give an exact result. It treats all the molecules as moving with a single velocity equal to $\langle v\rangle$ across a plane surface. The actual motion of the molecules is much more random. To do this calculation correctly requires averaging over many velocities and directions. If one does the complete careful calculation, one gets an additional numerical factor of 0.499 (which you can round off to 0.50 in calculations if you wish):

$$\eta = 0.499\, nm\langle v\rangle\ell,$$

or

$$\eta = 0.499\, \rho\, \langle v\rangle\ell.$$

This last equation shows us how to determine mean free paths from measurements of the viscosity. For example, for air at room temperature we found $\langle v\rangle$ to be around $500\,\mathrm{m\,s^{-1}}$ (see Eq. 6), and we know its density to be

$1.29\,\text{kg}\,\text{m}^{-3}$. Therefore, the mean free path must be

$$\ell = \frac{18 \times 10^{-6}}{0.499 \times 1.29 \times 500} = 56\,\text{nm}.$$

EXERCISES

17. What is the mean free path in a chamber evacuated to 1 Torr at room temperature?

18. Ultrahigh vacuum is considered to be around 10^{-11} Torr. What is the mean free path of a molecule in such a vacuum?

It is useful to substitute the value of ℓ from Eq. 12 into the expression for viscosity, for then we get the dependence of viscosity upon atomic parameters:

$$\eta = \frac{0.499 m \langle v \rangle}{\sqrt{2}\pi(2r_m)^2}. \tag{15}$$

Equation 15 is very interesting. From it you can tell how the viscosity of an ideal gas depends upon its density. What would you expect? The answer is—not at all. Are you surprised? James Clerk Maxwell, a very great physicist, was the first person to derive the result. He was so surprised that he built an apparatus and tested the prediction. It is true as long as the mean free path does not become comparable to the dimensions of the container or of the molecules.

EXERCISES

19. What will happen to the viscosity of a gas as you increase the pressure? Justify your answer.

20. Suppose you cool the gas. What will happen to its viscosity? [Hint: Equation 15 shows that η is proportional to $\langle v \rangle$. How does $\langle v \rangle$ depend upon temperature?] Would you expect motor oil to become more viscous or less as you cooled it? Compare this with the behavior of the viscosity of gases.

21. On a hot summer day air temperatures may rise to 40°C. By what percentage will the viscosity of this air change relative to room temperature?

5.7 ATOMIC SIZES

Using the viscosity and the simple theory relating it to the atomic model of gases, Loschmidt in 1865 made the first estimate of atomic sizes. We can use a slightly simplified version of his method to do the same.

Atomic Diameter

Rewrite Eq. 15 as

$$\eta = \frac{0.5}{\sqrt{2}}\frac{M}{N_A}\frac{\langle v\rangle}{\pi d^2},$$

where d is the atom's diameter ($d = 2r_m$), M is the molecular weight, N_A is Avogadro's number, and we have used the fact that $m = M/N_A$. Now substituting in for $\langle v\rangle$ from Eq. 11, we obtain

$$N_A d^2 = \frac{0.50\,M}{\sqrt{2}\pi\eta}\sqrt{\frac{3RT}{M}}. \tag{16}$$

Notice that all the quantities on the right-hand side are measurable; therefore, with this equation we can calculate a value for the combination $N_A\, d^2$ using measured quantities. Remember that thanks to Gay-Lussac and others, M was known without knowing N_A and that R/M is the same as k_B/m.

Because Eq. 16 only gives a value for the product of two unknowns, if we want to find d alone, we must find another independent relation between N_A and d. One way to do this is to consider a gas condensed into a solid. Then we would expect the gas molecules to be in contact with each other and the solid to become approximately a set of cubes of length d on a side. Then, as shown schematically in Fig. 5.5, a mole, i.e., a mass M, of the *solidified* gas would have a volume of $N_A\, d^3$, which is simply related to the molecular weight M and the density of the *solid* ρ_s:

$$N_A\, d^3 = V = \frac{M}{\rho_s}. \tag{17}$$

We can eliminate one of the unknowns by dividing Eq. 17 by Eq. 16, and then we get

$$d = \frac{2\pi\eta}{\rho_s}\sqrt{\frac{2M}{3RT}},$$

where ρ_s is the density of the solid. Remember that T is the temperature of the gas at which the viscosity was determined; it has nothing to do with the temperature at which the density of the solidified gas was measured.

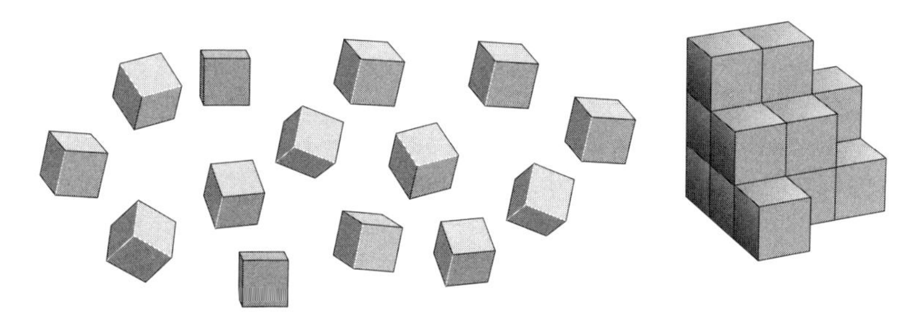

FIGURE 5.5 N blocks each of volume d^3 add up to a total volume of Nd^3.

Solid CO_2 at $-79°C$ has a density of $1.53\,g\,cm^{-3}$. Gaseous CO_2 at $0°C$ has a viscosity of $13.8\,\mu Pa\,s$. Substituting these values into the expression for d, we get $d = 0.2\,nm$ or $2\,Å$ where Å stands for the Ångstrom, a unit of length equal to 10^{-10} meters.[2] In other words, the radius of an atom is about $1\,Å$.

Avogadro's Number

Once we have an estimate of the atomic diameter, we can go one step further than Loschmidt and find N_A. Substituting the value of d back into Eq. 16 or into Eq. 17, we obtain $N_A = 30 \times 10^{23}$. This value is a factor of five larger than the currently accepted value of 6.02×10^{23}, but given the crudeness of our model of the solid, the value is pretty good.

EXAMPLES

6. If N_A is off by a factor of 5, how much error is there in our estimate of d? From Eq. 17, we see that if d were $1.7 = 5^{1/3}$ larger, N_A would come out just right. This means that our estimate of the diameter of the atom is too small by a factor of about 1.7. Not bad at all for a first attempt.

EXERCISES

22. At $-273°C$ the density of solid nitrogen is $1.14\,g\,cm^{-3}$, that of solid oxygen is $1.568\,g\,cm^{-3}$, and at $-260°C$ that of solid hydrogen is $0.0763\,g\,cm^{-3}$.

---

[2] It is named for a Swedish physicist who made the first precise measurements of atomic and molecular sizes. It is not an officially sanctioned SI unit, but it is still used by physicists too old to change their habits.

(a) What are the molecular radii of these three substances estimated by Loschmidt's method?
(b) What are the values of N_A that you obtain in each case?

23. Compare the mean free path of nitrogen at STP with the molecular diameter. What is the mean free path measured in molecular diameters?

5.8 CONCLUSION

This chapter is intended to give you the flavor of physical reasoning. It shows a few of the arguments and experimental data that established the atom as a physical entity. Much of what we have studied has been replaced by more complete and more sophisticated arguments, but the conclusions remain solidly founded.

Atoms exist. Their energy depends upon temperature. At room temperature, $T = 293$ K, gas atoms have an average kinetic energy of 0.038 eV, and they move with speeds of hundreds of meters per second. At atmospheric pressure they travel tens of nanometers between collisions. Atoms have diameters of the order of 0.2 nm. At STP a cubic centimeter contains 2.6×10^{19} of them. Over a large range of pressures and temperatures the ideal gas law describes real gases quite well. The model of atoms as tiny hard spheres explains a variety of physical properties of real gases. It predicts that viscosity (and diffusion, thermal conductivity, and the speed of sound) will be independent of gas density, will depend on the square root of temperature measured on the Kelvin scale, and will vary with the inverse of the square root of the molecular weight. These predictions are accurate over a range of pressures and temperatures where the mean free path is large compared to the molecule's size and small compared to the dimensions of the container.

There is a great deal more to be said, because the model of hard-sphere atoms is too simple. The physical extent of real atoms affects their behavior in ways other than the ones described above, and the assumption that atoms interact only when in contact is only approximate, as becomes especially apparent at high pressures. One serious omission is the absence of a discussion of the distribution of velocities of atoms. At any instant the gas atoms have a range of velocities, and it is possible to predict how many atoms will have which velocities. The deduction and prediction are well supported by experimental evidence.

However, it is time to examine the next advance in our understanding of the atomic nature of matter. Toward the end of the nineteenth century it became clear that matter is electrical in nature. Our understanding of that electrical nature led to very precise values for N_A. It also led deeper into the atom and revealed the internal structure of atoms. We need to study electrical properties of atoms next, starting with a review of some of the basic ideas of electricity and magnetism.

5.9 APPENDIX: AVERAGES OF ATOMIC SPEEDS

Introduction

In the main part of this chapter we argued that gas pressure P is the result of the average effect of many collisions of many different atoms with the walls containing the gas. The argument assumed that all the atoms had the same speed; this is extremely implausible. Here we give an argument that takes into account the fact that the atoms have a wide range of velocities. This more complete argument also is an opportunity to show you how to compute averages in complicated situations. The computation of averages will be important later in the book.

Sums and the $\sum$ Notation

We are often interested in adding up long strings of numbers. For example, suppose you wanted to know the average age of the students in a laboratory section. Suppose there are 16 students in the class with the names and ages shown in Table 5.2.

To find the average age we just add all the ages and divide by the number of students:

$$\text{average age} = \frac{18 + 17 + 19 + 19 + 18 + 18 + 18 + 18 + 20 + 19 + 17 + 21 + \cdots}{16}$$

TABLE 5.2 Names, ID Numbers, and Ages of 16 Hypothetical Students

Name	ID Number	Age	Name	ID Number	Age
Aimée	1	18	Brian	2	17
Ty	3	19	Kevin	4	19
Mannie	5	18	Doug	6	18
Yoko	7	18	Max	8	18
Sean	9	20	Laura	10	19
James	11	17	Ann	12	21
David	13	20	Jan	14	19
Kate	15	18	Luis	16	18

$$= 18.56\,\text{y}.$$

This is too tedious to write out. We need a notation that is both compact and general.

We start by giving a symbolic name to the quantity we are averaging. It is "age," so let's call it a. To talk about the age of a particular student we can use the ID number as a subscript, often called an "index." Then Aimee's age is a_1, while Jan's is a_{14} and Sean's is a_9. We can write the average in terms of these symbols as

$$\text{average age} = \frac{a_1 + a_2 + a_3 + a_4 + \cdots + a_{15} + a_{16}}{16}.$$

This form has the advantage of generality; it represents the average of any set of 16 ages. It is, however, just as tedious to write out as the previous form.

The $\sum$ notation provides a compact representation of a sum like the one above. We write

$$\sum_{i=1}^{16} a_i \equiv a_1 + a_2 + a_3 + a_4 + \cdots + a_{15} + a_{16}.$$

The idea is that whenever you see

$$\sum_{i=1}^{N} a_i$$

you know that it represents a sum of N quantities with the names a_1, a_2, $a_3, \ldots, a_{N-2}, a_{N-1}, a_N$.

In this notation the average age of the students of Table 5.2 can be written

$$\text{average age} = \frac{1}{16}\sum_{i=1}^{16} a_i.$$

The notation enables us to say that the average of any N quantities a_1, $a_2, \ldots, a_N$ is

$$\text{average of } a = \frac{1}{N}\sum_{i=1}^{N} a_i. \qquad (18)$$

Equation 18 is the general definition of an average.

There are many other useful applications of the $\sum$ notation. Sometimes the quantities being summed may be expressed algebraically in terms of the indices. For example, if you want to write down the sum of all the odd integers from 1 to 17 you can write

$$\text{sum of odd integers} = \sum_{i=1}^{9}(2i - 1) = 81.$$

TABLE 5.3 Distribution of Ages of 16 Hypothetical Students

ID Index i	Age Group	Number of Students in Age Group
1	17	2
2	18	7
3	19	4
4	20	2
5	21	1

Distributions and Averages

The average is not always a useful number for describing a set of quantities. It conveys only limited information. For example, a class consisting of 8 one-year-olds and 8 thirty-five-year-olds has the same average age as a class of 16 eighteen-year-olds. Knowing the average age would not tell you which of these two classes you would prefer to be in. Knowledge of the "distribution" of a set of quantities gives much more complete information about the set of data.

Let's reorganize the data of Table 5.2 to exhibit its distribution. We group the data by age. Then we tabulate the number of students in each group. Table 5.3 shows the result of this grouping. Because there are five different ages in this particular case, we created five groups or "bins." We have labeled each bin by assigning sequential identification numbers from 1 to 5.

Now we can describe the set of numbers by saying that in bin 1 there are 2 occurrences of the number 17; in bin 2 there are 7 occurrences of the number 18; in bin 3 there are 4 occurrences of the number 19; in bin 4 there are two occurrences of the number 20; and in bin 5 one occurrence of the number 21. Such sets of data are called "distributions" because they show how a property is distributed over a set of entities. Our example shows how the property of age is distributed over the set of students in a class.

Implicit in our example is the important idea that a bin has width. None of the students is exactly 17 or 18 or 19, etc. Indeed, we need to say what we mean when we say that someone is 18. Most likely, we mean that person has passed her eighteenth birthday and not yet reached her nineteenth, but life insurance companies usually mean that you are somewhere between 17.5 and 18.5 years old. In either case, we are grouping our data into bins that are 1 year in width. We could use a finer grouping, say in terms of months, or we could use a coarser grouping, e.g., two-year wide bins. Choosing the size of the bin for a distribution is an important decision based largely on common sense about how much information there is in the data set and on your judgment about how much of that information will be useful.

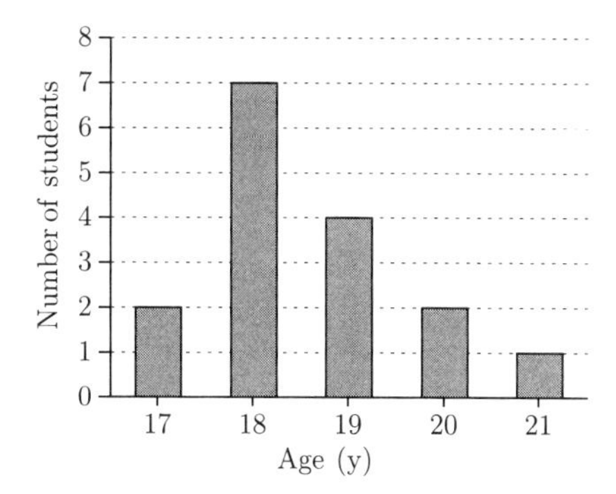

FIGURE 5.6 Distribution of student ages in our hypothetical class.

EXERCISES

24. In the example above why wouldn't you choose bins one month wide? Why not choose bins two years wide?

It is very helpful to represent a distribution graphically. This is usually done as a vertical bar graph. Each bar represents a bin. The width of the bar corresponds to the width of the bin, and the height of the bar represents the number of entities with values in the bin. Such a graphical representation is called a "histogram." You will use them frequently in this book. Figure 5.6 presents a histogram of the distribution of the data in Table 5.3.

EXERCISES

25. Find the ages of the students in your class and plot their distribution histogram.

We often want to calculate an average from a distribution. It is much easier to do this from the distribution data of Table 5.3 than from the complete data set in Table 5.2. In long form we can compute

$$\text{average age} = \frac{2 \times 17 + 7 \times 18 + 4 \times 19 + 2 \times 20 + 1 \times 21}{2 + 7 + 4 + 2 + 1} = 18.56\,\text{y}.$$

In the numerator and in the denominator there are as many terms as there are bins. In the numerator each term is the product of the quantity being averaged and the frequency with which it occurs. In the denominator each term is just the frequency of occurrence. We can put this in a compact and general form using the $\sum$ notation if we properly organize the names and labeling of the quantities involved.

You can see by referring to Table 5.3 that you must keep track of three different things when describing a distribution using the summation notation. First, you must know which bin you are talking about, so you need a bin number i. Second, you need the numerical value X_i that corresponds to bin i, this is the age in column 2 of Table 5.3. Third, you need the number of times that each value of X_i occurs; this is designated n_i here and is often called the frequency. In terms of these symbols we can rewrite the average as

$$\langle X \rangle = \frac{\sum_{i=1}^{i_0} n_i X_i}{\sum_{i=1}^{i_0} n_i}, \tag{19}$$

where i is the bin number; i_0 is the total number of bins; and n_i is the number of occurrences of the quantity X_i. Notice that $\sum_{i=1}^{i_0} n_i = N$ will always be the total number of quantities, 16 in the above example.

By convention the angle brackets denote "average value." Other notations used for average value are X_{ave} or $\bar{X}$ (read as "ex-bar"). In this book we usually use the angle brackets.

Equation 19 is the general form for the average of any distribution of quantities. When you see summations that look like this, you know they are averages.

A Distribution of Velocities

Let's apply these ideas to the data in Table 5.4. The table presents a distribution of velocities of some nitrogen molecules. There are really 29 bins in the distribution, numbered from -14 to $+14$. Each bin is 50 m/s wide. The bins from -14 to -1 are not shown because they are identical to the bins from 1 to 14. The velocity given in the table is the velocity at the midpoint of a bin.

EXERCISES

26. Fill in the blanks in Table 5.4.

TABLE 5.4 Distribution of x-Component Velocities of 1000 N_2 Atoms

x-velocity		number per cm^3 in the velocity interval $\delta v = 50$ m/s	
v_{xi}	(m/s)	n_i	bin number i
v_{x0}	0	68	0
v_{x1}	50	67	1
v_{x2}	100	64	2
v_{x3}	150	60	3
v_{x4}	200		
	250	51	5
v_{x6}		41	
	350	34	7
v_{x8}	400		8
v_{x9}	450	21	9
v_{x10}	500	16	10
v_{x11}	550	12	11
v_{x12}	600	9	12
v_{x13}	650	6	13
v_{x14}	700	4	14

Momentum Transfers by Collision

Figure 5.3 represents an atom rebounding from a wall. Assume that it has an x-component of velocity v_{xi} and that in a unit volume of the gas there are n_i atoms that have this same velocity. For these atoms the amount of momentum transferred to the wall in a time Δt is just

$$\delta p_i = (2mv_{xi})(n_i/2)(A\, v_x i \Delta t) = n_i A \Delta t v_{xi}^2.$$

This is the same result worked out in the text, because the atoms in a single bin are all taken to have the same magnitude of velocity in the x-direction. The difference is that we now have to add together the contributions of atoms that have different velocities, i.e., atoms in other bins. In other words, we propose to find the total change of x-momentum Δp by summing δp_i over all the different velocity bins indicated by i. Formally, this is

$$\Delta p = \sum_{i=1}^{i_0} \delta p_i = mA\Delta t \sum_{i=1}^{i_0} n_i v_{xi}^2,$$

where factors common to every term have been taken outside the sum. $\Delta p/\Delta t$ is the total force F exerted by the wall on the gas. By Newton's third law of motion, the gas must exert an equal but opposite force on the wall. Hence, F is the force on the wall, and F/A is the pressure P of the gas on the wall. We get

$$P = m \sum_{i=1}^{i_0} n_i v_{xi}^2.$$

This looks almost like an average. If it were divided by $\sum_{i=1}^{i_0} n_i$, the sum of the number of occurrences of each velocity, it would be an average. We multiply and divide by that sum, recognizing that the sum of the number of atoms per unit volume with a particular velocity v_{xi} must be the total number of atoms per unit volume, which we have been writing as n, i.e.,

$$P = mn \frac{\sum_{i=1}^{i_0} n_i v_{xi}^2}{\sum_{i=1}^{i_0} n_i}.$$

Referring back to Eq. 19, we see that

$$P = nm\langle v_x^2 \rangle,$$

where $\langle v_x^2 \rangle$ is the average of the squares of the x-components of the atomic speeds.

The argument in the text shows that $\langle v^2 \rangle = 3\langle v_x^2 \rangle$, which enables us to obtain the final result

$$P = \frac{nm\langle v^2 \rangle}{3}.$$

Velocity Bins

The above calculation assumes that the atoms have been sorted into i_0 bins. We never said how big i_0 is, and in fact, since no two atoms have exactly the same velocity, i_0 could be infinite. Calculus enables us to deal with that problem, but a little fudging does just as well. We define a bin width, a range of speeds, and we count as being in a particular bin all those atoms that have speeds within the specified range. For example, in Table 5.4 the bin width is $\delta v_{xi} = 50\,\text{m/s}$. The table says that in a collection of 1000 atoms in a cubic centimeter, about 64 of them have velocities between 75 and 125 m/s.

EXERCISES

27. Calculate the average velocity of these atoms.

28. If there are 1000 atoms in a cubic centimeter, what should be the sum of n_i in Table 5.4? What is it? Explain your answer.

29. What is the average of the square of these velocities? Calculate it directly from Table 5.4 (you might use a spreadsheet), and calculate it assuming that the gas is at room temperature. How do the two answers compare?

30. What will be the total change in the average momentum of all of these atoms that strike a wall in a time interval of 1 s?

31. For Table 5.4 evaluate $\displaystyle\sum_{i=-14}^{14} n_i$.

32. For Table 5.4 evaluate $\displaystyle\sum_{i=-14}^{14} n_i v_i$

33. For Table 5.4 evaluate $\displaystyle\sum_{i=-14}^{14} n_i m v_i^2$.

34. Make a graph of n_i vs. v_i. How would the values of n_i change if you used $\delta v_i = 25\,\text{m/s}$ instead of 50 m/s? How would the graph change under these circumstances? How many velocity groupings would you need now? Can you see that if instead of plotting n_i you plotted $n_i/\delta v_i$, you would get graphs that were nearly the same in the two cases? And can you see that the smaller you made δv_i, the smoother the curve would become?

The limiting ratio of the number in a velocity bin to the width of the bin is called the distribution function. In the limit of vanishingly small δv_i the ratio is a continuous function of v and is often written $n(v)$, although it is really just the derivative of the occurrence frequency with respect to the velocity. Every distribution function is a derivative. The notation $n(v)$ is also confusingly close to the n_i in Table 5.4, but it is different. For one thing, the units of $n(v)$ are number per m^3 per m/s (which is m^{-4} s); the units of n_i are just m^{-3}.

PROBLEMS

1. What is meant by the "mean free path" of a molecule?

2. How is viscosity η defined? What is a typical value of the viscosity of a gas?

3. **a.** Derive a relation between number density n, average speed $\langle v \rangle$, and mean free path ℓ. Take $\langle v \rangle \approx v_{\text{rms}}$.

b. What is the mean free path of a nitrogen molecule in air at room temperature?

4. In a lab experiment, N_2 molecules escape into a vacuum through a hole 0.051 cm in diameter. At what pressure will the mean free path of nitrogen molecules become comparable to the size of the hole?

5. Describe Loschmidt's method for estimating Avogadro's number, N_A.

6. Given that solid CO_2 at $-79°C$ has a density of 1.53 g cm^{-3} and that CO_2 vapor at STP has a viscosity $\eta = 13.8\ \mu$Pa s, estimate Avogadro's number using Loschmidt's method.

7. Given that $N_A = 6.02 \times 10^{23}$, estimate the size of an atom.

8. Suppose N_2 molecules are known to have an rms velocity of 500 ms^{-1}. What would be the rms velocity of hydrogen molecules at the same temperature and pressure? Explain your reasoning.

9. We found that the viscosity of an ideal gas is $\eta = 0.499\rho\langle v \rangle \ell$, where ρ is the density of the gas and the mean free path

$$\ell = \frac{1}{\sqrt{2}\pi(2r_m)^2 n},$$

where n is the number per unit volume of molecules or atoms of radius r_m. At STP, nitrogen gas in a $10\,\mathrm{m}^3$ container has a density of $1.25\,\mathrm{kg\,m}^{-3}$, a viscosity of $17\,\mu\mathrm{Pa\,s}$, and a mean free path of 60 nm. Suppose that without adding gas the volume of the container is doubled while keeping the temperature T constant. What then is

a. the density?

b. the rms velocity?

c. the mean free path?

d. the viscosity?

10. Assume you can prove that for our simple atomic model of an ideal gas the pressure $P = \frac{1}{3}\rho\langle v^2\rangle$, where ρ is the density of the gas and $\langle v^2\rangle$ is the average of the square of the velocity of the gas molecules.

a. Use the ideal gas law to derive a relation between the temperature of this gas (measured in kelvins) and the average kinetic energy of a gas molecule.

b. If the gas is helium ($M = 4\,\mathrm{u}$), what is the value of the average kinetic energy of a helium atom at room temperature?

c. If the gas consisted of oxygen molecules ($M = 32$), what would be the average kinetic energy of a molecule at room temperature? Explain your answer.

11. Consider the following table of measurements of pressure vs. volume for air in a closed volume at room temperature

Pressure	Volume
(kPa)	(cm^3)
90.9	24.89
73.6	30.73
59.5	37.93

a. Show that these data obey Boyle's law.

b. How many molecules are there in this volume? Explain your answer.

c. What is the rms velocity of the oxygen molecules in this sample of air? Explain your answer.

d. If the temperature is increased by 15%, by how much does the rms velocity of the oxygen molecules change? Explain your answer.

12. A quantity of oxygen gas is contained in a vessel of volume $V = 1\,\mathrm{m}^3$ at a temperature of $T = 300\,\mathrm{K}$ and a pressure of P. The vessel is connected to

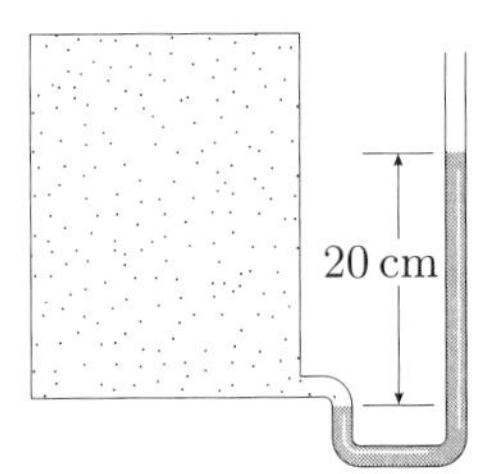

FIGURE 5.7 Apparatus for Problem 12.

a mercury-filled tube as shown in Fig. 5.7. Note that the upper end of the tube is *open* to the atmosphere.

a. Is P greater or lesser than 1 atm? Calculate P in units of *Torr* and also *pascals*.

b. If the temperature of the gas is doubled, keeping V constant, by what factor does each of the following change?

- **i.** density (g/cm^3)
- **ii.** average kinetic energy of a molecule
- **iii.** rate of molecular collision with the walls
- **iv.** v_{rms}
- **v.** mass of 1 mole of gas

c. By means of a small pump the gas pressure is reduced to 100 Torr, while the temperature and volume remain fixed at 300 K and 1 m^3. What then is the average kinetic energy of 1 molecule of the gas? Express your final answer in eV.

13. A box of H_2 gas is at STP

a. What is the pressure of the gas in

- **i.** Torr
- **ii.** pascals
- **iii.** atmospheres

b. What is the root-mean-square velocity of the H_2 molecules?

c. The temperature of the gas of H_2 is tripled and the number of molecules is halved. All else remains the same.

- **A.** What is the new pressure in the box?
- **B.** How does each molecule's average kinetic energy change from its value at STP?

14.a. If 1 liter of oxygen gas at 0°C and 101 kPa pressure contains 2.7×10^{22} molecules, how many molecules are in 1 liter of hydrogen gas at the same temperature and pressure? How do you know?

b. If at room temperature the average kinetic energy of an oxygen molecule ($M_{O_2} = 32$) is 0.04 eV, what is the average kinetic energy of a hydrogen molecule at the same temperature? Why?

c. When hydrogen and oxygen gas combine to form water, about 2.5 eV of energy is released as each water molecule forms. Assume that the hydrogen and oxygen combine inside a closed tank and that 10% of the released energy goes into kinetic energy of the molecules. Estimate the rise in temperature of the water vapor.

15. Suppose you have a small cube with "3" painted on 3 sides, "4" painted on 2 sides, and "1" painted on the sixth side.

a. What is the frequency distribution of the numbers?

b. What is the average value of the **squares** of the numbers?

16. What is the average kinetic energy of one mole of oxygen molecules at 27°C?

17. A beach ball with a radius of 0.5 m is rolling along the sidewalk with a speed of $2\,\mathrm{m\,s^{-1}}$. There is a light mist consisting of 3.1 small droplets of water in every cubic meter of air. Estimate to within a factor of 2 the mean distance the ball travels between collisions with droplets of the mist.

18. If the mean free path of a nitrogen molecule in a bell jar is 100 nm at atmospheric pressure, what is its mean free path at 0.76 Torr? Show how you get your answer.

19. If in the previous question you raise the temperature of the gas from 300 K to 600 K,

a. By how much does the collision time change?

b. Explain why the mean free path does not change when the temperature goes up.

20. Nitrogen gas (N_2, M = 28 u) at room temperature (300 K) and atmospheric pressure has a density of $1.25\,\mathrm{kg\,m^{-3}}$. The measured root mean square velocity of a N_2 molecule is $v_{rms} \approx 500$ m/s.

a. What is the average energy of a N_2 molecule under these conditions? Express your answer in J as well as eV.

b. At the same temperature and pressure, what is the average energy of a helium atom ($M = 4.0$ u)?

c. Under these conditions, what is v_{rms} for a helium atom?

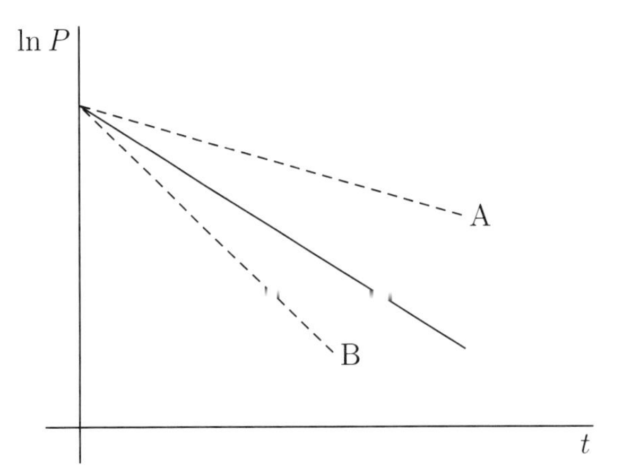

FIGURE 5.8 Logarithm of pressure vs. time for gases leaking into a vacuum (Problem 20).

d. Figure 5.8 shows a plot of the logarithm of pressure ($\ln P$) in a jar vs. time for an experiment in which gas leaks out of the jar through a tiny hole into a vacuum. The heavy solid line shown is for N_2. Which of the two dotted lines (A or B) best approximates the behavior of helium under the same experimental conditions? Explain briefly.

21. We showed that the pressure of a gas is related to its density ρ (mass/volume) of the gas and the average of the square of the velocity of the gas molecules:

$$P = \frac{1}{3}\rho\langle v^2\rangle.$$

a. Starting from the above equation, prove that the total kinetic energy of a gas is

$$KE_{\text{tot}} = \frac{3}{2}PV.$$

b. Starting from the last equation, prove that the average kinetic energy of a single molecule is

$$KE(\text{one molecule}) = \frac{3}{2}k_B T,$$

where k_B is Boltzmann's constant.

c. If the gas is H_2, what is the value of the average kinetic energy of a molecule at room temperature?

d. Find the ratio between the rms velocities (v_{rms}) for O_2 ($M = 32$) and H_2 molecules at the same temperature.

22. A spherical balloon with negligible elastic force is filled at room temperature (20°C with enough helium gas that it just floats in the surrounding air at standard atmospheric pressure. If the balloon is 2 m in diameter,

a. What is the pressure of the gas in the balloon?

b. What is the mass of the gas it contains?

c. What is the density of the gas in the balloon?

d. What is the density of the air outside, taking it to be 20% O_2 and the rest N_2?

e. What is the mass of the balloon without the gas?

C H A P T E R

Electric Forces and Fields

6.1 INTRODUCTION

By imagining a gas to be a collection of tiny spheres, we were able to explain many features of the behavior of gases and estimate the number and size of molecules. The results gave further credibility to the idea that atoms exist, but the numbers were imprecise, yielding estimates of Avogadro's number and atomic sizes accurate only to within an order of magnitude, i.e., only to within a factor of ten.

Greater precision came as physicists better understood electricity and magnetism. More important, along with increasing precision came a growing understanding of the inner structure of the atom. During the two decades bracketing the start of the twentieth century the electrical nature of the atom was discovered, the electron's minute electric charge was measured to better than 1%, and its behavior inside the atom began to be mapped out.

Before you can understand these advances, you need to review some electric and magnetic phenomena and the rudiments of how they are described and understood. In particular you need working ideas of

- electric charge,
- electric field and electric force,
- electric potential,
- electric current,
- magnetic field and magnetic force.

6.2 ELECTRIC CHARGE

Like mass, length, and time, electric charge is defined with great precision and care, but this definition is not very helpful for understanding the concept. Therefore, as for mass, length, and time, so for charge, we ignore the precision and care, and try to introduce the idea of charge in a more intuitive way.

The concept of electric charge is needed to explain the curious but simple experiments described below.[1]

Experiments with Electroscopes

First hang a thin, folded piece of gold or aluminum foil from a hook as shown in Fig. 6.1. The foil should be metallic, light in weight, and flexible. In fact, the foil should be so flexible that it is a good idea to put it inside a glass case

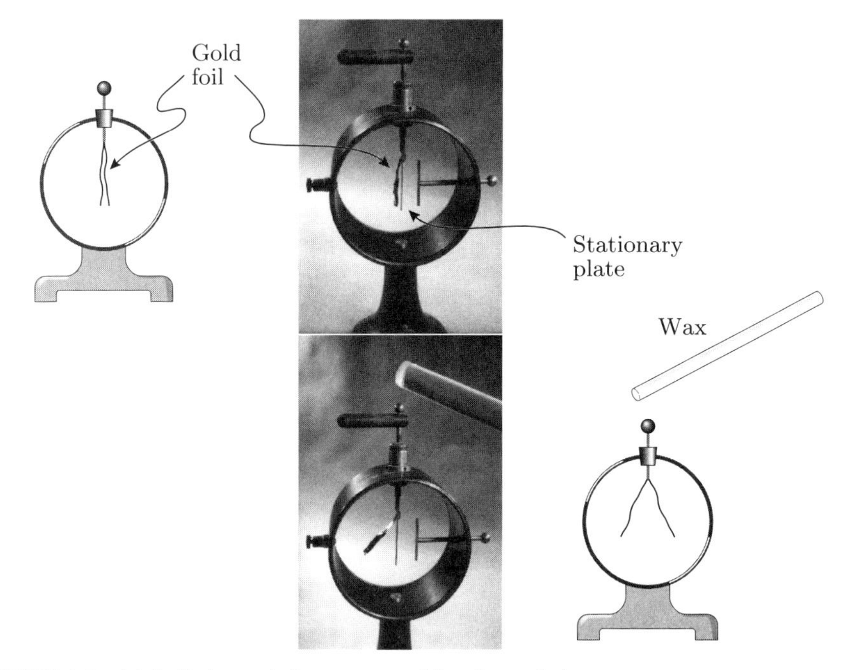

FIGURE 6.1 (a) A discharged electroscope; (b) a charged electroscope.

[1]Albert Einstein and Leopold Infeld in their fine book *The Evolution of Physics*, Simon & Schuster, NY 1938, point out that these experiments exhibit especially well the interconnection of theory and experiment. No one would dream of doing these experiments deliberately unless one already had a reasonably clear idea of the theory that is to be used to interpret them. Only then does the experimental plan make any sense.

to protect it from air currents. (Foils of gold leaf can be made so thin that they transmit light.) Then the foil must be connected by a metal path to a knob on top of the glass case. This device is called an "electroscope."

Next take a rod made of sealing wax or hard rubber and wipe it firmly with cat fur. Now hold the rod near the electroscope without touching it. If you have performed these operations correctly, the folded piece of foil will spread apart. The two leaves of the foil repel each other as shown in Fig. 6.1, and we say that the leaves of the electroscope have become "charged."[2]

If you move the rod away from the electroscope, the two leaves fall back together. The leaves become "discharged." If you bring the rod back near the electroscope, the leaves again spread apart.

If you touch the sealing wax to the metal knob connected to the electroscope's leaves, something new happens. Now when you move the sealing wax away, the leaves do not come entirely back together. Some mutual repulsion remains.

If you repeat the experiment using a glass rod that has been rubbed with silk, exactly the same sequence of events is seen. The leaves spread when the rod comes near; they fall back together when the rod is moved away. If the electroscope knob is touched with the rod, the leaves remain apart when the rod is moved away. This sequence of events is shown schematically in Fig. 6.2.

Now here is something new and curious: Suppose you charge the electroscope by touching it with the glass rod and then bring the sealing wax near. As the sealing wax previously rubbed with cat fur comes near the electroscope charged by the glass rod, the leaves of the electroscope come back

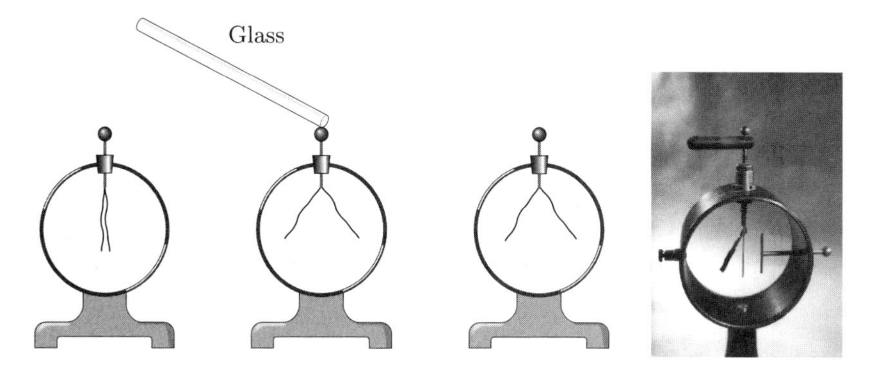

FIGURE 6.2 An electroscope charged by a glass rod rubbed with silk.

[2] Originally, the word "charged" was used simply in the sense that the electroscope was being loaded with something, as we say a gun is charged and refer to its load as a "charge."

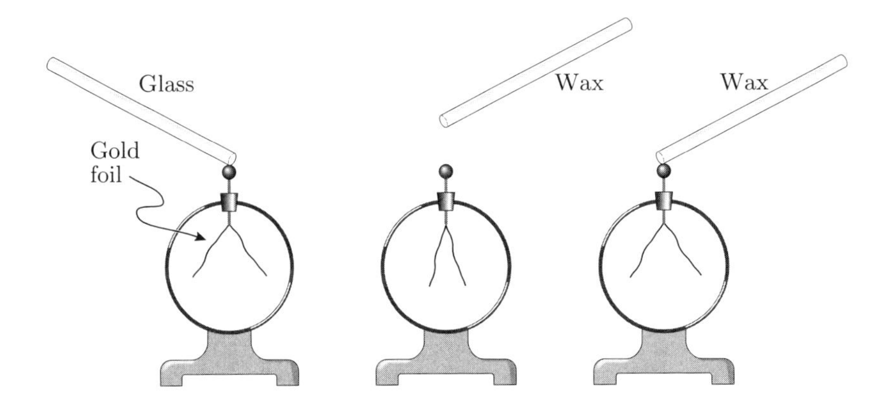

FIGURE 6.3 Sequence of events as rubbed sealing wax is brought nearer and nearer to an electroscope charged by a rubbed glass rod.

together. The electroscope appears to discharge! If the sealing wax has been rubbed vigorously, you may see the electroscope leaves close up as the wax is brought near and then spread apart as the wax comes closer yet!

Apparently, the property of the glass rod that causes the electroscope leaves to separate is different from the property of the sealing wax that causes a similar separation of the leaves.

We can make sense of these experiments by the following model. Assume there exist two "fluids"—"electric fluids" we might call them. In combination they exert no effect, and we say they are "neutral." When separated, the two fluids attract each other. They appear to strive to combine, each fluid tending to draw enough of the other to achieve neutralization.

The two fluids are called "positive" and "negative" partly because arithmetically a given amount of + will cancel an equal amount of − and produce a null value that corresponds to neutrality. The + and − kinds of electric fluid attract each other, but + repels + and − repels −. Unlike kinds attract; like kinds repel.

With this model we can interpret the above experiments as follows. Rubbing the sealing wax with cat fur removes some of one of the "fluids" and leaves an unneutralized remnant of the other. Similarly, the glass rod rubbed with silk gets left with an unneutralized amount of one of the fluids. Our experiments with the electroscope show that the sealing wax and the glass rod have been given excesses of unlike fluids.

Let's stop calling them "electric fluids." That terminology was acceptable in the eighteenth century,[3] but nowadays we call them "electric charge."

[3]When the internationally recognized American physicist Benjamin Franklin worked out these ideas.

In this terminology we say that positive and negative charges attract each other, but positive charges repel positive charges, and negative charges repel negative charges. In other words, *unlike charges attract; like charges repel.*

Which charge is on the glass rod? Which is on the sealing wax? The answer depends on how we choose to name them, and that is a matter of convention.

The official and universally accepted definitions of the names are as follows.

positive charge: any charge repelled by a glass rod that has been rubbed with silk;

negative charge: any charge repelled by sealing wax that has been rubbed with cat fur.

Thus, the proper designation of + and − on batteries or electrical power supplies throughout all the world's great electrical industries depends ultimately on cat fur.

Now we can explain the behavior of the electroscope when the sealing wax comes near as in Fig. 6.4. This is the same drawing as Fig. 6.1, except that the distribution of charges is shown. The sealing wax with its negative charge repels negative charge on the electroscope. The charge moves down onto the leaves of the electroscope; these now repel each other, and they spread apart. When the sealing wax is moved away, the negative charge moves back to the portions of the electroscope from which it had been displaced; the leaves become neutral again and collapse together.

If the sealing wax touches the electroscope, some of the negative charge on the sealing wax transfers to the electroscope, leaving it with a net negative

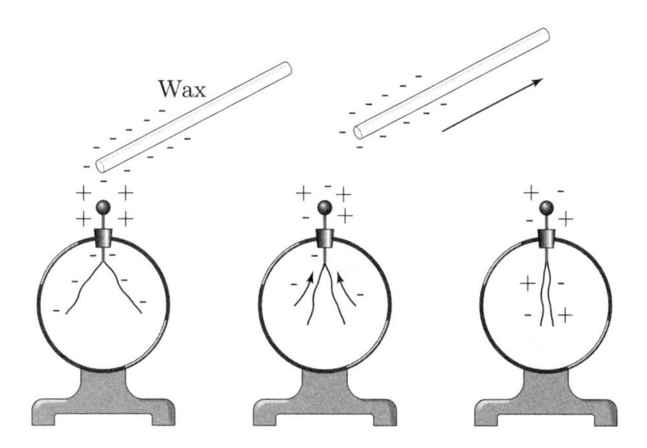

FIGURE 6.4 Negative charge on wax rod induces negative charge on electroscope leaves. As the rod is withdrawn, the charges redistribute themselves, and the foil leaves collapse.

charge. Now when the sealing wax is moved away, the electroscope cannot become neutral, and the leaves remain separated.

Notice that we cannot tell whether negative charge has been added to the electroscope or whether positive charge has been taken away. These experiments will not allow us to determine what kind of charge goes onto the electroscope or in what combination. From other experiments we now know that in metals it is the negative charge that is free to move while the positive charge is quite fixed. In liquids both positive and negative charges move.

Exactly the same description of the above three paragraphs applies to the glass rod and the electroscope except that now the charge is positive.

EXERCISES

1. If you understand this model of charge and its properties, you should be able to explain why an electroscope charged by a glass rod is first discharged and then charged as a piece of sealing wax is brought closer and closer. (See Fig. 6.4.) Explain it.

Conductors and Insulators

The electroscope's metal rod and leaves are particular examples of substances called electrical "conductors." An electrical conductor (or, more often, simply a conductor) is any substance on or in which electric charge moves easily. The outstanding example of conductors are the metals, but salt water and other solutions are also conductors of electricity.

The wax and the glass illustrate the opposite property. They are called "insulators," and charge does not move easily on or in them. You can see why the foils of the electroscope must be conductors if the device is to work well. Why must the wax and the glass rod be insulators? One reason is that if they were not insulators, the charge that you managed to rub onto them would immediately flow along them into your body and on out to the ground. The human body, as a bag of salt water, is a pretty good conductor.

The existence of materials with these two quite different properties is basic to all of electrical technology. Wires are possible only because they can be made of conducting material such as copper or aluminum. But control of the flow of electricity along a wire is possible only because you can wrap the wires in a material, such as rubber, fiberglass, or plastic, that does not conduct electricity.

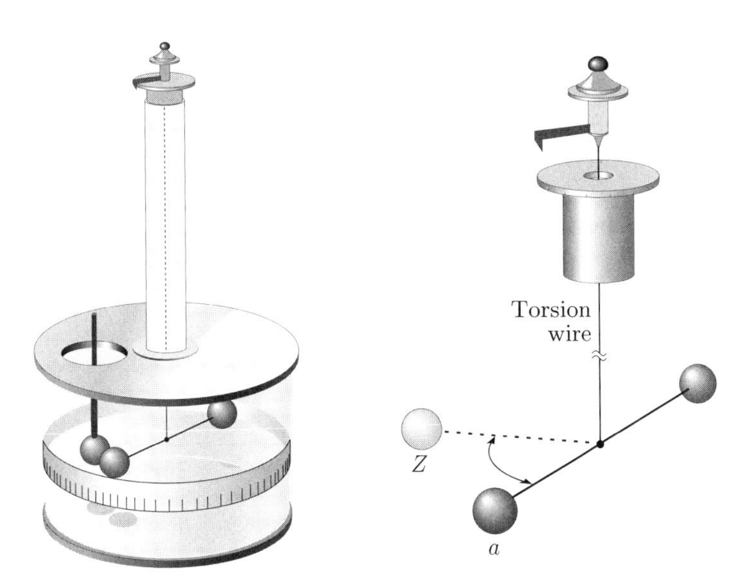

FIGURE 6.5 Coulomb's torsion balance.

Quantitative Measures of Charge

As always in physics, we want to make this new concept quantitative. To do this we arrange pieces of charged matter so that their attractive or repulsive forces on each other produce some observable mechanical effect such as motion, compression of a spring, deflection of a lever, or the twist in a fiber. The amount of the mechanical effect can then be used as a measure of the amount of charge present. One early method used the angle of separation between the leaves of the electroscope as a measure of the amount of charge on them.

The French engineer and physicist Charles Augustin Coulomb[4] developed a more precise measure. He built a device called a "torsion balance," shown in Fig. 6.5, that used the amount of torsion (twist) induced in a very fine wire or fiber to measure forces smaller than 10^{-8} N.

Coulomb set up a glass chamber consisting of a tall, narrow cylinder sitting on a short, large-diameter cylinder about 30 cm across and 30 cm high. The tall, narrow cylinder housed a very fine, delicate torsion wire that was clamped to a rotatable cap at the top of the housing. A lightweight, thin insulating rod, about 20 cm long, was hung by its midpoint on the lower end of the fine wire.

[4]Born in Angoulême, France on 14 June 1736; died in Paris on 23 August 1806. Had two sons, one born in 1790, the other in 1797. He married in 1802.

At one end of the rod was a small ball (ball a in Fig. 6.5). Another ball (ball Z) could be inserted through a hole in the cover of the larger cylinder and positioned near ball a. In this arrangement when balls a and Z were charged, ball Z would exert a force on ball a and cause the dumbbell-like object to rotate and twist the fine wire. The amount of twist in the wire measures the amount of force between ball a and ball Z.

To do the experiment, Coulomb first turned the cap that twisted the torsion wire so that the rod lined up where he wanted it. Then he put a charge on ball Z and inserted it so that it touched ball a and transferred some charge to it. The balls then repelled each other, and the torsion wire twisted through 36°. To see how the force between Z and a depended on their separation, he turned the cap at the top of the housing through 126° in order to move ball a to 18° from ball Z. He observed that by turning the cap through 567° the balls moved to within 8.5° of each other. This meant that the amount of twist in the fiber was 36°, 144°, and 575.5°, respectively, as the separation between the balls was halved, and then halved again.

EXERCISES

2. Show that these results imply that the force between the two balls depends inversely on the square of the distance between them.

With this device he showed that the magnitude of the force between charges small enough to be considered point charges was inversely proportional to the square of the distance between them:

$$F \propto \frac{1}{r^2},$$

where r is the distance between the centers of the two charged balls. The force was also found to be proportional to the product of the amount of the two interacting charges.[5] If we call those charges q_1 and q_2, then we have

$$F \propto q_1\, q_2.$$

Combining these two equations and letting k_c be some constant of proportionality, we have what is known as Coulomb's law for the force between two spherical charges:

$$F = k_c \frac{q_1\, q_2}{r^2}. \qquad (1)$$

[5] These are rather casual sounding statements. A deeper insight into Coulomb's accomplishments can be gotten if you ponder how he might have been able to arrive at such generalizations without having ways of measuring amounts of charge.

We can use this equation to define a unit of charge. By convention we agree that two identical charges separated by 1 m and exerting a force of 8.98755×10^9 N on each other each have a charge of 1 coulomb. This statement is a definition of the "coulomb" as a unit of charge. You might object that this is a strange way to do things. Why not just let $k_c = 1$? In fact, that was once the way it was done. However, some awkwardness in combining electric and magnetic units led to the creation of the currently used definition.[6] The abbreviation of the coulomb is C. All the SI multiples can be used, but the small ones like pC, nC, and μC are especially common.

Adopting this definition for our unit of charge means that we have agreed to assign to k_c, the constant in Eq. 1, the value $8.987552 \times 10^9\,\mathrm{N\,m^2\,C^{-2}}$. In this book, as in most practical work, we round off k_c and use the value $k_c = 9 \times 10^9\,\mathrm{N\,m^2\,C^{-2}}$.

For various reasons the constant k_c is often written as $k_c = 1/(4\pi\epsilon_0)$, where ϵ_0 has the mystifying name of "permittivity of free space" and has the precise value of $8.854187818 \times 10^{-12}\,\mathrm{C^2\,N^{-1}\,m^{-2}}$. We will be using the k_c notation for now, but some of the equations you will see later may be written in this other form.

EXERCISES

3. What is the force between a charge of 1.1 μC and one of 1.3 μC when they are separated by 1 cm? By how much does the force change if they are moved to a separation of 1 μm?

4. Two point charges of equal magnitude are placed 2 m apart. They are observed to exert a force of 9 μN on each other. What is the amount of each charge? What is the sign of each charge?

5. Suppose the two charges of the previous problem are moved until they are 1 m apart. What now is the force between them?

6. Coulomb calibrated his torsion fiber and found that it required a force of 1.87 μN to twist it through 360°. Assuming that the pith balls were equally charged, how much charge was on each?

---

[6]The official SI definition of the coulomb is actually cast in terms of an experimental method that is capable of much more precision than is possible by an experiment derived from Coulomb's law, but the results are the same. The official coulomb is defined as the amount of charge corresponding to the passage of 1 ampere of current for 1 second. The ampere of current is defined in terms of the force between two wires carrying that amount of electric current. We don't need to go into those details now.

6.3 ELECTRIC FIELD: A LOCAL SOURCE OF ELECTRICAL FORCE

Coulomb studied the force exerted by one charge on another. In this description, one charged object pushes or pulls another one some distance away. This is called "action at a distance." Many people don't like this idea. They feel intuitively that the force on an object should come from something located right where the object is. To satisfy this view, physicists invented the idea of a "field of force." We say that one charge produces something called an "electric field" that extends throughout all space. When another charge is placed anywhere in this electric field, it experiences a force because of the electric field. In other words, q_1's electric field exerts a force on q_2. This relationship is reciprocal, i.e., q_2 also produces an electric field at the location of q_1 and so exerts a force on it.

In principle it is easy to test for the presence of an electric field E without knowing how it was produced. Place a small neutral particle wherever you wish to test for an electric field. Observe its acceleration, that is, the effect of gravity. Now put a charge q_t (t for test) on the particle. If the charged particle now has a different acceleration from what it had when it was neutral, there is an electric field acting on it. The force due to the electric field acting on q_t is used to define the strength (magnitude) of the field:

$$E = \frac{F}{q_t}. \tag{2}$$

An important feature of E is that it is independent of q_t; the ratio F/q is the same for any charge q. In other words, the force exerted by an electric field on a charge q is strictly proportional to q. Therefore, Eq. 2 says that if you know E at some point in space, you can immediately calculate the force that will be exerted on any charge Q placed at that point in space. It will be

$$F = QE.$$

You just multiply the field strength E times the charge you are observing; that gives the force on it.

Equation 2 shows that the units of electric field are N C^{-1}, "newtons per coulomb."

Two Useful Electric Fields

An electric field is produced by electric charges, and different arrangements of charge produce different spatial distributions of electric field. There are two special electric fields that are important for you to know: (1) the electric field produced by a point charge or a sphere of charge; (2) the electric field produced in the space between two parallel, oppositely charged conducting plates.

Electric Field Outside a Point or Spherical Charge

EXAMPLES

1. A point charge produces an electric field everywhere in space. What is the magnitude of the electric field that a point charge of 0.5 C produces at a distance 2 m away from itself?

To answer this question place a small charge, let's say 1 μC, at a point 2 m from the 0.5 C charge. Then calculate the force on it. From Eq. 1 the force on this tiny charge will be

$$F = 9 \times 10^9 \frac{0.5 \times 1 \times 10^{-6}}{4} = 1.125 \times 10^3\,\mathrm{N}.$$

What then will be the value of the electric field at the position of the charge? If you just use the definition of Eq. 2, you get

$$E = \frac{F}{q_1} = \frac{1.125 \times 10^3}{1 \times 10^{-6}} = 1.125 \times 10^9\,\mathrm{N\,C^{-1}}.$$

In general, a point charge q_s (subscript s for "source") produces an electric field E in space a distance r away given by the expression

$$E = k_c \frac{q_s}{r^2}. \tag{3}$$

If q_s is positive, the field points radially outward; if negative, it points radially inward.

EXERCISES

7. Using Eq. 1 and the definition of electric field, show that the electric field, of a point charge is as written in Eq. 3.

Constant Electric Field

Another electric field that turns out to be useful is the constant field. When the electric field E has the same value everywhere, the acceleration of a

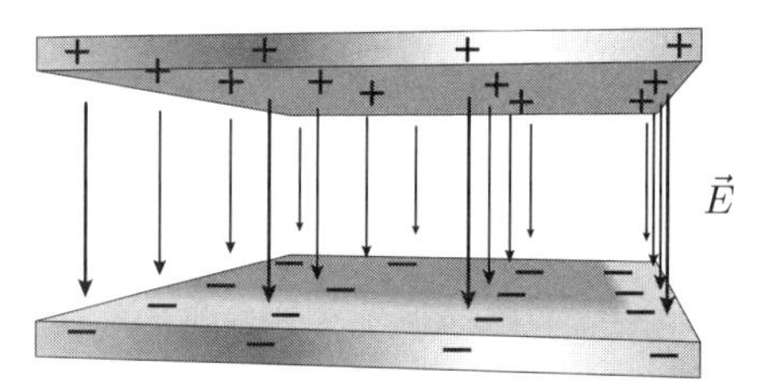

FIGURE 6.6 A common arrangement for producing a uniform electric field. You are looking edge on at a pair of parallel plates with inner surfaces uniformly covered with charge as indicated.

charged particle is constant everywhere. This greatly simplifies predicting the particle's motion.

It is also fairly easy to make an electric field that is nearly constant over a usable region of space. If you place opposite signs of charge on two parallel conducting plates, the charges will spread over the inside surfaces of the plates and produce a nearly constant electric field in the volume between them. If the separation between the plates is small compared to the lengths of the plates' edges, the electric field is quite uniform everywhere within this volume. Such an arrangement is shown schematically in Fig. 6.6.

EXAMPLES

2. Suppose you have a charge of $1\,\mu$C on a small sphere of mass $1\,\mu$g. When this charged mass is placed in a uniform (constant) electric field of $3\,\mathrm{N\,C^{-1}}$, how much acceleration does the electric field produce? Because we know that

$$F = E\,q = m\,a,$$

it follows that *anywhere* in the volume occupied by the electric field E

$$a = \frac{E\,q}{m} = \frac{3 \times 10^{-6}}{10^{-9}} = 3 \times 10^{3}\,\mathrm{m\,s^{-2}}.$$

EXERCISES

8. An electron has a charge of -1.60×10^{-19} C and a mass of 9.11×10^{-31} kg. Suppose it is placed in an electric field that points downward and has a magnitude of $1000\,\mathrm{N\,C^{-1}}$. What will be the magnitude and direction of its acceleration?

Direction of an Electric Field

An electric field has a direction. By convention it always points in the direction that a positive charge would accelerate if placed in the electric field.

EXERCISES

9. Explain why in Fig. 6.6 the direction of the electric field is as shown.

6.4 ELECTRIC POTENTIAL: V

The magnitude of an electric field at a point in space is equal to the magnitude of the force that would act on a unit charge if one were placed there, and the field exerts a force proportional to the amount of charge on any particle placed in it. If this force is the only one acting on it, the charged particle will accelerate. If the acceleration changes the magnitude of the particle's velocity, then the particle's kinetic energy changes. In general, changes in a particle's kinetic energy can be calculated from details of its acceleration, which is usually different at different points in space, and from its path of motion, which can be quite complicated.

However, if all you need to know is how much a particle's kinetic energy changes as it goes from one point in space to another, you can use the conservation of mechanical energy to avoid all the details. As discussed in Chapter 2, changes in a particle's kinetic energy ΔK are related to changes in its potential energy ΔU by the conservation of mechanical energy, which guarantees that $\Delta K = -\Delta U$. Of course, to apply this relationship to a charged particle in an electric field, you must know the particle's potential energy at every point in the space filled by the electric field.

The concept of "electric potential" gives a convenient way to specify the potential energy of a charged particle at any point in a given electric field independent of the amount of charge on the particle. But before we describe electric potential, let's review how changes in the potential energy of a body give information about its velocity by considering a body attracted by gravity near Earth's surface.

EXAMPLES

3. As the body falls, its gravitational potential energy is transformed into kinetic energy. Suppose a 2 kg mass m is held at height $h = 10$ m above the floor and then dropped.

- What is the total energy of the mass (relative to the floor)?
- By how much has the potential energy of the mass changed when it is 4 m above the floor?
- By how much has the kinetic energy of the mass changed when it is 4 m above the floor?
- What is the velocity of the mass when it is 4 m above the floor?

The total mechanical energy of this mass is conserved. Initially, it is at rest, so its kinetic energy is 0 J, and it has only potential energy, which is

$$mgh = 2 \times 9.8 \times 10 = 196\,\mathrm{J},$$

where $g = 9.8\,\mathrm{m\,s^{-2}}$ is the acceleration due to gravity near Earth's surface, and the zero of the potential energy is chosen to be the floor. Since this is the only mechanical energy it has, this must also be the *total* mechanical energy.

When the mass has fallen to where it is 4 m above the floor, its potential energy has become only

$$mgy = 2 \times 9.8 \times 4 = 78.4\,\mathrm{J}.$$

Because energy is conserved, the total energy must remain the same as before, 196 J. Where is the missing $196 - 78.4 = 117.6$ J of energy? It must be in kinetic energy, so

$$\frac{1}{2}mv^2 = 117.6\,\mathrm{J}.$$

Knowing the kinetic energy and the mass, you can find the velocity. Call the kinetic energy K. Then show that

$$v = \sqrt{\frac{2K}{m}} = \sqrt{\frac{235.2}{2}} = 10.8\,\mathrm{m\,s^{-1}}.$$

The example reminds you how to use conservation of mechanical energy to find the change in velocity that results from a change in potential energy.

Electric Potential Energy

A similar example also applies to charges in electric fields. Mechanical energy turns out to be constant for these systems too, and the sum of kinetic and potential energy stays the same.

But what is the potential energy of an electric charge? If the electric field E is constant, the answer is simple. Just as in the case of gravitational potential energy, the zero is arbitrary. If we set $U = 0$ at the starting point, the potential energy U of a charge q that has been pushed against a constant field E for a distance y is

$$U(y) = -q\,E\,y. \tag{4}$$

What will be the change in potential energy of a 2 C charge pushed 10 m against an electric field of 9.8 N C^{-1}? It will be $2 \times 9.8 \times 10 = 196$ J.

EXERCISES

10. What will happen to the charged particle if you let it go after having pushed it 10 m against this electric field?

11. Suppose you now release the charged particle. If its mass is 0.20 kg, how fast will it be moving after it has traveled 6 m?

12. Incidentally, it would be quite a trick to collect 2 C on a reasonably sized object. To see why this is so, calculate the force exerted by a 1 C point charge on an identical charge 6 cm away. Your answer should be a huge number.

It is nice that the potential energy function is so simple for a constant electric field. What is the potential energy function like for other kinds of electric fields? One other simple case is the potential energy function of a small particle carrying a charge q a distance r away from a point charge Q. The potential energy function of q is

$$U(r) = k_c \frac{q\,Q}{r}. \tag{5}$$

The potential energy of a point charge q in a *constant* E field is given by Eq. 4. The potential energy of q in the E field of a *point charge* Q is described by Eq. 5. You should know these simple formulas.

Although for other combinations of charge and field the formula for potential energy—if there is one—can be quite complicated, you can often do without any formula at all. If you know the *numerical values* of the potential energy at two points in space, you can find the change in potential energy when a charge moves from one point to the other just by subtracting one value of the potential energy from the other. Thus, if you measure that the potential energy of a 2 C charge is 196 J at a point A in space, and that at point B it is 78.4 J, then obviously, when the charge moves from A to B its potential energy changes by 117.6 J. Since the total energy of the charged particle does not change, the potential energy must have become kinetic energy, and the kinetic energy must have increased by 117.6 J.

Do you see that this electrical case works like the gravitational one? Suppose you have a map of the electric potential energy of a charged particle at different places. Then if you know the particle's mass, its electric charge, and its initial velocity, you can find its velocity anywhere that you know its potential energy. The new kinetic energy equals the original kinetic energy plus the change in energy the charge acquired by moving through the potential difference, i.e.,

$$\Delta K = -\Delta U.$$

EXAMPLES

4. Suppose you have a 2 mg mass with a charge of 1 μC and an initial kinetic energy of 1 J. If the charge moves so that its potential energy decreases by 1 J, what is its change in velocity?

Start by finding the particle's initial velocity v_1 from your knowledge of its initial kinetic energy (1 J). Since the potential energy drops by 1 J, the kinetic energy must increase by 1 J. Then find the final velocity v_2 from your knowledge that the final kinetic energy is 2 J. The change in velocity is the difference between the initial and final velocities, $v_2 - v_1$.

$$\frac{1}{2}mv_1^2 = 1,$$

$$v_1^2 = \frac{2}{m} = \frac{2}{2 \times 10^{-6}} = 10^6,$$

$$v_1 = 1.00 \times 10^3 \,\mathrm{m\,s^{-1}}.$$

Since the final kinetic energy is

$$\frac{1}{2}mv_2^2 = 1 + 1 = 2\,\mathrm{J},$$

the new velocity v_2 is

$$v_2 = \sqrt{\frac{4}{2 \times 10^{-6}}} = 1.414 \times 10^3 \,\mathrm{m\,s^{-1}},$$

from which it follows that the *change* in velocity is

$$v_2 - v_1 = (1.414 - 1.000) \times 10^3 = 414 \,\mathrm{m\,s^{-1}}.$$

Electric Potential—Volts

Now let's consider the idea of "electric potential." Although electric potential is related to electric potential energy, it is not the same thing. This means you have to be alert to whether the word potential is being used as a noun—electric potential—or whether it is being used as an adjective—potential energy.

We introduce the idea of electric potential to distinguish between two separate aspects of the electric potential energy of a particle carrying a charge q. The electric potential energy of the particle always arises from these two separate features: (1) the amount of electric charge q on the particle; (2) some property of the space in which the charge q is situated. In fact, electric potential energy U is always strictly proportional to q; if you double q, the particle's electric potential energy doubles.

Because the potential energy of a charge q is proportional to its own magnitude, we can separate the two factors just by dividing the potential energy U by q. What remains is then a property of the space alone; it has units of joules per coulomb, $\mathrm{J\,C^{-1}}$. It is this property of space that we call "electric potential." It is usually denoted by $V(x, y, z)$ or frequently just V. Do you see that $V(x, y, z)$ will always have a numerical value equal to the electric potential energy $U(x, y, z)$ of a unit electric charge?

In effect, the electric potential gives a map of the electrical condition of space. If you know the value of the electric potential V at a point in space, you can find the electric potential energy of any charged particle carrying a charge Q simply by multiplying Q times V. Electric potential enables us to separate the electric properties of space from the electric properties of the charged particle.

The units of electric potential, i.e., joules/coulomb, or $\mathrm{J\,C^{-1}}$, are given their own name—the "volt," abbreviated V. All the SI multiples are used, e.g., picovolts, or pV, or 10^{-12} V; nV, or nanovolts or 10^{-9} V; megavolts, or MV, or 10^{6} V; gigavolts, or GV, or 10^{9} V; teravolts, or TV, or 10^{12} V.

It is common to refer to electric potential as "voltage." Instead of asking what the electric potential is at some point in space, people will ask, What is the voltage there? Often you will hear people refer to differences of electric potential simply as "potential differences" or, equivalently, as "voltage differences." These mean the same thing.

The distribution of voltage in space can be quite complicated, so once again we consider only the two special, simple, cases that come up often: the electric potential associated with a constant electric field and the electric potential produced by a point charge.

For the first case, set up an x-y coordinate system with the x-axis lying in the bottom plate and the y-axis going from the bottom plate toward the top one. If we choose the bottom plate to be at zero volts, then the electric potential function between the charged plates is

$$V(y) = -E\,y. \tag{6}$$

The minus sign appears in this equation because E has direction, and electric potential increases in the direction opposite to E.

For the second case, the electric potential function at any radial distance r from a point-charge source Q is

$$V(r) = k_c \frac{Q}{r}, \tag{7}$$

where the zero of the potential has been chosen to be at very large values of r, i.e., $r \to \infty$. Know and understand these two equations.

EXAMPLES

5. Equation 7 tells you that 1 m away from a point charge of 1 μC the electric potential is

$$V = \frac{9 \times 10^9 \times 10^{-6}}{1.0} = 9000\,\text{V}.$$

6. How much energy would it require to move a 0.2 C charge from a position 1 m away from the microcoulomb charge of the previous example, to a position 0.4 m away?

From Eq. 7 you can show that the potential at 0.4 m is 22500 V; or you can use the fact that the voltage due to a point charge varies as $1/r$ and just divide the 9000 V of the previous example by 0.4 to get 22500 V. Therefore, the change in potential is 13500 V, and the energy required to move 0.2 C in closer is $0.2 \times 13500 = 2700$ J.

EXERCISES

13. Suppose that a charge of -0.2 C were placed 1 m away from a point charge of 1 μC and released. What would happen? Assume that the 1 μC charge is firmly fixed at its location. What would be the kinetic energy of the -0.2 C charge when it reached a distance 0.4 m from the other charge?

Notice from the above examples that given the potential difference between any two points in space you can immediately calculate the change in energy associated with the movement of any charge Q from one of the points to the other. The change in potential energy ΔU will be Q times the change in electric potential ΔV, i.e.,

$$\Delta U = Q\,\Delta V. \tag{8}$$

You will repeatedly use Eq. 8 and the fact that $\Delta K = -\Delta U$ to find the change in kinetic energy of an electric charge.

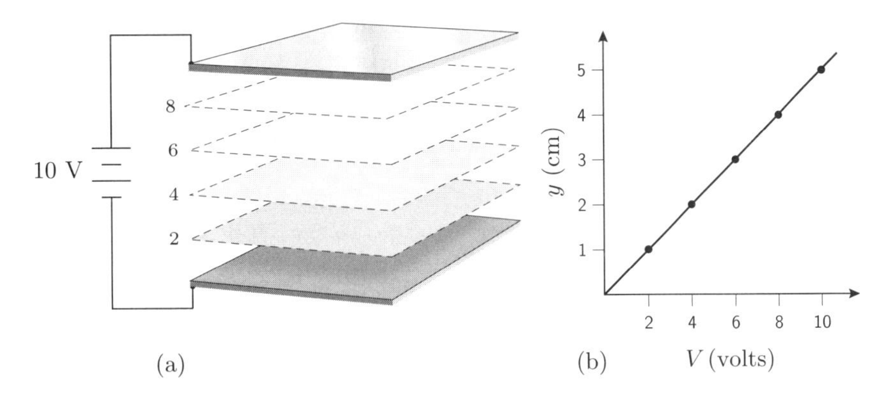

FIGURE 6.7 (a) Equipotential surfaces in the constant electric field between two oppositely charged parallel conducting plates; (b) a plot of the potential function along any line perpendicular to the plates.

EXERCISES

14. An electron is known to have a kinetic energy of 0 J. The electron (mass 9.11×10^{-31} kg) then goes from -10 V to $+10$ V. What is its change of velocity?

15. Suppose the electron of the previous problem initially has a kinetic energy of 16×10^{-19} J. After going from -10 V to $+10$ V, what is its new velocity?

EXAMPLES

7. The change in kinetic energy of a charge of 3 C that moves through a potential difference of 7 V is $3 \times 7 = 21$ J.

EXERCISES

16. How much energy does a 1.5 V "D" battery expend when it pushes $10\,\mathrm{mC\,s^{-1}}$ through a wire for 20 minutes?

Visualizing Electric Potential

Most physicists carry in their heads some kind of picture of $V(y)$ or $V(r)$. Such mental images are often useful, so we offer here some suggestions for how you might visualize the electric potential functions of our two special cases.

Keep in mind that the potential function assigns a number to each point in space; thus it costs you four numbers to specify the potential—three coordinates, x_0, y_0, and z_0, to describe the point about which you are talking, and one more number, which is the value $V(x_0, y_0, z_0)$ at that point.

It is hard to graph such a function, but you can think of the electric potential as a cloud of water vapor. Where the values of the potential are large, the cloud is thick and opaque. Where V is small, the cloud becomes more transparent. The cloud corresponding to the electric potential between the parallel plates that produces a constant electric field E has a density that increases linearly from the bottom plate up to the top plate. The cloud corresponding to the electric potential produced by a point charge would have spherical symmetry centered on the charge. The cloud's density would be very great near the center (infinitely dense at $r = 0$) and then get more and more tenuous away from the center, falling off as $1/r$.

It is probably more useful to visualize the electric potential in terms of surfaces of constant potential. The idea is that the potential will have the same value at many different points in space, and these points will define extended surfaces. For example, in the case of Fig. 6.6 the surfaces of constant potential, often called "equipotential surfaces," are planes parallel to the surfaces of the plates. Figure 6.7(a) illustrates the equipotential surfaces for values of $V = 2$, 4, 6, and 8 volts.

The equipotential surfaces around a point charge are concentric spherical shells. Figure 6.8(a) shows such surfaces for $V = 2, 4, 6$, and 8 volts. Notice that around a point charge equipotential surfaces separated by the same ΔV are not uniformly spaced the way they are between the parallel plates.

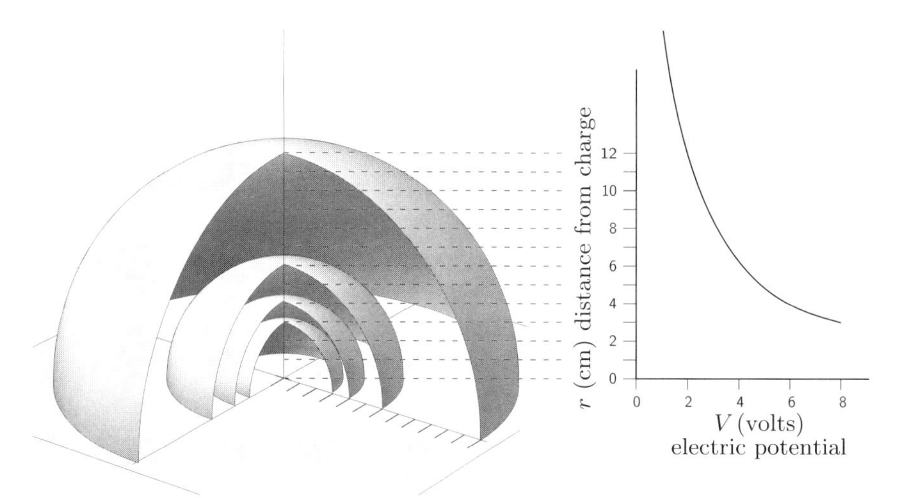

FIGURE 6.8 (a) Spherical equipotential surfaces in the electric field around a point charge; the portions of these surfaces below the plane are not shown. (b) A graph of the potential function along any line extending radially out from the source charge.

It is often helpful to plot a graph of the electric potential function. This is easy to do for our two special cases, because V is in each case a function of only one variable. Such plots are shown in Figs. 6.7(b) and 6.8(b). They are marked to show the locations and values of the equipotential surfaces illustrated in the adjacent Figs. 6.7(a) and 6.8(a).

Keep in mind that no energy is required to move a charged particle between any points on the same equipotential surface.

EXERCISES

17. How much energy will be required to move a $-2\,\mu\text{C}$ particle from the 8 V equipotential surface to the 4 V equipotential surface around a $Q = 24\,\mu\text{C}$ source charge?

18. How much energy will be required to move a $-2\,\mu\text{C}$ particle from the 8 V equipotential surface to the 4 V equipotential surface between a pair of oppositely charged parallel conducting plates?

There are instruments called "voltmeters" with which you can measure the electric potential difference between any two points. With such devices it is possible to map experimentally the electric potential over an extended region of space. Such experimentally determined maps can be used to predict the change in potential and kinetic energy of a charged particle in quite complicated electric fields. For example, Fig. 6.9 shows the equipotential lines that might occur in a plane containing a sphere of charge and a charged needle.

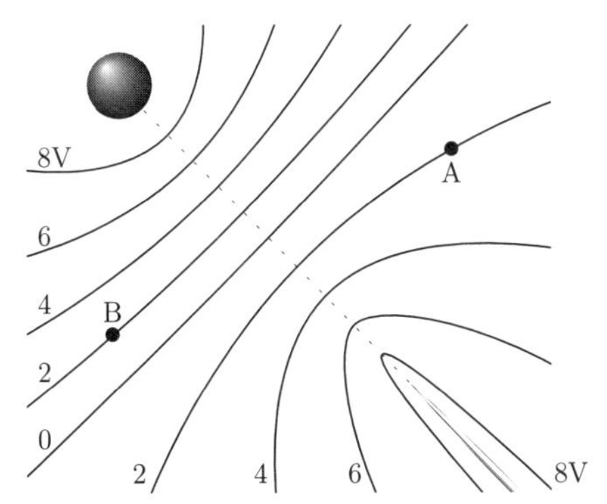

FIGURE 6.9 Equipotential lines such as might be found around the combination of a positively charged needle and a sphere of charge.

EXAMPLES

8. What is the sign of the charge on the needle? Imagine a small positive charge moved toward the needle. Because the potential rises and because the charge is positive, its potential energy will increase. This means that the charge is being repelled by the needle. Therefore, the needle is positively charged.

EXERCISES

19. Use a similar argument to show that the charge on the sphere is also positive.

EXAMPLES

9. How much energy would it take to move a 2 μC charged particle from point A to point B?

Because the electric potential is the same at both points, the net energy needed to move the particle from A to B is zero. Notice that this does not mean that the kinetic energy of the particle did not change as it moved from A to B. Its kinetic energy increased as the particle moved from the 2 volt equipotential line to the 0 volt line; but then it decreased by the same amount as the particle moved from the 0 volt line up to point B on the second 2 volt line.

The Electron Volt

Earlier, when discussing energy, we told you that the unit of energy commonly used in discussing atomic phenomena is the electron volt or eV. It has the seemingly peculiar definition of

$$1\ \text{eV} = 1.6 \times 10^{-19}\ \text{J}.$$

Now we can show you where that unit comes from. The charge on the electron is -1.6×10^{-19} C. The charge on the proton is $+1.6 \times 10^{-19}$ C. In fact, every charge we have ever found in nature is some integer multiple of 1.6×10^{-19} C. This number is called the "elementary charge" and is so important that it is given its own symbol e, where

$$e = 1.6 \times 10^{-19}\ \text{C}.$$

The charge on an electron is $-e$. The charge on a proton is e. (People also carelessly call e without the negative sign "the electronic charge," and then you are supposed to remember to put in the minus sign when you do any calculations.)

By how much does the energy of an electron change when it moves through a potential difference of $V = 1$ volt? From Eq. 8 it follows that

$$e\,V = 1.6 \times 10^{-19} \times 1 = 1.6 \times 10^{-19}\,\text{J},$$

which is the same as 1 electron volt, or 1 eV.

The unit eV, i.e., electron volt, makes calculations of the change in kinetic energy of an electron accelerated through a potential difference a trivial act. In these units the change in energy is numerically equal to the change in electric potential.

EXERCISES

20. What is the change in energy of an electron accelerated through a potential difference of 1 kV? Give your answer in eV.

Of course, you still have to do some figuring to find the velocity.

EXAMPLES

10. What is the velocity of an electron accelerated from rest (initial kinetic energy = 0) through a potential of 1 kV?

Clearly, its energy changes by 1 keV, or 1.6×10^{-16} J. Since it was at rest initially, its kinetic energy is now

$$\frac{1}{2}mv^2 = 1.6 \times 10^{-16}\,\text{J},$$

so we find that

$$v = \sqrt{\frac{3.2 \times 10^{-16}}{9.11 \times 10^{-31}}} = 1.85 \times 10^{7}\,\text{m}\,\text{s}^{-1}.$$

This is 6.2% of the speed of light, so this electron is moving right along.

6.5 ELECTRIC CURRENT

A stream of moving charges is called "electric current." For historical reasons the direction of the current is taken to be the direction positive charges would move to produce the observed transport of electricity, even though quite often the transport is caused by the motion of negative charges in the opposite direction. Charge flow is measured as the rate at which charge passes any given point. It is, therefore, $\Delta q/\Delta t$, and the units are coulombs per second ($\mathrm{C\,s^{-1}}$). There is a special name for this group of units. It is called the "ampere" and is abbreviated "A." All the standard SI multiples are used: e.g., pA, nA, μA, mA, A, kA, MA.

EXAMPLES

11. If 1 μC passes a certain point every second, the current will be 1 $\mu\mathrm{C\,s^{-1}}$ or, equivalently, 1 μA.

EXERCISES

21. Two carbon rods are placed in a solution of H_2SO_4. A current of 12 A is passed between the two rods for 2 hr 14 min. This results in the passage of an amount of charge that releases 1.008 g of H_2. How much charge is transported in this time?

Speed of Charges in a Current

If electric current involves the motion of electric charges, how fast do they move? Let's consider electric current flowing in copper, a metal easily drawn out into wire. It has an atomic mass of 63.55 u and a density of 8.9 $\mathrm{g\,cm^{-3}}$. Copper wire is sold in standard sizes specified by a number called its gauge. In the U.S. we use American Standard Gauge. As the gauge number gets larger, the wire gets thinner. For example, #18 copper wire has a diameter of 1.024 mm; #20 has a diameter of 0.812 mm; and #22 has a diameter of 0.644 mm.

EXAMPLES

12. Show that a 1 m length of #18 wire has a mass of about 7 g.

This wire is a cylinder. To find the mass of a 1 m long cylinder of #18 wire, first find its volume and then multiply by the density of copper. The volume of a cylinder is its length L times its cross-sectional area A.

The cross-sectional area A can be found from the diameter $D = 0.1024$ cm using the fact that $A = \frac{\pi}{4}D^2$. This gives

$$A = \frac{\pi}{4} \times 0.1024^2 = 8.23 \times 10^{-3}\,\text{cm}^3,$$

and the volume V of a 1 m length is then

$$V = 8.23 \times 10^{-3} \times 10^2 = 0.823\,\text{cm}^3.$$

The mass m_1 of a 1 meter length of wire is then its volume V times the density of copper. This gives $m_1 = 0.823 \times 8.9 = 7.33$ g.

Metallic copper is a good conductor of electricity because each atom has essentially one electron free to move around among the other atoms. These electrons move quickly and erratically from atom to atom, randomly exchanging places with one another. This happens whether the metal is a chunk of copper or whether it is drawn out into a wire.

If a difference of electric potential is set up, for example by connecting a battery to the ends of the wire, then in addition to their rapid, random motion, the electrons acquire a small average velocity that carries them from high potential energy to low potential energy. (For electrons this is from low electric potential to high electric potential.) The motion of the negatively charged electrons means that an electric current flows. Because of the convention that electric current flows in the direction that positive charge would move, i.e., from high electric potential to low electric potential, the electric current flows in the direction opposite to the motion of the electrons.

EXAMPLES

13. How fast are the electrons moving in a piece of #22 wire when the electric current is 1 A?

To answer this question you will need to know the cross-sectional area of the wire and how many electrons pass across that area each second. Do you recognize that this is essentially the same problem as finding how many molecules pass through a hole of area A in 1 s, or how many molecules hit an area A of a container's wall in 1 s?

Do you remember the argument? If a container holds n particles per unit volume and each of them is moving toward A with a speed $\langle v \rangle$, then in a time interval Δt the number ΔN that pass through the hole (or strike the area) A is the number of particles in a cylinder of base area A and length $\langle v \rangle \, \Delta t$. We can write this as

$$\Delta N = A\, n \, \langle v \rangle \Delta t,$$

so the rate at which the molecules strike A is

$$\frac{\Delta N}{\Delta t} = n \, \langle v \rangle \, A. \tag{9}$$

As previously noted, the flow of current is the rate at which electric charge passes through an imaginary surface that cuts across the wire. Thus an amount of charge $\Delta Q = 1\,\text{C}$ passing through such a surface in a time $\Delta t = 1\,\text{s}$ constitutes a current of 1 A; in other words, $I = \frac{\Delta Q}{\Delta t} = \frac{1\,\text{C}}{1\,\text{s}}$. If you could find a connection between $\frac{\Delta Q}{\Delta t}$ and $\frac{\Delta N}{\Delta t}$, you could rewrite Eq. 9 in terms of the electric current and solve for $\langle v \rangle$.

The connection is quite simple. If $\frac{\Delta N}{\Delta t}$ particles pass a point along the wire each second and if each of these particles has a charge e, then the amount of charge passing must be

$$I = \frac{\Delta Q}{\Delta t} = e \, \frac{\Delta N}{\Delta t}.$$

In other words, you can replace $\frac{\Delta N}{\Delta t}$ with I/e. Then solving for $\langle v \rangle$ you will get

$$\langle v \rangle = \frac{I}{e\, n\, A}.$$

The current I is just 1 A. What is n; what is A? For #22 wire the diameter is 0.0644 cm, so the area A is $\pi/4 \times 0.0644^2 = 3.26 \times 10^{-3}\,\text{cm}^2$.

To find n use the fact that 63.55 g of copper contains 6.02×10^{23} copper atoms and the fact that the density of copper is $8.9\,\text{g}\,\text{cm}^{-3}$. This last fact tells us that 1 mole of copper occupies $63.55/8.9 = 7.14\,\text{cm}^3$. From this it follows that in copper the number density of the electrons that are free to move around and cause an electric current is

$$n = \frac{6.02 \times 10^{23}}{7.14} = 8.43 \times 10^{22}\,\text{cm}^{-3}.$$

Now you can compute the value of the average velocity of the electrons in a #22 wire carrying 1 ampere of current.

$$\langle v \rangle = \frac{I}{e\, n\, A} = \frac{1}{1.60 \times 10^{-19} \times 8.43 \times 10^{22} \times 3.26 \times 10^{-3}} = 0.0227\,\text{cm}\,\text{s}^{-1}.$$

The electrons move along the wire at a fraction of a millimeter per second!

How long will it take them to travel a distance of 1 cm along a piece of #22 wire?

6.6 SUMMARY OF ELECTRICITY

Here is what you need to know about electricity.

There is something called electric charge. There are two kinds of it, + and −. Like kinds of charge repel; unlike kinds attract. Small point charges q_1 and q_2 exert a force on each other proportional to the magnitude of each charge and inversely proportional to the square of the distance r between them:

$$F = k_c \frac{q_1 q_2}{r^2}.$$

This equation is Coulomb's law. Choosing the constant k_c in the above equation equal to $9 \times 10^9\,\mathrm{N\,C^{-2}\,m^2}$ defines the unit of charge called the "coulomb." In other words, if two identical point charges separated by 1 meter exert a force of 9×10^9 N on each other, they each carry a charge of 1 coulomb, i.e., 1 C.

It is convenient to attribute the electric force to two separate factors: to a property of space and to a property of the body to be placed in that space. The property of space we call "electric field." The property of the body is its "electric charge." The magnitude of the electric field E at a point in space equals the ratio of the force exerted on a small test charge q placed at that point to the magnitude of the charge:

$$E = \frac{F}{q}.$$

Electric field has a direction. It points in the direction that a small, positive charge will move when placed in the field.

The simplest electric field is the constant field $E =$ constant. A charge q placed in such a field is subjected to a constant force

$$F = Eq$$

everywhere in the field. The constant electric field is of practical importance because it is a good approximation to the electric field between two metal sheets close together and charged with opposite signs of charge. This is an arrangement easy to make in the laboratory.

It follows from the definition of electric field and Coulomb's law that a point charge Q will create an electric field E a distance r from itself given by the following expression:

$$E = k_{\mathrm{c}} \frac{Q}{r^2}.$$

Electric potential V is often easier to work with than electric field. Electric potential is the potential energy of a *unit* charge, where this energy arises from the charge's location in an electric field. The change in potential energy of a nonunit charge q is simply related to the change in electric potential:

$$\Delta U = q\, \Delta V.$$

We often calculate the change in kinetic energy of a charged particle from its change in potential energy:

$$\Delta K = -\Delta U,$$

$$\frac{1}{2}mv_2^2 - \frac{1}{2}mv_1^2 = q(V_1 - V_2).$$

The units of electric potential are "volts"; 1 volt = 1 J/C.

When the electric field is constant, there is a particularly simple relationship between electric field and electric potential. Two points separated by a distance y along the direction of a constant electric field have a potential difference V given by the expression

$$V = -Ey.$$

For this special case the potential difference is the product of the electric field E and the distance between the points. If you know the potential difference V between two points separated by a distance d, and if you know that the electric field E is constant, then you can find the strength, i.e., the magnitude, of the field by the inverse of the above argument:

$$E = \frac{V}{d}.$$

This is a very useful relation. You can find the direction of E by seeing in which direction V decreases most steeply.

The electric potential that a point charge Q produces a distance r away from itself is

$$V(r) = k_{\mathrm{c}} \frac{Q}{r}.$$

An electric field always points in the direction from high electric potential to low electric potential. (Why? You should be able to explain this from your knowledge about the electric field.)

Moving streams of charge are called electric currents. The direction of flow of electric current is always in the direction that positive charge would move even if the actual current is negative charge flowing in the opposite direction. The magnitude of electric current is the time rate of passage of charge: $\Delta Q/\Delta t$. The SI unit of electric current is the "ampere." It is the current of $1\ \mathrm{C\,s^{-1}}$ passing a given point. The abbreviation for ampere is "A."

PROBLEMS

1. The force of gravity due to a spherical mass falls off proportionally to $1/r^2$ just like the electrical force due to a charge of 5 C. What would be the value of g (acceleration due to gravity) one Earth radius above Earth's surface?

2. The force between two charges 2 m apart is measured to be 3 N.

What will be the force between the charges if they are moved to be

a. $\frac{2}{3}$ m apart?

b. 6 m apart?

3. Coulomb measured the force between two charges by the amount of twist in a fiber from which a small charged pith ball hung at the end of a 10 cm long balanced arm, as shown in Fig. 6.10.

a. If $q_1 = -10\,\mathrm{pC}$ and $q_2 = -20\,\mathrm{pC}$, what is the electric force between the two "point" charges when the fiber has twisted through 0.2 radians, as shown in Figs. 6.10 and 6.11?

b. How much twist in the fiber will be needed to make the angle between the two lines from the fiber to the two charges become 0.1 radians, as shown in Fig. 6.11?

c. In part (a) what is the magnitude of the electric field produced by q_2 at the location of q_1?

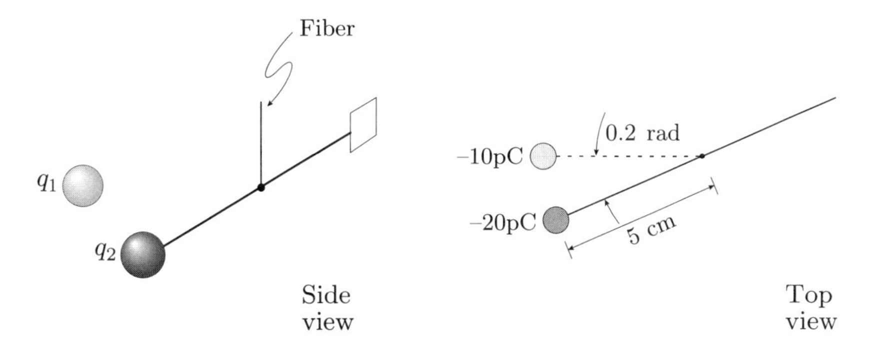

FIGURE 6.10 Positions and charges of pith balls (Problem 3).

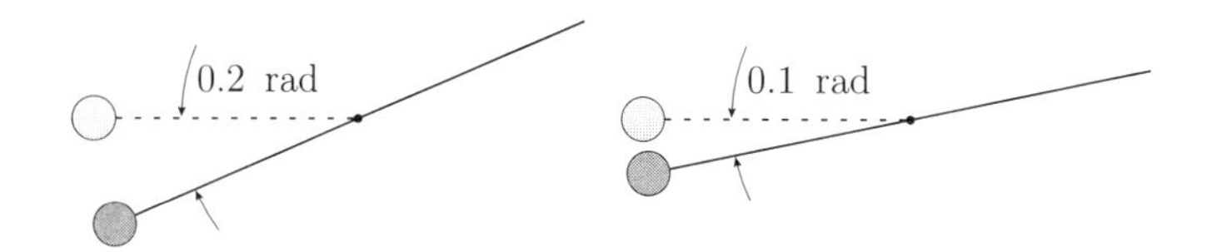

FIGURE 6.11 Before and after twisting the fiber (Problem 3).

d. By what factor will the electric field at q_1 due to q_2 change when the angle goes from 0.2 to 0.1 radians?

4. Looking down onto Coulomb's torsion balance, we see a needle 2 cm long suspended from a thin filament, as shown in the diagrams of the following table:

Some Data from Coulomb's Torsion Balance

View from above	Separation between charges (cm)	Rotation of needle (deg)	Twist of cap (deg)	Total twist of filament (deg)
(i)	0.62	36	0	36
(ii)	0.31	18	120	138
(iii)	0.15	8.5	567	575.5

a. Write down Coulomb's law and define the symbols you use and their units.

b. Explain and show in what way Coulomb's data given in the table support his law. Be quantitative.

c. If the charge on each tiny sphere was 1 nC, what would be the force between the two charges in diagram (i) of the table?

d. What would be the electric field acting on the charge at the end of the needle in (i)?

e. By how much would that field change if we doubled the charge at the end of the needle but left the other charge unchanged?

5. Suppose you have two pointlike charges $q_1 = 0.5$ C and $q_2 = 1$ C separated by 0.5 m. What is the magnitude of the force on the smaller of the two charges?

6. In the previous question, if q_1 is removed, what is the value of the electric field at the position where it formerly had been located?

7. What property of matter gives rise to electrical fields?

8. A charge of 1 pC and mass 1 pg is put at a certain point in space. As soon as it is released it begins to accelerate at 10^8 m s^{-2}. What is the magnitude of the electric field?

9. **a.** How much kinetic energy does an electron gain when it moves through a potential difference of 250 V?

b. Also give your answer in units of electron volts.

10. The diagram in Fig. 6.12 shows two charged parallel plates separated by 5 mm. A device, which might be a battery or a power supply, has been attached to the two plates and has established a difference of electrical potential of 10 V between the plates.

The symbol ⏚ for "ground" or "common" shows that the potential of the bottom plate is 0 V.

a. What is the electrical potential anywhere in the plane parallel to and halfway between the two metal plates?

b. What is the electric potential energy of a +2 μC charge when it is halfway between the two plates?

c. What is the electrical potential energy of a -2 μC charge when it is halfway between the two plates?

d. For the previous two cases:

i. Which charge has the larger electric potential energy?

ii. Which charge is at the larger electric potential?

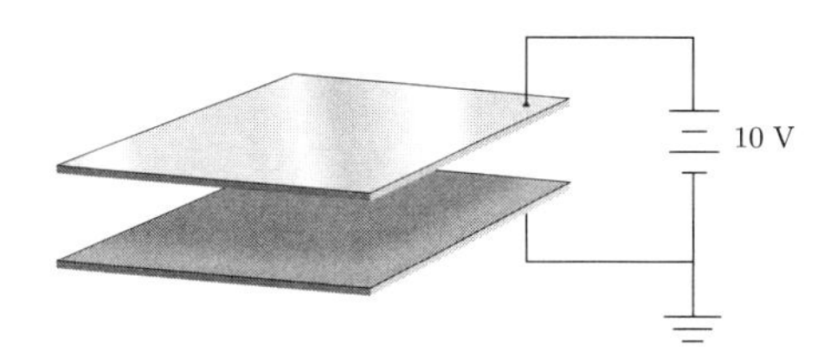

FIGURE 6.12 A pair of metal plates connected to a 10 V battery (Problem 10).

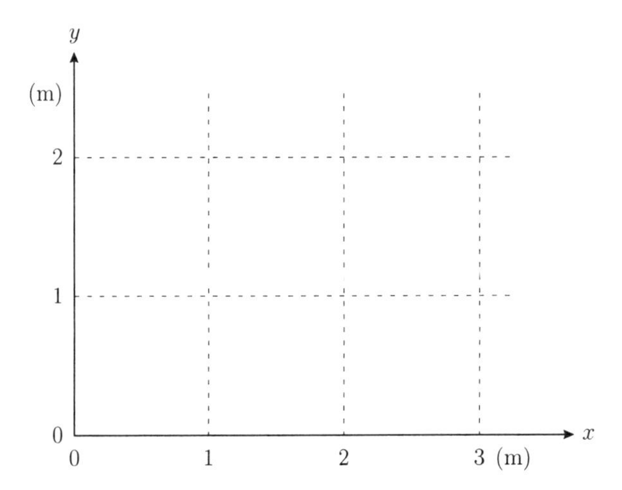

FIGURE 6.13 Coordinate grid for problem 11.

e. If the two charges are moved to the bottom plate, which of them undergoes the largest increase in potential energy? How much?

f. For the negative charge, what is the direction in space in which its potential energy will decrease most steeply? Why or why not is this surprising?

g. What is the value of E between the plates shown in Fig. 6.12?

11. Figure 6.13 shows an x-y coordinate system. The table gives values of the electric potential measured at the indicated points.

(x, y) (m)	Potential (V)
$(0, 0)$	0
$(0, 2)$	4
$(1, 0)$	2
$(1, 1)$	8
$(1, 2)$	8
$(3, 1)$	10
$(3, 2)$	20
$(2, 2)$	10

a. Mark the points $(1, 0)$, $(2, 2)$, $(3, 2)$, and $(0, 2)$ on the graph

b. Label each point with the value of the electric potential at that point

c. A droplet containing a net charge of $-7e$ ($e = 1.6 \times 10^{-19}$ C) is released at $(0, 0)$ and moves freely to $(3, 1)$. By how much does its kinetic energy change? Give your answer in units of electron volts.

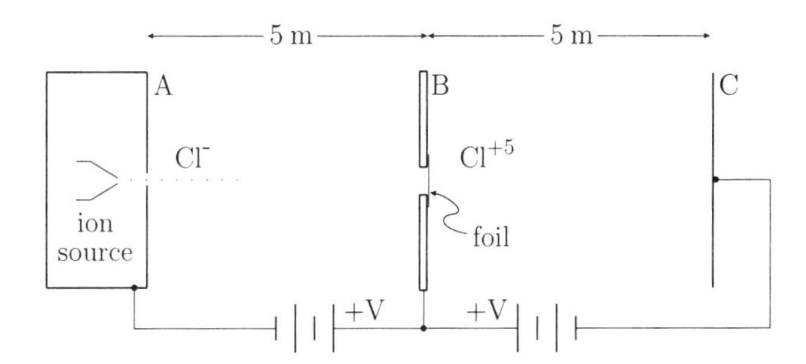

FIGURE 6.14 Problem 12: tandem accelerator.

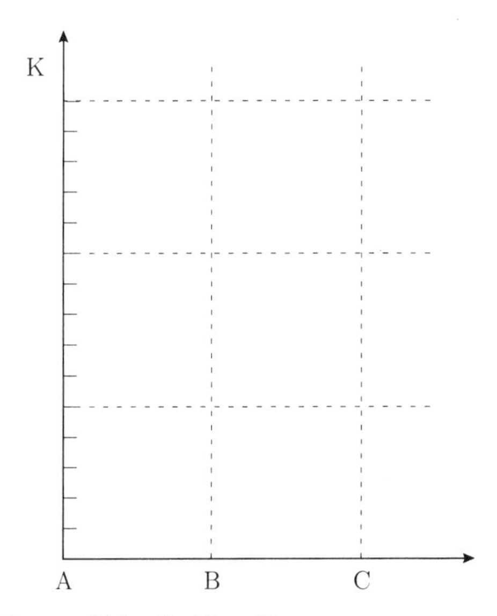

FIGURE 6.15 Coordinate grid for Problem 12.

d. The particle then continues moving and goes from $(3, 1)$ to $(1, 2)$. Does its kinetic energy increase or decrease? How do you know?

12. The apparatus diagrammed in Fig. 6.14, called a "tandem accelerator," is used to accelerate ions to high kinetic energies.

At the left, labeled A, is a source of negatively charged chlorine ions Cl^- (i.e., each ion carries a charge $-e$). A 1 MV potential difference is applied between plates A and B, as shown.

a. Find the electric field in the space between A and B. Specify both direction and magnitude.

b. Find the kinetic energy of the ions when they reach plate B. Use appropriate units.

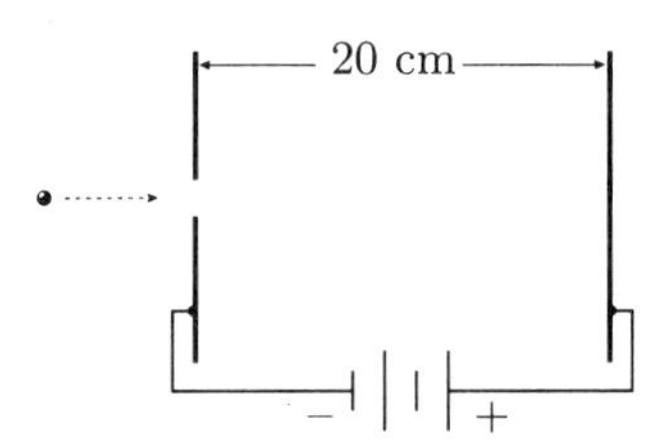

FIGURE 6.16 Region of electric field for Problem 13.

In the center of plate B is a hole covered by a very thin foil. Each Cl^- ion passing through this foil is stripped of six electrons while its velocity hardly changes (really!). As a result, chlorine ions of charge $+5e$ emerge from plate B and are accelerated further in the region between B and C. The potential difference between B and C is 1 MV.

c. On a coordinate system like that in Fig. 6.15 graph the kinetic energy vs. position from A to C. Scale the vertical axis in appropriate energy units. Your graph should be as quantitatively correct as possible.

d. The atomic mass of Cl is 35. Show that the mass of a chlorine ion is 5.8×10^{-26} kg. What is the velocity of the ions reaching plate C?

13. A voltage of 400 V is applied to the plates shown in Fig. 6.16.

a. If the plates have a separation of 20 cm, calculate the magnitude and direction of the electric field.

b. We now send a particle of charge $q = +2e$ and mass $m = 6.4 \times 10^{-27}$ kg through the hole in the left plate, as shown in the figure. What would be the particle's initial velocity such that it slows down and stops right at the other plate (without crashing)?

c. Describe qualitatively the motion of the particle if we change the voltage to 200 V (same polarity).

d. We now restore the voltage to 400 V. Describe qualitatively the motion of the particle if we decrease its initial velocity from the value found in (b).

C H A P T E R 7

Magnetic Field and Magnetic Force

7.1 MAGNETIC FIELD

Electric fields produce forces on electrical charges whether they are moving or sitting still. There is another quite different field that exerts forces on electric charges *only* if they are moving. Equally strange, the strength of the exerted force depends upon the direction of the charge's motion. This peculiar behavior reveals the presence of a *magnetic field*.

Magnetic Force on a Moving Charge

Figure 7.1 shows different states of motion of an electric charge in a magnetic field B. When the charge is at rest, no force acts. This behavior tells you that no electric field E is present. If you now put the charge in motion, you will see that a force acts on it! If you experiment by moving the charge in different directions, you will observe some curious things.

- There is a direction of motion along which no force occurs.
- When the charge moves in a direction such that a force occurs, the force is in a direction perpendicular to the charge's line of motion *and* perpendicular to that line of motion along which no force occurred.
- The force varies in strength but becomes a maximum when the motion of the charge is perpendicular to the line along which no force was seen.

Clearly, some new kind of force is present. We attribute this force to another kind of field that we call the "magnetic field." It is this kind of field that one finds near a bar magnet. It is this kind of field that surrounds the

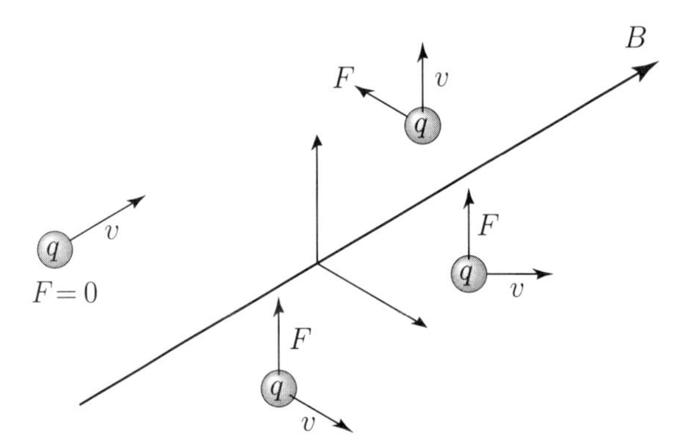

FIGURE 7.1 A set of pictures showing how the magnetic force on a positive charge q varies in magnitude and direction for various directions of motion with respect to the line along which there is no force. The direction of the magnetic field is taken to be parallel to that line.

Earth and makes the little magnet that is a compass needle point northwards. *A magnetic field also appears around any electric current or moving charge.*

If you do a series of experiments that measure the force on a succession of different charges all moving in the same direction at the same location in space, you find that the force due to a magnetic field is proportional to the size of the charge:

$$F \propto q.$$

If you find a direction of motion such that the moving charges experience a maximum force F, and then vary the speed v along this direction, you find that the force is proportional to v, so

$$F \propto q\,v.$$

As with the electric field, the force arises from properties of the particle—its charge and its velocity—and from this new property of the space through which the charge is moving—the magnetic field. It is customary to represent the magnetic field by the symbol B. In a manner analogous to the case of the electric field, the magnitude of the magnetic force is proportional to the product of the particle properties and the field strength

$$F \propto qvB.$$

The units of magnetic field B are defined to make the constant of proportionality in this equation be 1. Then the numerical value of B, i.e., the strength of the magnetic field, is defined by the equation

$$F = q\,v\,B.$$

Thus, if a charge of 1 C moving with a speed of 1 m/s experiences a maximum force of 1 N, the magnetic field must have a value of 1 unit of field strength.

The SI units of B are $\mathrm{N\,C^{-1}\,m^{-1}\,s}$, and they are given the name "tesla" which is abbreviated "T." A magnetic field of 1 tesla (1 T) can exert a maximum force of 1 N on a charge of 1 C moving with a speed of 1 m/s. Earth's magnetic field is a bit less than 10^{-4} T. In the northeastern United States the field strength is about 55 μT. Historically, the unit of magnetic strength was something called a "gauss" (abbreviated G). This unit is still often used: 1 gauss = 10^{-4} tesla.

As mentioned above, the magnetic field B has direction as well as magnitude. The direction is parallel to the line of motion along which a moving charge experiences no force. This means that the line of B is perpendicular to the direction of motion that gives rise to a maximum force, and, as indicated in Fig. 7.1, both of these directions are perpendicular to the force. Which way B points along this line is a matter of convention. The agreed-upon convention is to say that B points in the direction that the northward-pointing pole of a compass needle will point when placed in the magnetic field B. Thus direct observation shows that in the Northern Hemisphere, Earth's magnetic field points downward at a fairly steep angle.

The directions of motion and field are important because they determine the direction of the force. The diagram in Fig. 7.2 shows how to determine the direction of the force on a positively charged particle moving with velocity v and at some angle θ relative to B. The force is always perpendicular to the plane determined by the directions of v and B. To decide which way F acts relative to that plane, imagine that you stand at a point from which diverge lines in the directions of v and B. If the line of v is to the right of B, the force is in the direction of your head. Otherwise, the force is toward your feet. If the charge is negative, the forces are reversed. This way of determining the direction of the force on a moving charge is called "the right-hand rule."

We must quantitatively describe one more aspect of the magnetic force on a moving charge q. The force on q will be less when it moves in a direction that is not perpendicular to B. It turns out that the force is proportional to the sine of the angle between v and B. Therefore, the full statement of the

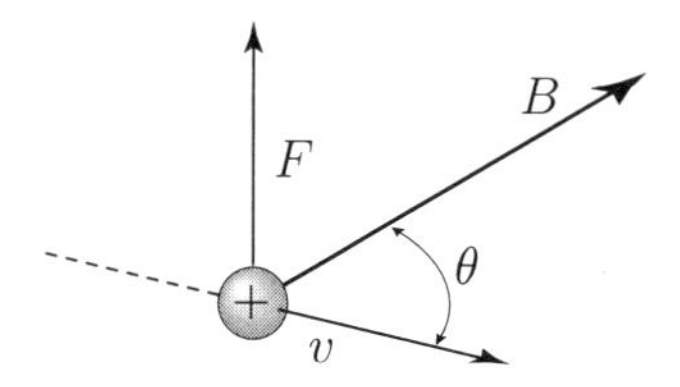

FIGURE 7.2 The force on a positively charged particle moving with velocity v at an angle θ relative to B.

magnetic force on a charge q moving with velocity v at an angle θ relative to B is

$$F = qvB \sin\theta. \quad (1)$$

This force of a magnetic field on a moving particle is often called the "Lorentz force."

EXERCISES

1. The three diagrams in Fig. 7.3 show a charge q moving with velocity v in different directions relative to the magnetic field B. The diagram uses the customary convention of dots representing B pointing out of the plane of the page (tips of arrowheads), and x's representing B pointing into the plane of the page (tail feathers of arrowheads). For each case, what is the direction of the force?

Another way to describe how the magnitude of the force depends on the angle between the directions of v and B is to observe that $v \sin\theta$ is the component of the velocity that is perpendicular to B. Therefore, F is proportional to $v_{\text{perp}} = v_{\perp} = v \sin\theta$. For most of the cases of interest to us, v will be perpendicular to B and $\sin\theta$ will equal 1.

EXAMPLES

1. Suppose a particle with a charge of -1.6×10^{-19} C moves with a speed of 2×10^{6} m/s perpendicular to Earth's magnetic field of 55 μT. What is the force on this particle? This is a straightforward calculation:

$$F = 1.6 \times 10^{-19} \times 2 \times 10^{6} \times 55 \times 10^{-6}$$
$$= 1.76 \times 10^{-17}\ \text{N}.$$

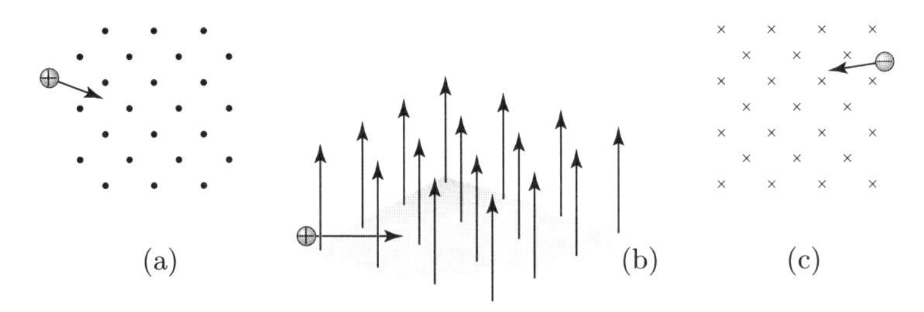

FIGURE 7.3 Three examples of charged particles moving in a uniform magnetic field. The dots mean that the magnetic field B points out of the page; the x's mean that the magnetic field B points into the page.

Suppose the particle in the example is an electron. Knowing that an electron has a mass of 9.11×10^{-31} kg, you can find the electron's kinetic energy:

$$\begin{aligned}\frac{1}{2}mv^2 &= 0.5 \times 9.11 \times 10^{-31} \times 4 \times 10^{12}\\ &= 1.82 \times 10^{-18}\,\mathrm{J}\\ &= 11.4\,\mathrm{eV}.\end{aligned}$$

From this you see that because of its very small mass, it takes only about 11 V to get an electron moving as fast as $2 \times 10^6\,\mathrm{m\,s^{-1}}$.

EXERCISES

2. What is the acceleration of the electron by the force in the above example? How does your answer compare to g, the acceleration due to gravity?

A Moving Charge in a Constant Magnetic Field

The motion of a charge in a constant magnetic field will be quite important to us. Such magnetic fields are frequently used to measure the charged particle's momentum, which can, in turn, be used to find its energy and velocity. If the mass is not known, combinations of electric and magnetic field can be used to find the velocity of a charged particle and also to measure the ratio of its charge to its mass.

Figure 7.4 shows a positive charge q moving with velocity v perpendicular to a constant magnetic field B. As before, the little dots in the figure indicate that B is pointing up out of the plane of the page. From the rules for finding the direction of the force on q, you should be able to show that the force on q is initially toward the bottom edge of the page.

Notice that the direction of the force changes. Because the force acts in a direction perpendicular to the velocity of the charge, the force changes the direction of the velocity of the charge. As soon as the direction of v shifts, so does the direction of the force. These directions remain at right angles to each other, with the result that the direction of v shifts steadily around in a circle, and the charge moves in a circle of some radius R at a steady speed v. In other words, in a constant field B *the charge moves with uniform circular motion*.

We know from elementary mechanics that an object moves in a circle at a steady speed v only if it is accelerated toward the center of the circle with a magnitude of acceleration equal to v^2/R. Therefore, the magnitude of the

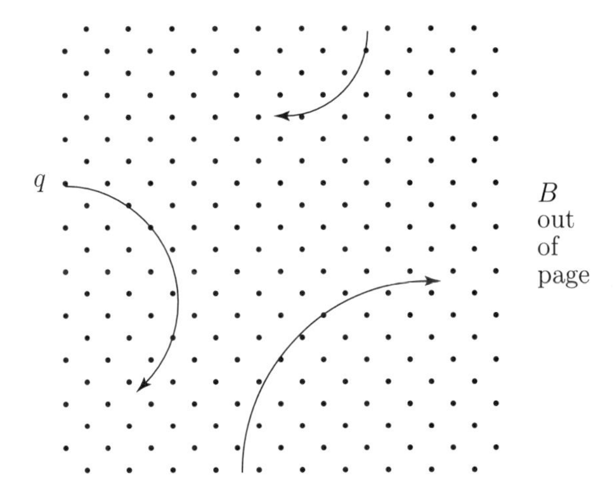

FIGURE 7.4 Circular trajectories of positively charged particles entering a uniform magnetic field pointing out of the page.

force making the particle move in a circle is

$$F = m\,a = m\frac{v^2}{R}.$$

Because this force is produced by the Lorentz force, you can equate the two expressions,

$$qvB = \frac{mv^2}{R},$$

and rearrange them to get

$$mv = p = q\,B\,R, \tag{2}$$

where, as previously, the symbol p represents momentum mv.

An important consequence of Eq. 2 is that knowing the magnetic field and the particle's charge, you can measure its momentum by measuring its radius of curvature as it moves through the uniform magnetic field. Or if you know the velocity and field and charge, you can predict the radius of curvature of the charge as it moves in the uniform (i.e., constant) magnetic field.

EXAMPLES

2. If electrons are accelerated from rest through 200 V, they acquire a kinetic energy of 200 eV. Assume that a beam of these electrons enters a region of uniform magnetic field of $B = 5$ mT. If the beam enters perpendicular to B, what is its radius of curvature?

First, find the velocity of the electron.

$$\frac{1}{2}mv^2 = 200 \times 1.6 \times 10^{-19}\,\text{J};$$

$$v = \sqrt{\frac{64 \times 10^{-18}}{9.11 \times 10^{-31}}} = 8.38 \times 10^6\,\text{m}\,\text{s}^{-1}.$$

From this result find the momentum

$$p = mv = 9.11 \times 10^{-31} \times 8.38 \times 10^6 = 7.64 \times 10^{-24}\,\text{N}\,\text{s},$$

from which you can get

$$R = \frac{p}{qB} = \frac{7.64 \times 10^{-24}}{1.6 \times 10^{-19} \times 5 \times 10^{-3}} = 9.54 \times 10^{-3}\,\text{m} = 9.54\,\text{mm}.$$

3. Consider an electron (mass 9.11×10^{-31} kg) in a uniform magnetic field of 0.1 mT. If the electron's path is observed to bend with a radius of 10 cm, what is its momentum? What is its kinetic energy?

Its momentum is

$$p = qRB = 1.6 \times 10^{-19} \times 0.1 \times 10^{-4} = 1.6 \times 10^{-24}\,\text{N}\,\text{s},$$

and its kinetic energy is

$$\frac{1}{2}mv^2 = \frac{p^2}{2m} = \frac{2.56 \times 10^{-48}}{18.22 \times 10^{-31}} = 1.41 \times 10^{-18}\,\text{J} = 8.78\,\text{eV}.$$

EXERCISES

3. What are the momentum and kinetic energy of an electron that bends through a radius of 5 cm in a magnetic field of 2 mT?

Sources of Magnetic Fields

The sources of magnetic fields and magnetic forces exhibit a remarkable reciprocity: A magnetic field exerts a force on moving charges; moving charges,

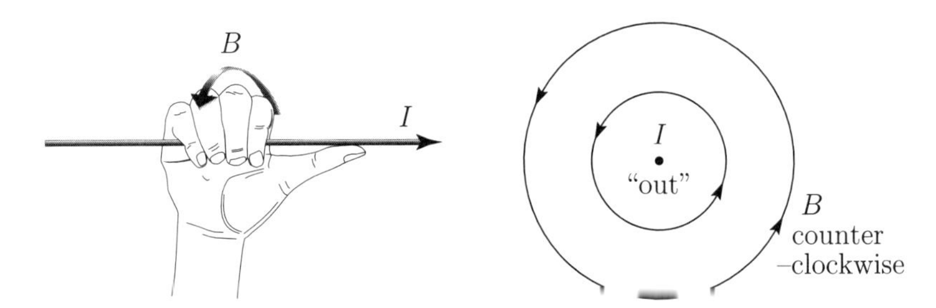

FIGURE 7.5 The right-hand rule for finding the direction of the magnetic field surrounding a long, straight, current-carrying wire.

i.e., an electric current, produce a magnetic field. Just as electric charges produce electric fields that act on other electric charges, *electric currents produce magnetic fields* that exert forces on other currents.

An electric current I flowing in a long straight wire will produce a magnetic field around the wire. Its direction is always tangent to circles centered on the wire. To find the direction tangent to these circles, imagine that you grasp the current-carrying wire with your *right hand* so that the direction of your thumb points in the direction of the flow of the current. Then your fingers curl around the wire in the direction of the magnetic field. This rule, also called the "right-hand rule," is illustrated in Fig. 7.5; it is *not* the right-hand force rule.

The strength of the magnetic field B a distance r from a long, straight wire is

$$B(r) = \frac{\mu_0 I}{2\pi r},$$

where μ_0 is a constant equal to $4\pi \times 10^{-7}\,\mathrm{T\,A^{-1}\,m}$ and I is the current flowing in the wire.

The magnetic field produced by the current in one wire will add (vectorially) with the magnetic field produced by a current in another wire. By using many wires or one wire bent into many loops it is possible to design different arrangements and strengths of magnetic field. In particular, it is possible to produce a magnetic field that is essentially constant in a limited volume of space. Such a constant B field is often used in practice as well as for generating physics homework problems and exercises.

7.2 MAGNETIC FIELDS AND ATOMIC MASSES

Physicists used magnetic deflection of charged particles to make two major advances in our understanding of atoms.

First, magnetic deflection provided conclusive evidence that many chemical elements have more than one kind of atom. As early as 1912, J.J. Thomson used magnetic deflection to show that there are at least two kinds of neon atoms occurring in nature, one of atomic mass 20 and another of mass 22. (There is also a third naturally occurring neon atom of mass 21.) Atoms that behave chemically the same but differ in atomic weight are called "isotopes." Thus, chemically pure neon gas is a mixture of three different isotopes. Throughout the 1920s and 1930s Francis W. Aston used magnetic deflection to discover that most elements have more than one isotope. Physicists and chemists have now identified 327 naturally occurring isotopes of the approximately 90 elements found on Earth.

For example, the two familiar chemical elements carbon and oxygen each have different isotopes. Most carbon atoms have mass 12, but about 11 out of every 1000 carbon atoms have mass 13. Most oxygen atoms have mass 16, but out of every 10^5 oxygen atoms 39 have mass 17, and 205 have mass 18. In 1931 Harold Urey discovered that familiar, much studied hydrogen has an isotope of mass 2 (called "deuterium"). It had gone unnoticed because only 15 out of every 10^5 hydrogen atoms have mass 2.

The second dramatic advance was in the measurement of atomic masses. There are two parts to this story. For one thing, every isotope was found to have a nearly integer mass. For example, chlorine, which has a *chemical* atomic weight of 35.4527 u, turns out to be 75.4% atoms of mass 35 and 24.6% atoms of mass 37. Its noninteger chemical atomic weight is just the average of two (almost) integer isotopic masses. The existence of integer masses strongly suggests that there is an atom-like building block within the atom, another level of atomicity.

EXERCISES

4. Show that the above relative abundances of the two chlorine isotopes explain the observed noninteger chemical atomic weight.

However, when measured precisely enough, the mass of every isotope turns out to be slightly different from an exact integer. And these differences are very important. Using magnetic deflection it became possible to measure these differences with accuracies of parts per ten thousand and in this way find atomic masses of isotopes to a precision of parts per million. This enormous advance over the traditional chemical methods of determining atomic weights (Chapter 2), which even when pushed to their limits yielded values accurate to no better than a few tenths of a percent, had dramatic consequences well beyond advancing our understanding of atoms.

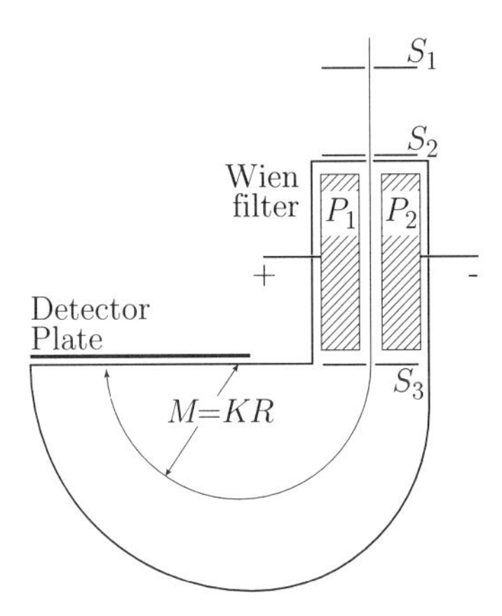

FIGURE 7.6 Bainbridge's apparatus for measuring the ratio of the charge to mass of ions; B is out of the plane of the page.

The new knowledge helped to explain how stars generate their energy, and it contributed significantly to the development of nuclear weapons.

Magnetic Mass Spectrometry

Both the existence of isotopes and the precise measurement of their atomic masses were made possible by devices called "magnetic mass spectrometers." These devices bend ions in a magnetic field, and the diameters of their orbits provide a measure of their masses. Let's look at one such device used by K.T. Bainbridge in the 1930s that uses electric and magnetic forces in simple, interesting ways. Its design is shown schematically in Fig. 7.6. There are two parts: a device that produces ions of a given charge and sends them down a channel in which there are both electric and magnetic fields exerting forces in opposite directions on the ions; and a region that contains only magnetic field where the ions then move in circular orbits.

The channel is an example of a "Wien velocity filter," named after the physicist who invented it. Because the magnetic field pushes the ions to one side of the channel while the electric field pushes them to the other, only those ions for which the two forces are equal and cancel can get through the channel without hitting the walls. The electric force is qE; the magnetic force is qvB. The two forces will be equal only for ions moving with a particular velocity v such that

$$qE = qvB,$$

which reduces to

$$v = \frac{E}{B}.$$

In other words, only those ions with velocity $v = E/B$ can get through the channel. The other ions are filtered out. This assures that all the ions entering the magnetic field have the same velocity.

If the ions entering the magnetic field all have the same charge and the same velocity, they will bend with the same radius of curvature R in the magnetic field only if they have the same mass M. If a photographic plate is placed as shown in Fig. 7.6, the ions will curve around and strike it. When the plate is developed, there will be a dark line where the ions struck. The distance from the entrance slit to the dark line is the diameter of the ions' orbit. Because $Mv = qRB$ and $v = E/B$, it follows that

$$M = \frac{qB^2}{E} R = K R, \tag{3}$$

i.e., the mass of the ion M is directly proportional to R, which can be measured from the dark lines on the photographic plate.

To operate this mass spectrometer, you set up B and E for some particular velocity. Then you vary the accelerating voltage from, say, 5 kV to 20 kV. If in that range of accelerating voltages any ions acquire the correct velocity to pass through the filter, they will enter the magnetic field and be bent proportionally to their masses. Let's see how this might work for isotopes of neon.

EXAMPLES

4. Suppose you want neon ions of mass 20 that have been accelerated through 10 kV to enter a magnetic field of 1.5 T and bend and strike the photographic plate.

First find their velocity using the fact that they will have kinetic energy

$$\frac{1}{2}Mv^2 = 1.6 \times 10^{-19} \times 10^4 \,\text{J},$$

from which it follows that their velocity will be

$$v = \sqrt{\frac{1.6 \times 10^{-19} \times 10^4 \times 2}{20 \times 1.66 \times 10^{-27}}} = 3.11 \times 10^5 \,\text{m/s},$$

where 1.66×10^{-27} kg is the mass of an atomic mass unit.

Now use the fact that

$$R = \frac{Mv}{qB}$$

$$= \frac{20 \times 1.66 \times 10^{-27} \times 3.11 \times 10^{5}}{1.5 \times 1.60 \times 10^{-19}}$$
$$= 0.041 \text{ m}.$$

You see that these ions will strike the plate 8.2 cm from where they exited the Wien filter.

EXERCISES

5. Now show that if the neon isotope of mass 22 is present, its ions will enter the bending region of the spectrograph when the acceleration voltage is 11 kV and strike the photographic plate 9.0 cm from the exit of the Wien filter.

6. Where would you look on the photographic plate for evidence of a mass-21 isotope of neon?

It is always important to understand the precision of measurements. For this case you need to ask, How precisely can mass be measured with Bainbridge's mass spectrometer? Clearly, the answer will depend on how precisely you can measure E, B, and R. It is not difficult to measure R to a fraction of a millimeter, say ± 0.1 mm. Out of 8.2 cm this would be a fractional precision of

$$\pm \frac{\Delta R}{R} = \pm 1.2 \times 10^{-3} \approx 0.1\%,$$

or a few parts per thousand. This is not good enough to show that mass-20 neon's mass deviates slightly from being an integer, because that deviation is only $\approx \Delta M/M = 4 \times 10^{-4}$.

There is a maneuver often used by experimental physicists to improve the precision of their measurements. Rather than measure the value of a quantity directly, measure its difference from something that is already known accurately. And if you don't have an accurately known reference, you can set one up. You have already seen one example of this technique: When chemists and physicists did not know how to determine the actual masses of atoms of the chemical elements, they made up a scale by assigning a mass of 16 to oxygen (today we use carbon) and measuring all other atomic masses relative to this standard. The magnetic mass spectrometer led to another version of this technique based on measurements of "mass doublets."

The idea is to send through the mass spectrometer two different ions composed of different types of atoms so that they have nearly the same molecular

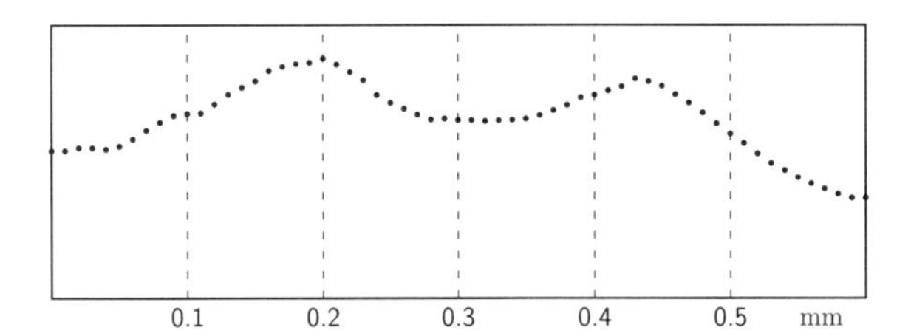

FIGURE 7.7 Two lines on a photographic plate of a mass spectrometer. The separation between the lines corresponds to the mass difference between an ionized hydrogen molecule H_2^+ and an ionized atom of mass-2 hydrogen.

weight. For example, Aston was able to measure the difference in mass between the mass-2 isotope of hydrogen and two atoms of the more common mass-1 isotope by sending through his mass spectrometer single-atom ions of mass-2 hydrogen and diatomic hydrogen ions, H_2^+. They each have a total mass of nearly 2 u, but their masses are slightly different. Figure 7.9 shows a tracing of the darkening of the photographic plate that then occurred in Aston's spectrometer. The separation of the two peaks is quite clear and is measurable to within ± 0.01 mm. This was found to correspond to a mass *difference* of 0.00155 u.

EXERCISES

7. Show that the fractional uncertainty in this mass difference is $\approx 4\%$ and that the uncertainty in the mass difference is therefore $\pm 6 \times 10^{-5}$ u.

Referring back to Example 7.4 you can see that this result is better than would have been obtained from two independent measurements with fractional uncertainties of $\pm\frac{0.01}{82} = 1.2 \times 10^{-4}$, because then you would measure $2\,\text{u} \pm 2.4 \times 10^{-4}$ u and $2.0016\,\text{u} \pm 2.4 \times 10^{-4}$ u. And when you took the difference of these two numbers you would get $2.0024 \pm 0.00024 - 2.0000 \pm 0.00024 = 0.0016\,\text{u} \pm 0.0003$ u, which is a result with a 19% uncertainty.

Further advance in precision comes when you relate these measurements of relative differences to some wisely chosen scale. Aston's measurements can easily be related to the scale of atomic masses based on assigning to the mass-12 isotope of carbon an atomic mass of *exactly* 12, because he also showed that one could measure the mass difference between C^{++} and a 3-atom molecule of mass-2 hydrogen. Notice from Eq. 3 that a doubly charged mass-12 C ion bends with the same radius of curvature as a singly charged mass-6 ion, and therefore it will strike the same part of the plate as a singly charged ion consisting of 3 atoms of mass-2 hydrogen. The separation between the two peaks corresponded to a mass difference between the two ions of 0.0423 u.

These results yield a very precise value of the mass of the hydrogen atom. To see how this works, write the two mass differences as linear equations:

$$2M_1 - M_2 = 0.00155\,\text{u},$$
$$3M_2 - \frac{M_{12}}{2} = 0.0423\,\text{u},$$

where M_1 is the mass of mass-1 hydrogen, M_2 is the mass of mass-2 hydrogen, and M_{12} is the mass of mass-12 carbon. By agreement, this last value is chosen as the standard of the atomic mass scale and is assigned the value of exactly 12.000.

To solve the two equations, multiply the top one by 3 and add it to the bottom one. Solve for M_1. You should get $M_1 = 1.00782\,\text{u}$, and because the above mass differences are modern values, this is the currently accepted result. It is confirmed in many ways.

EXERCISES

8. Use these mass differences to find the mass of mass-2 hydrogen.

Because of mass spectrometry and other techniques, we now know the atomic masses of all naturally occurring isotopes accurately to six decimal

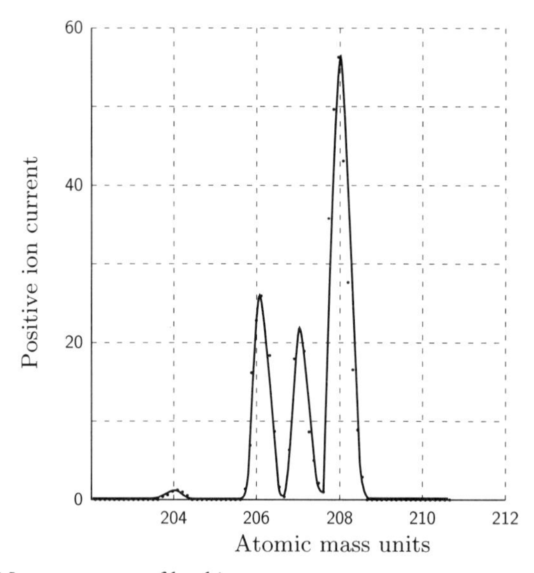

FIGURE 7.8 Mass spectrum of lead isotopes.

places. Mass spectrometry also tells us the relative abundances of the different isotopes. For example, comparison of the areas of the peaks in Fig. 7.8 (taken by A.O. Nier in 1938) shows the relative abundance of the different isotopes of lead found in nature.

7.3 LARGE ACCELERATORS AND MAGNETIC FIELDS

We have described the forces exerted by magnetic fields on moving charged particles and shown that when a particle of charge q moves with momentum p perpendicular to a magnetic field B, the magnetic field will bend the particles in a circle of radius R such that

$$p = qBR.$$

When engineers and physicists design particle accelerators, they often use this property of magnetic fields to bend and guide fast-moving charged particles around a closed path.

The following two examples show how the strength of the magnetic field determines the size of the accelerator. As the momentum of the accelerated particles becomes greater and greater, it is necessary to make the accelerator very large.

EXAMPLES

5. With modern superconducting materials it is feasible to build large magnets that operate reliably at magnetic fields as high as 6.6 T. Such magnets are at the edge of what is technically possible. The Superconducting Supercollider (SSC), which was to have been built in Waxahachie, Texas, until the project was abandoned in 1994 as too expensive, was intended to produce beams of protons with momentum of $1.1 \times 10^{-14}\,\mathrm{kg\,m\,s^{-1}}$.

What is the smallest possible circumference of a ring of 6.6 T magnets that could guide such protons around a closed loop?

You know that the momentum p is connected to the magnetic field B, particle charge q, and circle's radius R by the relationship $qBR = p$. For the SSC, $q = e = 1.6 \times 10^{-19}$ C, $B = 6.6$ T and $p = 1.1 \times 10^{-14}$. Solving the above equation for R gives

$$R = \frac{p}{eB} = \frac{1.1 \times 10^{-14}}{1.6 \times 10^{-19} \times 6.6} = 1.04 \times 10^{4}\,\mathrm{m},$$

which is the same as 10.4 km. The circumference of a circle of this radius is about 65 km.

The actual distance around the SSC was to have been 87 km because its shape would have been that of a race track rather than a circle. There needed to be straight sections between the magnets in which to place apparatus for studying the behavior of the high-momentum particles.

6. On eastern Long Island at Brookhaven National Laboratory is an accelerator called the Relativistic Heavy Ion Collider (RHIC for short). Its 3.5 T magnets are designed to bend into a closed loop gold ions that have a charge of $79e$ and a momentum of 1.1×10^{-14} kg m s^{-1}.

What is the circumference of the smallest possible ring of magnets for RHIC?

This is the same problem as the previous one, and you can work it out in exactly the same way. You can also notice that the momentum of RHIC's gold ions is the same as that of SSC's protons and just scale the answer of 65 km.

How do you scale the answer? Notice that the charge of the gold ions in RHIC is 79 times that of the charge of protons in the SSC, and RHIC's magnetic field is $3.5/6.6 = 0.53$ that of SSC's. Therefore, the smallest possible circumference of RHIC will be

$$2\pi R_{\text{RHIC}} = \frac{65}{0.53 \times 79} = 1.6\,\text{km}.$$

The actual circumference will be around 3.8 km. Figure 7.9 shows you that this is because of the six long straight sections between the bending regions of the ring.

The ring is very prominent from the air. Look for it if you are flying over eastern Long Island.

7.4 A SUMMARY OF USEFUL THINGS TO KNOW ABOUT MAGNETISM

In the chapters to come you will need to know the following things about magnetic fields. A magnetic field exerts a force only on a moving charge. The magnetic, or Lorentz, force is proportional to the charge, q, to that part of the charge's velocity, $v_\perp$, that is perpendicular to the magnetic field, and to the strength of the magnetic field, B. These relations are expressed in the equation

$$F = q\, v_\perp\, B = q\, v\, B\, \sin\theta.$$

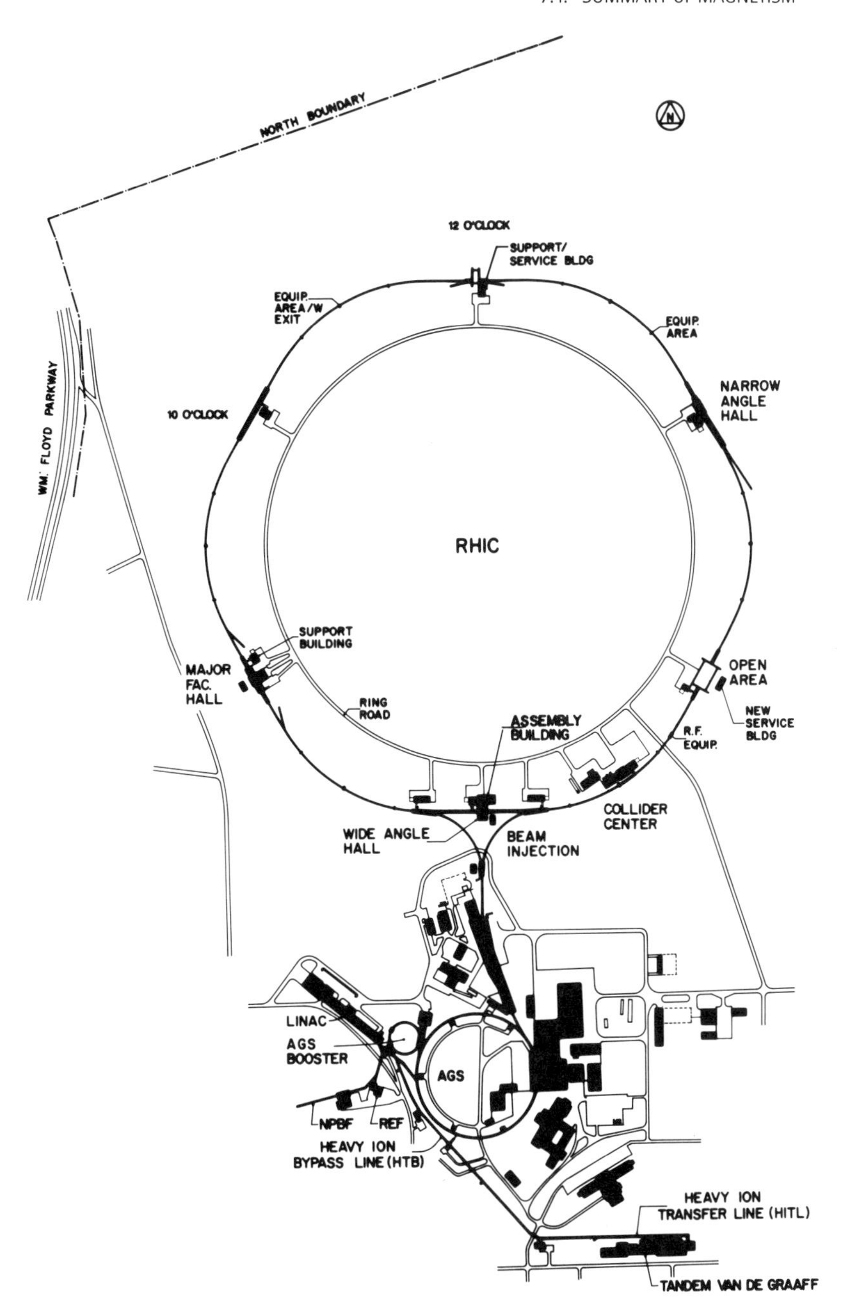

FIGURE 7.9 A map of RHIC. The large hexagonal ring is the main accelerator and storage ring containing the 3.5 T bending magnets. *Drawing provided courtesy of Brookhaven National Laboratory.*

This equation defines the unit of magnetic field strength to be $1\,\mathrm{N\,C^{-1}\,(m\,s^{-1})^{-1}}$, which is called a tesla, or T. The direction of a magnetic field is the direction that the northward-pointing end of a compass needle points when placed in the field. This direction is parallel to the line of motion where no force acts on a moving charge.

The force exerted by a magnetic field on a moving charge is perpendicular both to the velocity of the charge and to the magnetic field. To find the direction of the force, imagine two arrows, one pointing in the direction of the velocity and the other in the direction of the magnetic field. Now imagine the two arrows put tail to tail and that you are standing at the point where the tails join. Point your right arm in the direction of v and your left arm in the direction of B. If your arms are crossed, then the force is down, toward your feet; if your arms are not crossed, the force is up, toward your head.

Here is an important special case used often in this book. If a charge q moves with a velocity v perpendicular to a uniform magnetic field B, it will move in a circle of radius R such that

$$mv = p = qRB.$$

Magnetic fields are produced by charges that are moving, e.g., currents in wires or beams of charged particles. They have been used to measure atomic masses to great precision. Such measurements show the existence of isotopes, different-mass atoms of the same element. Magnetic fields are basic to the guidance and control of beams of charged particles whether they are the electrons that make the picture on the screen of your television set or the beams of high-energy particles that circulate inside accelerators.

PROBLEMS

1. A proton is accelerated from rest through a potential difference of 0.5 MV.

- **a.** What is its energy after this acceleration?
- **b.** What is its velocity?
- **c.** If it enters, as shown in Fig. 7.10, a uniform magnetic field of 0.1 T, pointing into the plane of the paper, which way does it curve, toward the top edge of the page or toward the bottom edge?
- **d.** What is the radius of its curvature in the magnetic field?

2. An electric potential of 1000 V is placed across two plates separated by 20 mm, as shown in Fig. 7.11. A beam of particles each with charge q and mass

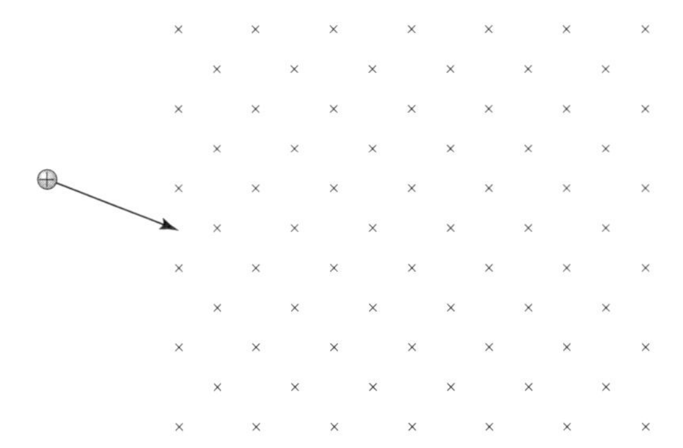

FIGURE 7.10 A positive charge enters a uniform magnetic field (Problem 1).

m travels at $v = 3 \times 10^6$ m/s between the plates and is deflected downward. A uniform magnetic field applied to the region between the plates brings the beam back to zero deflection.

a. What is the intensity of the deflecting electric field E?

b. What is the intensity of the deflecting magnetic field B?

c. Prove that $v = E/B$ when the electric and magnetic forces balance.

d. Assuming that the particles have been accelerated through 25.6 V and have kinetic energy of 41×10^{-19} J, find the value of q/m. What are these particles likely to be?

3. Physics students often measure the ratio of the charge e of an electron to its mass m. In one experiment to measure e/m, electrons are boiled off a hot filament and accelerated through 150 V (See Fig. 7.12).

a. What kinetic energy do the electrons acquire as a result? Give your answer both in joules and in electron volts.

b. After they have passed through a potential difference of 150 V, what is the speed of the electrons?

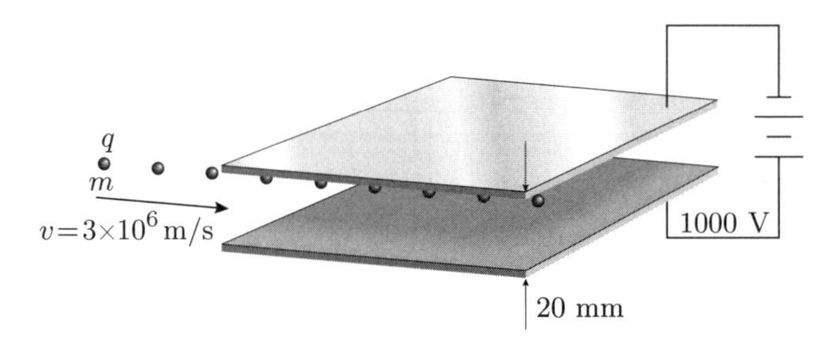

FIGURE 7.11 Arrangement for electric deflection in Problem 2.

c. After the electrons enter a uniform magnetic field of 0.5 mT traveling at right angles to the field as shown in Fig. 7.12, what force will the magnetic field exert on them? Give both the magnitude and direction of the force.

d. What will be the electrons' radius of curvature in the magnetic field?

4. A uniform magnetic field of 45.6 mT points in the z direction, as shown in Fig. 7.13. At time $t = 0$ a particle with a charge of -1.60×10^{-19} C is moving along the positive x-axis, as indicated in the diagram. It has a speed of 4.4×10^5 m/s and moves in a circle of radius $r = 10$ cm.

a. What is the direction of the force on the particle?

b. What is the numerical value of the force on the particle?

c. What is the momentum of the particle?

d. What is the particle's mass?

5. Suppose a particle made of equal amounts of negatively and positively charged matter with a total mass of 2 μg is put initially at rest somewhere in space.

a. You notice that although the two charges of matter remain evenly mixed, this neutral particle begins to accelerate. What is likely to be the cause of such acceleration?

b. Imagine you remove from the particle 1 μC of negative charge. You observe that now when the particle is placed at rest at the same point in space as before, it begins to accelerate much more than before. What is likely to be the cause of this additional acceleration?

c. You notice that once it is in motion, the acceleration of this charged particle remains essentially constant regardless of the direction in which the particle is moving. What does such behavior tell you about the presence of a magnetic field?

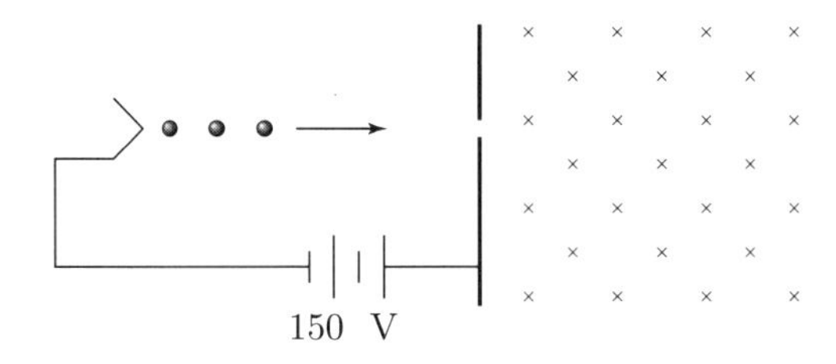

FIGURE 7.12 A beam of electrons accelerates through 150 V and then enters a uniform magnetic field of 0.5 mT pointing into the plane of the paper (Problem 3).

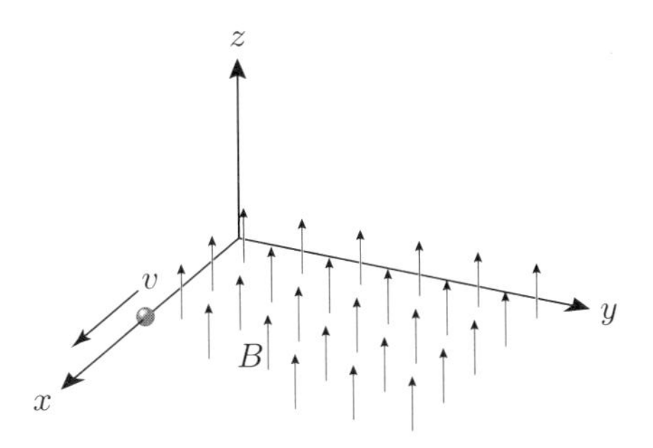

FIGURE 7.13 Direction of motion of a negatively charged particle in a uniform magnetic field at some instant of time (Problem 4).

6. Suppose you did an experiment in which electrons came off a cathode and were accelerated through 200 V. Suppose these electrons enter a magnetic field perpendicular to the field and bend as shown in Fig. 7.14.

a. What was the kinetic energy of the electrons? Give your answer in electron volts.

b. What was the direction of the force on them? Give your answer by drawing arrows on the diagram.

c. What must have been the direction of the magnetic field to bend the electrons as shown?

d. Explain how you figured out your answer to (c).

7. Figure 7.15 illustrates the operation of a mass spectrometer. The mass-to-charge ratio of an ion, the accelerating voltage, the magnetic field, and the radius of the ion's orbit are related as follows:

$$\frac{m}{q} = \frac{B^2 R^2}{2\,V_{\mathrm{acc}}}.$$

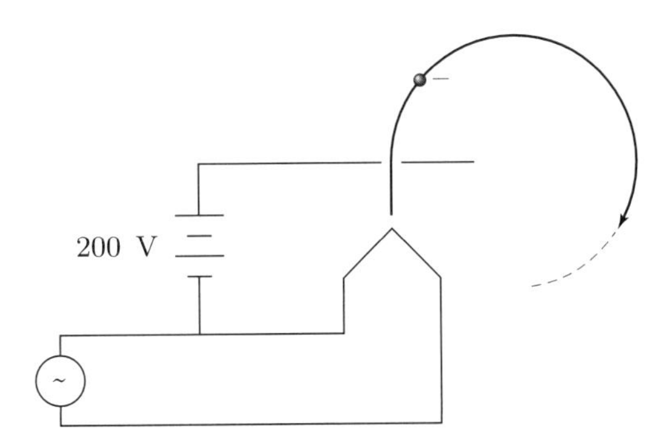

FIGURE 7.14 A beam of electrons accelerated through 200 V bends in a magnetic field (Problem 6).

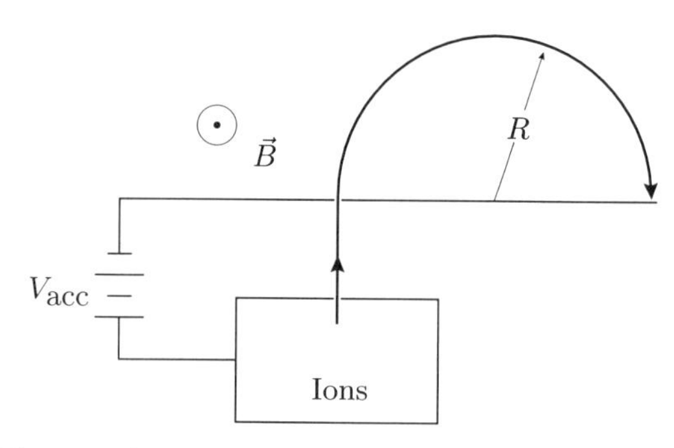

FIGURE 7.15 Schematic diagram of the mass spectrometer of Problem 7.

For an accelerating voltage of 1000 V and a magnetic field of 0.1 T, helium ions (He^+) have a circular orbit of radius 9 cm. (Note $m = 4$ u).

a. If V_{acc} is increased to 4000 V, what is the new orbital radius of He^+?

b. If V_{acc} is returned to 1000 V, and B is doubled, find R.

c. Let B once again equal 0.1 T. At what voltage would O_2^+ ions (m = 32 u) have a radius of 9 cm?

d. What is the kinetic energy (in convenient units) of the helium and oxygen ions when they enter the magnetic field?

8. An electric field is produced by applying a potential difference of 400 V between two parallel plates spaced 2 mm apart (see Fig. 7.16). A beam of 20 keV electrons is sent between the plates along the x-axis.

a. What is the momentum p_x of the electrons as they enter the space between the plates?

b. If the plates are 0.5 cm long, what is the change in momentum Δp_y between the time the electrons enter the space between the plates and the time they emerge?

c. With no voltage across the plates, the electrons travel on from the plates and strike some point on a screen 15 cm away. Turning on the voltage across the plates causes the electrons to strike a different point on the screen. What is the distance between the points that the electrons strike with voltage off and with voltage on?

9. Now the electric field in the previous problem is turned off and a uniform magnetic field is made to fill the space between the plates. The direction of B, the magnetic field, is out of the plane of Fig. 7.16.

a. In what direction will the electrons be deflected?

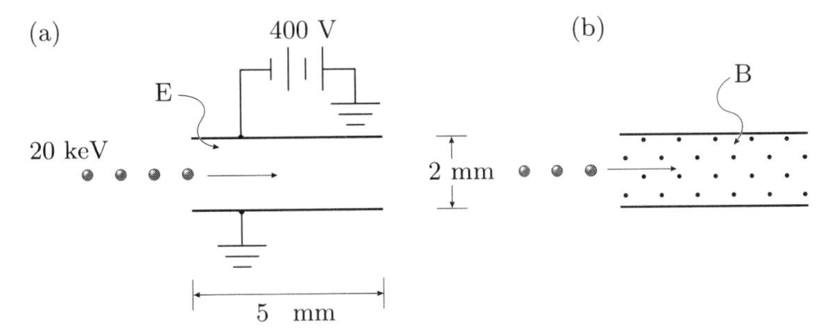

FIGURE 7.16 A beam of electrons passes (a) through an electric field (Problem 8) and then (b) through a magnetic field (Problem 9).

b. For what value of B will the magnitude of the change in momentum Δp_y of the electrons passing between the plates be the same as that produced by the electric field of the previous problem?

CHAPTER 8

Electrical Atoms and the Electron

8.1 INTRODUCTION

One important result of physicists' increased understanding of electricity and magnetism was the recognition that atoms are electrical in nature. In the 1830s Michael Faraday's[1] studies of the flow of electricity through solutions contributed evidence that electricity is itself "atomic," i.e., made up of small indivisible units of electric charge. In 1894, G.J. Stoney proposed the word "electron" as the name for such a natural unit of charge. In 1897 the British physicist J.J. Thomson used electric and magnetic deflection to establish the existence of the tiny particle of electricity that we now call "the electron." His work also showed that the electron is a fundamental component of every atom and intimately related to its chemical properties. By the end of the first decade of the twentieth century the American physicist Robert A. Millikan had measured the electron's mass and charge to within one percent.

The experiments that led to the discovery of the electron and revealed its properties are good examples of how physicists use physics to discover more physics, and they illustrate the kinds of physical argument that physicists find persuasive. They are also direct applications of the laws of electricity and magnetism laid out in the previous two chapters, and they turned out to be the seeds from which have grown pervasive, important contemporary technology that profoundly shapes our lives today.

[1] Michael Faraday, 1791–1867, born of a poor family near London, received only the equivalent of an elementary school education. Apprenticed to a bookbinder at thirteen, Faraday took to reading everything he could find, especially scientific books. Attendance at some lectures on chemistry by Sir Humphrey Davy led to his applying for and receiving a position as assistant to Davy. From this beginning, Faraday trained himself in science to the point where his discoveries in chemistry and electromagnetism, as well as his extremely popular public lectures, brought him renown and many honors.

8.2 ELECTROLYSIS AND THE MOLE OF CHARGES

As early as 1800, the passage of electric current through a solution was known to decompose it into elemental components. For example, Davy decomposed water into hydrogen and oxygen gas in this way. In the 1830s, Michael Faraday studied the passage of electric current through fluids that are good conductors of electricity. A brine solution, sodium chloride dissolved in water, is a good example.[2] Acid and base solutions also are good conductors.[3] From this work Faraday deduced that electrically charged atoms, "ions" as he named them, are the conductors of electricity in solutions. He measured the amounts of different elements that were released from the solution during the passage of a given quantity of electric charge, and he established a quantitative relation between this charge and the atomic masses of elements produced by the electric decomposition of molecules in the solution.

Faraday called solutions that could be thus electrically decomposed "electrolytes." He also coined the following words:

electrode. The (usually) metal pieces inserted into the electrolyte to pass the electrical current into and out of the solution.

electrolysis. The electrically induced breakup of molecules in the electrolyte when current flows through it.

anode and **cathode.** Respectively, the positively and negatively charged electrodes.

ions. The bodies, charged atoms or molecules, that carry current through the electrolyte between the anode and the cathode.

anions and **cations.** The ions that move to the anode and cathode, respectively.

Faraday varied the nature of the electrolyte and the size and shape of the electrodes. He tried different arrangements of electrodes to ensure that when gases were the product of decomposition, all the evolved gas was collected. When sufficient care was taken, the universal result was that *the measured*

---

[2] Ordinary tap water has enough dissolved material to be a pretty good electrical conductor, which is why you should not stand in a puddle of water during a lightning storm or become an electrolytic cell by touching a live wire.

[3] See *Great Experiments in Physics*. Morris H. Shamos, Ed., Holt and Co., New York, 1959, p. 128, for extensively annotated excerpts from the original publications of Faraday on electrolysis and electromagnetism, as well as from the works of other important physicists. Short biographies of the pioneers we are discussing in this chapter are also included.

amount of a substance decomposed was "in direct proportion to the absolute quantity of electricity [i.e., total charge] which passes."

EXAMPLES

1. In one experiment with zinc and platinum electrodes,[4] Faraday observed that hydrogen gas was generated from the platinum electrode, while zinc oxidized and left the zinc electrode. Faraday collected the hydrogen gas; there was 12.5 cubic inches of it at a temperature of 52°F and a pressure of 29.2 inches of mercury. He corrected this to 12.2 cubic inches of dry hydrogen at "mean temperature and pressure." This corresponds to the electrolysis of an amount of water that would yield 18.3 cubic inches of mixed hydrogen and oxygen, which he tells us would have a mass of 2.35 grains. Before electrolysis, the zinc electrode weighed 163.10 grains; after electrolysis it weighed 154.65 grains. This means that during the electrolysis of 2.35 grains of water, 8.45 grains of zinc were removed from the zinc electrode.

If for every molecule of H_2O that is electrolyzed, one Zn atom is removed from the zinc electrode, then the ratio of the masses of the electrolyzed water and zinc should equal the ratio of their molecular weights. To check this, let's assume that the molecular weight of water is 18 u and determine the atomic weight of Zn from Faraday's data.

$$M_{\mathrm{Zn}} = 18\frac{8.45}{2.35} = 64.7\,\mathrm{u},$$

which compares very well with the accepted value of 65.4 u.

Notice that in this example it is not necessary to know the SI equivalent of a grain. And a good thing too, because it is not clear what it is. Also, Faraday corrected the gas volume for temperature and pressure. This also is a good thing for him to have done because it is not clear now what "mean temperature and pressure" meant in 1834.

Even though Faraday had no way to measure the actual quantities of charge passed through the electrolyte, he could stabilize the current so that the amount of charge conducted was proportional to the amount of elapsed time. Thus he could know reasonably accurately the relative amounts of

[4] pp. 254–255 in *Experimental Researches in Electricity* (3 vols. bound as 2) by Michael Faraday. Dover Publications, Inc., New York, 1965. The three volumes were originally published in 1839, 1844 and 1855 respectively.

TABLE 8.1 Relative Masses of Elements Produced by Electrolysis Using the Same Amount of Charge

Element	Hydrogen	Oxygen	Chlorine	Iodine	Lead	Tin
Relative Mass	1	8	36	125	104	58
Atomic Weight	1	16	36	125	208	116

charge used in different experiments. Taken together, all his results showed a simple relation between atomic masses and the relative masses of elements electrolyzed by the same amount of charge. As Table 8.1 shows, these measured relative masses are in the ratio of the atomic masses or some simple integer multiple of them. For example, the amount of charge that releases one atomic weight of hydrogen releases only half an atomic weight of lead; similarly for tin. As for oxygen, the release of these atoms requires (exactly) twice as much electricity for electrolysis as for hydrogen.

When it became possible to measure charges accurately, the amount of charge required to electrolyze exactly one mole of hydrogen, or silver, or chlorine, was found to be 96485 coulombs. In other words, 96485 C will cause the electrolytic deposition of one gram atomic weight of silver, i.e., 107.88 g, which is one mole of silver atoms.

Therefore, it appears quite reasonable to identify this quantity of charge, 96485 C, as an Avogadro's number N_A of basic charges. This suggests that there exists some basic, elementary charge and that 96485 C is a mole of them. This quantity of electricity, the mole of elementary charges, is called "the faraday" and is often represented by the symbol F. If we designate the elementary atom of charge by the symbol e, then the Faraday is

$$F = N_A e = 96485\,\text{C}.$$

Figure 8.1 shows a typical setup for electrolysis along with a simplified modern interpretation of what goes on in the process.

EXAMPLES

2. Suppose that a cell arranged as in Fig. 8.1 carries a current of 15 amperes for 10 minutes. How much silver will plate out on the cathode?

The amount of charge, Q, delivered by this current during this amount of time is $15\ \text{C}\,\text{s}^{-1} \times 10\,\text{min} \times 60\,\text{s}\,\text{min}^{-1} = 9000\,\text{C}$. As a fraction of a faraday F this is

$$Q/F = 9000/96485 = 0.0933,$$

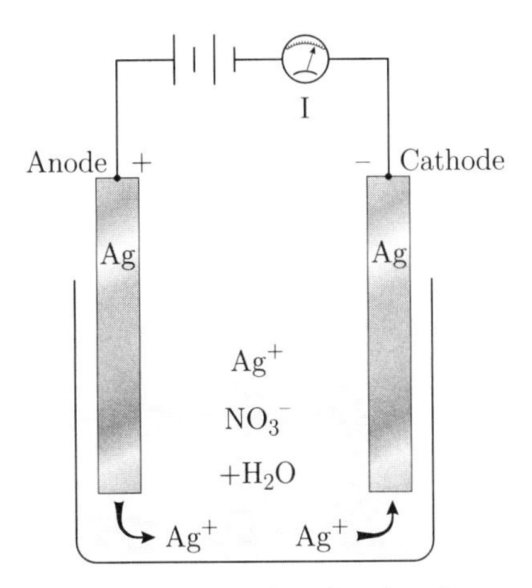

FIGURE 8.1 An electrolytic cell in which silver dissolves from a silver anode and plates onto the cathode.

which from the definition of F must be the same fraction of silver's mole atomic weight of 107.88 g. Thus:

$$\text{Deposited Wt. of Ag} = 107.88 \times 0.0933 = 10.06\,\text{g}.$$

Faraday's work strongly suggested that chemical reactions involve integer amounts of some basic quantity of electric charge. In conjunction with the chemical evidence for atoms, these results led to the idea of an integer number of these basic charges accompanying ions in solution. Faraday even was so bold as to suggest that it is the electrical charges that are the source of chemical "affinity," or bonding, as we would say now.

EXERCISES

1. Faraday's results are shown in Table 8.1 along with the known atomic weights. Use these data to deduce the number of faradays required to produce one mole of each of the elements he studied. Assume the possibility of different numbers of elementary charges on different ionic species (which Faraday apparently did not consider) and explain why a faraday of charge might not always generate a mole atomic weight. Explain each of the six results reported in the table.

Note that the faraday is intimately connected with Avogadro's number. If you know any two of the three quantities N_A, F, or the elementary atomic charge e, you can determine the third.

EXERCISES

2. In Chapter 5 you saw that a rough estimate of Avogadro's number ($\approx 30 \times 10^{23}$) could be obtained from kinetic theory and measurements of the viscosity of air. Using this result and the value of the faraday, estimate a value for the elementary charge e.

Notice that Faraday's law of electrolysis makes it possible to find the ratio of an ion's charge to its mass. In the case of hydrogen this is the ratio

$$\frac{F}{M_H} = \frac{N_A e}{N_A m_H} = \frac{e}{m_H},$$

where M_H is the atomic weight of a mole of hydrogen atoms and m_H the mass of a single hydrogen atom.

EXERCISES

3. From the known value of the faraday determine the value of the charge-to-mass ratio of the hydrogen ion. Of all the possible ions that one can imagine forming from the periodic table of the elements, which will have the largest value of its charge-to-mass ratio?

Keep this result in mind for comparison with the charge-to-mass ratio discussed in the next section.

8.3 THOMSON'S EXPERIMENTS AND e/m

Another line of inquiry leading to further insights into the electrical nature of atoms was begun some thirty-five years after Faraday's work. By then it was possible to generate large electric potentials that could supply sustained currents. In the last third of the nineteenth century these were used to produce electrical discharges in gases. In such discharges rays emanated from the negative electrode, i.e., the cathode, and so they were named "cathode rays."

When cathode rays passed through low-pressure gas, the gas would glow. Certain materials—called "phosphors" (e.g., coatings on TV tubes)—when struck by cathode rays gave off visible light. Because magnets strongly deflected cathode rays, it seemed they might be electrically charged particles. However, an attempt by the German physicist Heinrich Hertz to deflect them by passing them through an electric field between two plates showed no effect, and so supported the belief that cathode rays were *not* electrical.[5]

In 1887 J.J. Thomson[6] began experiments that proved that cathode rays are electric particles. He not only improved on the experiments of others and convincingly showed the electrical nature of cathode rays, but he produced clear evidence that they had a well-defined, distinct ratio of charge to mass, a finding that could only be understood if the rays were streams of distinct particles. These particles were named "electrons." For being the first to exhibit their existence and for determining some of their properties, Thomson is considered the discoverer (in 1897) of the electron.

His work also provided evidence that the electron is an important part of the atom. That an atom has parts was a dramatic change in the atomic concept. It was no longer "not cuttable"; it had parts.

Determining the Nature of Cathode Rays

To show that cathode rays are charged particles, Thomson had to solve several problems. For one thing, he had to show that as charged particles they were deflected by electric fields. For another, he needed to demonstrate conclusively that electrical charges followed the exact path of the cathode rays themselves. He had to understand and explain the unusual behavior of the rays when a magnet was brought near. (You can see this effect by bringing a magnet close to the screen of your TV or computer monitor.)[7] This was not an easy task, because such gas-discharge experiments were the first observed instances of well-defined charged particles moving in magnetic fields, and most physicists did not yet have the convenient tool of vector representation or all the "hand rules" that now make it easy (?) to predict the motion of charged particles in a magnetic field.

Thomson tried several different configurations of electrodes inside evacuated tubes to rule out possible leakage of charges from elsewhere in the tube when the cathode-ray beam was deflected by a magnet; he showed conclusively that the observed charge was associated only with the beam. He repeated Hertz's experiment on electrostatic deflection and showed that in better vacuums the cathode rays were deflected by electrostatic fields. He also showed that when the beam struck an object, it deposited there an

[5] Hertz's experiment failed because of poor vacuum. He was working at the edge of what was technically possible in his time, and it was not good enough. Ionization of residual gas in his cathode-ray tube produced enough conductivity to short circuit the applied voltage and reduce the electric field to a value too low to produce any deflection.

[6] Joseph J. Thomson, 1856–1940. See Shamos's book, p. 216.

[7] If you overdo this, you can cause the color alignment on the screen to go out of adjustment, and you will become quite unpopular with other users.

amount of heat energy equal to the electrical energy given to the beam by the accelerating voltage.

EXAMPLES

3. Suppose that it took 0.02 joules to heat the beam detector of Thomson's tube enough to raise its temperature by 1 K, and that the beam was produced by a potential difference of 400 volts between the anode and cathode. If the cathode-ray beam delivered a current of 2.0 μA at the detector, how long would it take to raise the temperature of the detector by 2 K?

A 2 K rise in temperature would require 0.04 J of energy (assuming no serious losses by cooling). In time t the beam delivers an energy of VIt. For $V = 400$ V, and $I = 2$ μA, so the time t is $0.04/0.0008 = 50$ s.

Determination of e/m

After convincing himself that cathode rays were particles, Thomson investigated their dynamics under the assumption that they had definite charge and mass. He used a combination of electric and magnetic deflection to show that cathode rays have a distinct, well-defined ratio of charge to mass. He studied the ratio of charge to mass because that was all he could measure with the experimental techniques of his day. There are several ways to measure this ratio, and we will describe two of them. The first is close to the method used by Thomson; the second is an approach devised by K.T. Bainbridge, a modern version of which is now used in many undergraduate laboratories.

e/m by the Thomson Method

The key to Thomson's method of measuring e/m is deflection of charged particles by electric and magnetic fields. Electric deflection is commonly achieved by passing a beam of charged particles between oppositely charged parallel plates as shown schematically in Fig. 8.2. You have seen that in the volume between such plates the electric field E is constant. While the beam is passing between the plates, the electric field exerts a constant force on the beam, upward in the case shown in Fig. 8.2. The deflection is usually measured by looking at the luminous spot made when the beam strikes a screen a distance L from the deflecting plates. When the field is off, the beam strikes one point on the screen; when the field is on, the spot is deflected some measurable distance y.

We can predict the amount of deflection by calculating the angle at which the beam emerges from between the plates when E is on. That angle is going to be the ratio of the upward momentum of the beam to its horizontal

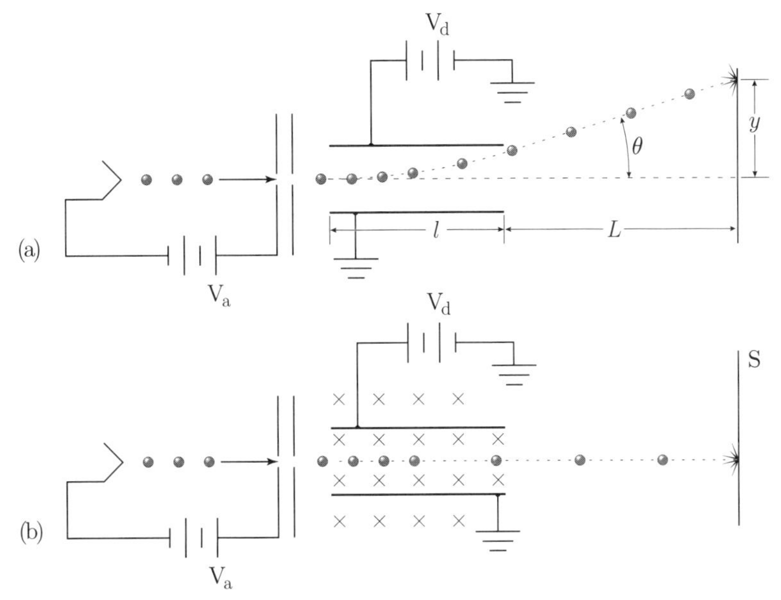

FIGURE 8.2 Thomson's arrangement of electric and magnetic fields to deflect cathode rays. The screen S on which the deflection was observed is at the extreme right.

momentum p_x. When the beam enters the region between the plates, it has no upward momentum. It acquires an amount Δp_y from the action of the field E. The angle of deflection will be

$$\theta = \frac{\Delta p_y}{p_x}.$$

From the definition of force as $\Delta p/\Delta t$ it follows that $\Delta p_y = F_y\, \Delta t$, where Δt is the time that any given electron spends between the deflecting plates, and $F_y = eE$. If the length of these plates is ℓ and the speed of the electrons is v_x, then

$$\Delta t = \frac{\ell}{v_x}.$$

From the fact that p_x is mv_x, the angle of deflection will be

$$\theta = \frac{\Delta p_y}{p_x} = \frac{eE\ell}{mv_x^2}.$$

To find the displacement on the screen assuming that the small-angle approximation is valid, just multiply θ by the distance to the screen:

$$y = L\theta = \frac{eE\ell L}{mv_x^2}. \tag{1}$$

Equation 1 shows a basic property of electric deflection: It is inversely proportional to the kinetic energy of the particles being deflected.

This calculation of displacement on the screen neglects y-displacement occurring while the beam particles are between the plates. It is usually negligible compared to the amount of displacement that occurs after leaving the plates, but it is not hard to include it if you need to. Electric deflection is a basic tool for controlling charged particles of all kinds; examples of practical applications will be given later.

Equation 1 was not enough for Thomson to find the value of e/m of cathode rays. He could measure y and E, but not v_x. You might think he could use the fact that he gave the rays their kinetic energy by accelerating them through a known voltage V_a (subscript "a" for acceleration). As a result

$$\frac{1}{2}mv_x^2 = eV_a,$$

but, as you can easily show, using this equation to eliminate v_x from Eq. 1 also eliminates e and m.

EXERCISES

4. Show that this last statement is true.

Thomson's solution was to use magnetic deflection. A simple way to do this is to produce in the region between the plates a B field with just the right strength to cancel the deflection due to E. The effects of the two fields cancel when

$$eE = ev_xB,$$

or

$$v_x = \frac{E}{B}. \qquad (2)$$

(This is the condition for the Wien velocity filter discussed in the previous chapter.) Thomson could determine both B and E, which meant that he could replace v_x in Eq. 1 with its experimentally determined value. Or he could substitute Eq. 2 into Eq. 1 and solve to get

$$\frac{e}{m} = \frac{Ey}{B^2\ell L}, \qquad (3)$$

where everything on the right side was measurable. These early measurements were not very precise; one value Thomson obtained was 0.71×10^{11} C/kg, about a factor of 2.5 less than the currently accepted value.

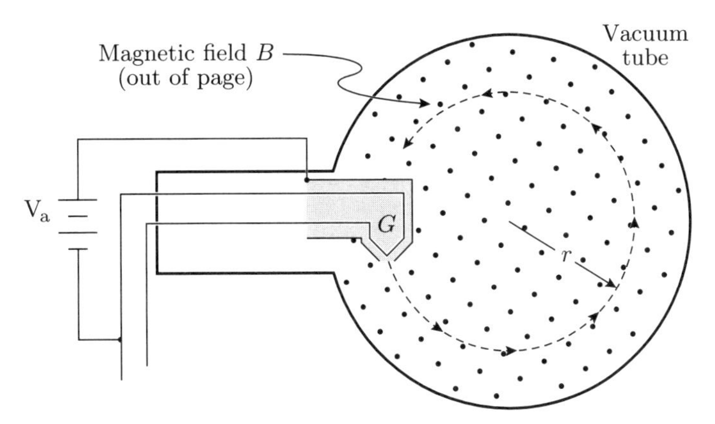

FIGURE 8.3 Diagram of an apparatus for measuring e/m. The source of electrons G directs a beam accelerated through a voltage V_a in a downward direction. An external magnetic field B bends the beam in the circle with a radius r as shown.

EXERCISES

5. If the separation between the plates was $d = 5$ mm and the deflecting voltage V_{d} (d for deflection) was 10^4 V, what was the value of the deflecting electric field?

e/m by the Bainbridge Method

All methods of measuring e/m use both electric and magnetic fields. In the Bainbridge method, as shown in Figure 8.3, a large tube is situated between two coils that produce a reasonably uniform magnetic field B. This field bends the cathode rays in a circle with a radius r that depends upon the particles' momentum and the strength of the magnetic field. The beam of cathode rays is produced by a small "electron gun" oriented to inject electrons tangent to the circumference of the tube so that the beam can make as large a circle as possible. The gun gives the electrons kinetic energy by accelerating them across a potential difference of V_{a} volts. With a charge e and mass m, each electron gains a kinetic energy of

$$\frac{1}{2}mv^2 = e\,V_{\mathrm{a}}. \tag{4}$$

In modern versions of the Bainbridge apparatus the tube is evacuated during its manufacture and then refilled to a low pressure with an inert gas such as helium. When electrons pass through this gas, they cause it to glow, and the path of the rays becomes visible so that an observer can measure the radius of curvature r of the electron beam.

The external magnetic field B produces a force on the moving charges perpendicular to their path. This force acts as a centripetal force to produce circular motion. Thus

$$\frac{mv^2}{r} = evB, \tag{5}$$

where r is the radius of the circle. Solve for v,

$$v = \frac{eBr}{m},$$

then square the result and substitute into Eq. 4 to get

$$\frac{2eV_a}{m} = v^2 = \frac{e^2B^2r^2}{m^2}$$

and

$$\frac{e}{m} = \frac{2V_a}{B^2r^2}. \tag{6}$$

Every quantity on the right side of Eq. 6 is measurable, so e/m can be determined. The modern value is

$$e/m = 1.759 \times 10^{11} \text{ C/kg}.$$

EXAMPLES

4. In Fig. 8.3, suppose that a radius of 8.0 cm is measured and the accelerating voltage is 400 V. What magnetic field is being used to bend the cathode rays?

Solving Eq. 6 gives

$$B^2 = \frac{2V}{(e/m)r^2}$$

from which it follows that

$$B = \sqrt{2 \times \frac{400}{1.76 \times 10^{11} \times 64 \times 10^{-4}}}$$

(where r^2 was converted to square meters). Evaluating this expression gives

$$B = \sqrt{7.10 \times 10^{-7}} = 8.43 \times 10^{-4} \text{ tesla}.$$

The significance of e/m

Thomson's value for e/m was remarkably large, or, to put it another way, the value of m/e was remarkably small. From experiments in electrolysis one finds that m/e for the hydrogen ion is about 10^{-8} kg/C (see Exercise

3). Thomson's value of m/e for electrons was one or two thousand times smaller, indicating that in terms of mass the electron was a very small part of the atom.

At least as important as its discovery was the recognition that the electron was a part of *every* atom. The fact that cathode rays were the same regardless of the kind of gas atoms in the electric discharge and entirely independent of what material the cathode (or anode) was made of suggested that electrons were to be found in every kind of atom. Furthermore, a year earlier W.A. Lorentz had shown that he could explain a peculiar change that occurred in the frequency of light emitted by neutral atoms placed in a magnetic field if there was inside the atom a charged particle with a charge-to-mass ratio of about 2×10^{11} C/kg, the same value to within experimental uncertainty as the value of e/m measured by Thomson for a free electron. This was strong evidence that the electron is an internal part of every atom and intimately involved in atomic behavior.

8.4 THE ELECTRON'S CHARGE

Introduction and Overview

Accurate knowledge of the electron's charge is important. It sets the scale of electric forces in atoms, and these forces determine atomic size. Moreover, knowledge of e would yield N_A from the faraday, and the mass of the electron from e/m. After Thomson's work, experimenters sought to measure the atomic charge directly. The most conclusive of the direct measurements were the experiments of the American physicist R.A. Millikan.[8] He obtained accurate results by refining a technique for producing and studying the movement of small electrically charged droplets in air under the influence of gravitational and electric fields. His work established that there is such a thing as an elementary indivisible unit of charge and that its value is $e = 1.6 \times 10^{-19}$ C.

Millikan used droplets of oil a few micrometers in diameter (a human hair is about 100 μm in diameter). When such a small droplet falls in air, it accelerates only very briefly—until resistance from the viscosity of air balances the force of gravity—after which it falls at a slow, steady speed. These droplets are too small to be seen directly, but when illuminated with a bright light against a dark background, an individual droplet scatters enough light to make a visible glint. Millikan could then measure its speed of fall

[8]Robert A. Millikan, 1865–1953. See Shamos, *op. cit.*, p. 238. The results of Millikan's experiment are reported in *Physical Review* **32**, 349 (1911).

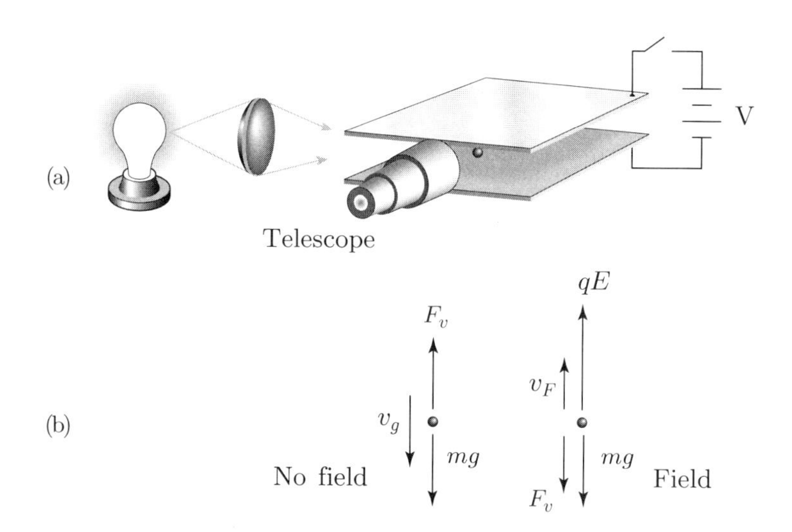

FIGURE 8.4 (a) Millikan apparatus for determining the charge of an electron. (b) A charged droplet of oil can be observed through a telescope to fall at terminal velocity v_g between parallel plates in the absence of any electric field and to rise with terminal velocity v_F when an electric field E is present.

very accurately. The apparatus to apply the necessary fields and illuminate the oil drops is illustrated in Fig. 8.4.

Measuring this velocity was the key to Millikan's experiment. A droplet's speed of fall in air depends on its shape and size and mass; and by measuring a droplet's speed he could figure out its mass. Similarly, an electric field acting on a charged droplet will accelerate it briefly until air resistance and gravity balanced the electrical force. After that the droplet will move at a steady speed. He measured that speed also. These two measurements gave him enough information to deduce the strength of the electrical force on the droplet, from which he could determine how much electric charge it carried. After measuring many different charges both on the same and on different drops, he showed that all his measured charges were integer multiples of a single, smallest amount—the elementary charge.

Droplet Size from Terminal Velocity

To understand the effects of electric field on his charged oil droplets, Millikan had to know their masses. He was able to find the mass, or, what was equivalent, the radius, of any droplet by an ingenious application of the mechanics of falling bodies in a viscous medium (air in his experiments). The idea is that when a small, spherical body falls freely through a viscous medium, it quickly stops accelerating and falls at a steady rate v_g called its "terminal velocity."

The terminal velocity v_g depends on the acceleration due to gravity, g, on the viscosity η of the medium, and the sphere's radius a and density ρ of the oil of which the droplet was made. Millikan measured the terminal velocity of a falling droplet and then from his knowledge of g, η, and ρ, he calculated a. Then he calculated the mass m of the spherical droplet from the fact that $m = \frac{4}{3}\pi a^3 \rho$.

To see how this works, think of an oil droplet falling freely through the air inside the chamber of Millikan's experimental apparatus (Fig. 8.4) with the electric field turned off. Two forces act on the droplet, gravitational attraction downward and viscous drag opposing the droplet's motion. The gravitational force is just mg. The retarding viscous force due to the air is proportional to its viscosity and to the droplet's velocity. The British physicist Stokes showed in the nineteenth century that the viscous force acting on a small sphere of radius a falling with velocity v through a fluid with viscosity η is

$$F_v = 6\pi\eta a v. \tag{7}$$

Consider what happens when the droplet falls. It starts off accelerating at g, but as its velocity increases, the retarding viscous force increases, balancing some of the gravitational attraction and thereby decreasing the acceleration. The value of v continues to increase and the acceleration continues to decrease until the droplet is nearly at the velocity v_g at which the force F_v retarding the droplet just equals mg, the force of gravity downward. As the velocity approaches v_g, the acceleration goes to zero, and the velocity stops changing. The velocity v_g at which this balance occurs is called the "terminal velocity."

We can find an expression for v_g in terms of the droplet radius a and density ρ. Begin by equating the viscous and gravitational forces, $F_v = mg$, and solving for v_g:

$$\begin{aligned} 6\pi\eta a v_g &= mg, \\ v_g &= \frac{mg}{6\pi\eta a}. \end{aligned} \tag{8}$$

Then eliminate m from this equation using the fact that $m = \frac{4}{3}\pi a^3 \rho$. Replacing m by this expression and doing some algebra, you will get

$$v_g = \frac{2}{9}\frac{\rho g a^2}{\eta}. \tag{9}$$

EXERCISES

6. Derive Eq. 9 from Eq. 8.

EXAMPLES

5. How slowly does a droplet fall through air if its radius is only 1 μm, i.e., what is its terminal velocity? Assume that the density of the oil is $\rho = 0.92\,\mathrm{g\,cm^{-3}}$ and the air temperature is 20°C.

We know from Chapter 5 that $\eta = 18.3\,\mu$Pa s. Then, using Eq. 9 we get

$$v_g = \frac{2}{9}\frac{0.92 \times 10^3 \times 9.8 \times 10^{-12}}{18.3 \times 10^{-6}} = 1.1 \times 10^{-4}\,\mathrm{m/s}.$$

A droplet 1 μm in radius will fall with a terminal velocity of 0.11 mm/s; it will take 9 s to fall 1 mm!

For a small droplet in air this equilibrium speed is reached almost immediately, and for all practical purposes the droplet would be seen to fall slowly at constant velocity.

EXERCISES

7. What would be the terminal velocity of an oil droplet if it were 10 μm in radius?

Millikan used the inverse of this argument to find droplet radii and masses. Looking through a telescope he would measure the time t_g it took a droplet to fall a known, small height h marked by two reference lines on his telescope. Then he could calculate the terminal velocity as $v_g = h/t_g$. Knowing v_g, he could use the inversion of Eq. 9 to find the droplet radius a. From this he could find the mass of the droplet.

EXAMPLES

6. What is the mass of an oil droplet with a radius of 1 μm? Assuming, as we have been doing, that the droplet is a sphere, we can calculate m because we know that the mass of the sphere will equal its volume times the density of the oil: $m = \frac{4}{3}\pi a^3 \rho$. Therefore,

$$m = \frac{4}{3}\pi(10^{-6})^3 \times 0.92 \times 10^3 = 3.9 \times 10^{-15}\,\mathrm{kg} = 3.9\,\mathrm{pg}.$$

The oil drop technique is a sensitive method for weighing a tiny mass as well as for determining its charge.

EXERCISES

8. Suppose a droplet has a terminal velocity of 0.43 mm/s. What is its radius and mass? Efficiency tip: Use the values worked out for a 1 μm radius drop and scaling arguments to avoid unnecessary calculation.

9. How big is the viscous force acting on the 1 μm-radius droplet when it is moving with terminal velocity? On the 10 μm droplet?

Finding the Charge on a Droplet

By applying an electric field Millikan could cause the charged droplets to rise. Turning the field repeatedly on and off, he could cause an individual drop to rise and fall over and over as he observed its motion for hours. From the speeds of fall and rise that he measured, Millikan could determine the amount of charge on a drop. He could also change the amount of charge on a given droplet, measure changes in its speed of rise, and calculate by how much the droplet's charge had changed. Let's see how he did these things.

Millikan produced an electric field E by applying a potential difference V to a pair of metal plates separated by a small distance d. As noted earlier, this creates a constant electric field $E = V/d$ between the plates. When the field is on, the droplet experiences a force qE upward in addition to the downward force mg, and under the combination of these forces it rises with a terminal velocity v_F (where the subscript F denotes the terminal velocity with the field turned on). When the field is turned off, the droplet falls with a terminal velocity, which we call v_g (the subscript g is intended to make you think "gravity only," i.e., field off).

There is a direct relation between the charge on the droplet and the measurable variables, the electric field E, and the two terminal velocities, v_F and v_g. From Eq. 8 it follows that for a freely falling droplet,

$$v_g \propto mg.$$

With the electric field on, the net force on the droplet is qE upwards minus mg downwards: $qE - mg$. As far as the droplet is concerned, this situation seems like "falling upwards," so the droplet will again reach a terminal velocity v_F proportional to the total force:

$$v_F \propto (Eq - mg).$$

Dividing the two proportionalites gives

$$\frac{v_F}{v_g} = \frac{Eq - mg}{mg},$$

from which it follows that

$$q = \frac{mg}{Ev_g}(v_F + v_g). \tag{10}$$

By expressing m in terms of a and ρ and then using Eq. 9 you can rewrite Eq. 10 to give the charge on the droplet solely in terms of quantities that can be determined from measurements. You will get

$$q = \frac{6\pi\eta^{3/2}}{V/d}\sqrt{\frac{9v_g}{2\rho g}}\,(v_F + v_g). \tag{11}$$

EXERCISES

10. Derive Eq. 11.

EXAMPLES

7. In studying one droplet Millikan used a potential difference of 5085 V between two plates 16 mm apart. He observed that when the field was off, the droplet fell with a speed of 0.08584 cm/s; when the field was on, the droplet rose with a speed of 0.01274 cm/s. What was the charge on this droplet?

Using $\rho = 0.92 \times 10^3\,\mathrm{kg\,m^{-3}}$, $\eta = 18.2\ \mu\mathrm{Pa\,s}$, and $g = 9.8\,\mathrm{m\,s^{-2}}$, we obtain from Eq. 11

$$q = \frac{6\pi}{5085/16 \times 10^{-3}}(18.2 \times 10^{-6})^{3/2} \tag{12}$$

$$\times \sqrt{\frac{0.08584 \times 10^{-2}}{9.8 \times 0.92 \times 10^3}}(0.08584 + 0.01274) \times 10^{-2}$$

$$= 1.40 \times 10^{-18}\,\mathrm{C}. \tag{13}$$

Notice that this is not really what we want to know. It is the charge on the given droplet; it is not the elementary charge. Only if a droplet had exactly one unneutralized elementary charge would the above equation give us the value of the elementary charge. Indeed, a singly charged droplet could not be made to rise in the electric field of Millikan's apparatus, so he could never observe such a droplet. To get around this problem Millikan changed the charge on a given droplet from time to time. He did this by slightly ionizing the air of the chamber with x-rays or radioactivity. A droplet would then pick up one or more ionized molecules and change its charge appreciably but not its mass. As a result, the electric force on the droplet would change and so would the droplet's terminal velocity v_F; v_g would stay the same. By measuring the changes in v_F, Millikan could calculate by how much the charge had changed. He observed that this change was always an integer multiple of some smallest amount, 1.602×10^{-19} C by modern measurements. He concluded that this smallest change corresponded to the elementary charge that we denote as e.

EXERCISES

11. How many elementary charges were there on the droplet referred to in Example 8.7?

Quantization of Electric Charge

It is interesting that Millikan did not need actually to measure the value of the elementary charge e in order to demonstrate experimentally that the electric charge on a droplet is always an integer multiple of e. His work is such a nice example of reasoning with ratios that it is worth looking at closely. Table 8.2 is a reproduction of data that Millikan published[9] to show the atomic nature of charge. (Other examples of Millikan's data can be found in Shamos's book.) Along with the data, Table 8.2 shows various parameters of his apparatus, such as voltage V, spacing between the plates d, the distance h of fall and rise of a droplet, etc. The data recorded by Millikan are the values of the quantities actually measured. These are the times of fall under gravity, t_g, and the times of rise when the field was on, t_F. You can see that when he caused the charge on a droplet to change, the time of rise with the field on changed because the electrical force Eq changed.

There is enough information in the table so that from these times you (and he), using Eq. 11, could calculate the charge on every droplet. But this would be neither efficient nor very insightful. Also, as noted in the previous section, Millikan concentrated on the *changes* in charge. He did this because these changes were smaller multiples of e than was the total charge on a droplet.

From Eq. 11 you can see that the change in charge Δq is

$$\Delta q = \frac{6\pi\eta^{3/2}}{V/d}\sqrt{\frac{9v_g}{2\rho g}}\left(v_F' - v_F\right), \tag{14}$$

[9]R.A. Millikan, *Electrons (+ and −), Protons, Photons, Neutrons, Mesotrons and Cosmic Rays*, revised edition, University of Chicago Press, Chicago, 1947, p. 75.

TABLE 8.2 Data from Millikan's Measurements on an Oil Drop (*Taken with permission from p. 75 of R. A. Millikan,* Electrons (+ and −), Protons, Photons, Neutrons, Mesotrons and Cosmic Rays, *University of Chicago Press, ©1937 and 1947 by the University of Chicago*)

t_g Sec.	t_F Sec.	$\frac{1}{t_F}$	$\left(\frac{1}{t'_F}-\frac{1}{t_F}\right)$	n'	$\frac{1}{n'}\left(\frac{1}{t'_F}-\frac{1}{t_F}\right)$	$\left(\frac{1}{t_g}+\frac{1}{t_F}\right)$	n	$\frac{1}{n}\left(\frac{1}{t_g}+\frac{1}{t_F}\right)$
11.848	80.708	.01236				.09655	18	005366
11.890	22.366		[illegible]	6	[illegible]			
[illegible]	22.390	.04470				.12887	24	.005371
11.904	22.368		.03751	7	.005358			
11.882	140.565	.007192	.005348	1	.005348	.09138	17	.005375
11.906	79.600	.01254				.09673	18	.005374
11.838	34.748		.01616	3	.005387			
11.816	34.762	.02870				.11289	21	.005376
11.776	34.846							
11.840	29.286					.11833	22	.005379
11.904	29.236	.03414	.026872	5	.005375			
11.870	137.308	.007268				.09146	17	.005380
11.952	34.638	.02884	.021572	4	.005393	.11303	21	.005382
11.860			.01623	3	.005410			
11.846	22.104					.12926	24	.005386
11.912	22.268	.04507	.04307	8	.005384			
11.910	500.1	.002000				.08619	16	.005387
11.918	19.704		.04879	9	.005421			
11.870	19.668	.05079				.13498	25	.005399
11.888	77.630		.03794	7	.005420			
11.894	77.806	.01285				.09704	18	.005390
11.878	42.302	.02364	.01079	2	.005395	.10783	20	.005392
11.880			Means		.005386			.005384

Duration of exp.	= 45 min.	Pressure	= 75.62 cm.
Plate distance	= 16 mm.	Oil density	= .9199
Fall distance	= 10.21 mm.	Air viscosity	= $1{,}824 \times 10^{-7}$
Initial volts	= 5,088.8	Radius (a)	= .000276 cm.
Final volts	= 5,081.2	$\frac{l}{a}$	= .034
Temperature	= 22.82° C.	Speed of fall	= .08584 cm./sec.

where it is necessary to write v'_F to distinguish the velocity of rise after the charge has changed from v_F, the velocity of rise before the droplet charge changed. Notice that for a given droplet, Eq. 14 means that

$$\Delta q \propto \left(v'_F - v_F\right),$$

or, since $v'_F = h/t'_F$ and $v_F = h/t_F$,

$$\Delta q \propto \left(\frac{1}{t'_F} - \frac{1}{t_F}\right),$$

because the distance h over which the fall and rise times were measured was the same for every droplet.

If the values of the various occurrences of Δq are multiples of some small, elementary value, then so must also be the values of the various occurrences of $\left(\frac{1}{t'_F} - \frac{1}{t_F}\right)$. In columns 4 and 5 of Table 8.2 you see Millikan's examination of this possibility. All the different values of $\left(\frac{1}{t'_F} - \frac{1}{t_F}\right)$ are convincingly close to being small integer multiples of the value $0.005386\,\mathrm{s}^{-1}$.

EXERCISES

12. How close to being exact integer multiples of $0.005386\,\mathrm{s}^{-1}$ are the first three entries in column 4 of Table 8.2?

13. Explain why Millikan probably would have found the same result in column 6 of Table 8.2 even if there had not been an instance in which the charge changed by a single elementary unit.

Of course, if there is a basic "atom" of charge, then the total charge q on any droplet should be some integer multiple of the same quantity as any change in charge Δq. From Eq. 10 it follows that

$$q \propto (v_g + v_F) \propto \left(\frac{1}{t_g} + \frac{1}{t_F}\right).$$

Consequently, the quantity $\left(\frac{1}{t_g} + \frac{1}{t_F}\right)$ should be an integer multiple of the same quantity as $\left(\frac{1}{t'_F} - \frac{1}{t_F}\right)$ is. Columns 7, 8, and 9 of Table 8.2 show this to be the case.

EXERCISES

14. Explain why column 7 of Table 8.2 represents the total charge on a droplet (as distinct from the change in charge).

15. As well as showing the times of fall and rise, Table 8.2 lists the parameters of Millikan's experiment.
(a) Use these to show that the radius a is correctly calculated from his values of viscosity, speed of fall, and oil drop density.
(b) From these parameters and Millikan's conclusion that the elementary charge corresponds to $\left(\frac{1}{t'_F} - \frac{1}{t_F}\right) = 0.005386\,\mathrm{s}^{-1}$ determine the value of the elementary charge in coulombs.

Millikan reported data on over 50 drops altogether. He excluded measurements on any trial when he thought the charge on the droplet changed while he was measuring its speed in the electric field. In every case both the charge on a droplet and any change of that charge were integer multiples of a single value. The repeatability of these measurements was strong evidence that a unique atomic unit of charge existed. That charge e is called the "elementary charge." Often, somewhat carelessly ignoring its sign, e is called the "electron's charge." The quantity e is also the magnitude of the charge on singly ionized atoms such as hydrogen ions and the like. The modern value for e is 1.602×10^{-19} C. The results you can obtain from Table 8.2 are slightly different from this value of e because they have not yet been corrected for inaccuracies that occur in Stokes's law when the size of the droplet is comparable to the mean free path of the molecules through which it is falling. Millikan made these corrections and obtained a value of 1.593×10^{-19} C for e.[10] Remember that the charge of an electron is negative, although the quantity e is usually given as a positive number. It is your job to remember the sign when it is needed.

Important Numbers Found from e

From accurate knowledge of e came accurate knowledge of Avogadro's number, the mass of the electron, and an order-of-magnitude estimate of the energy of interaction of an electron with its atom. Knowing a precise value of Avogadro's number, you can calculate the actual mass in kilograms of any atom or molecule.

EXERCISES

16. Calculate Avogadro's number from e and the faraday F.

17. Show how once Millikan had measured the elementary charge, he could determine that the mass of an electron is 9.11×10^{-31} kg.

Millikan's result permits us to estimate with moderate precision the energy of interaction of the electron with its atom. An electron a distance of $r_0 = 0.1$ nm—a reasonable number based on the sizes of atoms—from a singly charged ion will have a potential energy of $-ke^2/r_0 = -14.4$ eV. This number tells us roughly what energy will be involved when atoms interact with each

[10] Millikan claimed that his answer was precise to about a tenth of a percent. However, when other techniques became available to measure atomic spacings in solids to high precision, the value for Avogadro's number derived from these data disagreed with the value calculated from Millikan's e. The source of the discrepancy was ultimately traced to a slightly inaccurate value of the viscosity of air. Millikan used 18.240 μPa s; the currently accepted value is 18.324 μPa s. Scale Millikan's value for e by the ratio of these two numbers and see what you get.

other by means of their electrons. In other words, this number tells us that the energy scale of chemistry is of the order of 10 eV.

8.5 SUMMARY

This chapter has examined three important contributions to our understanding of atoms. Faraday's studies of electrolysis and the regularities of electrolytic deposition showed that atoms have an electrical nature. The results hinted that electricity, like matter, has an atomic nature and that there exists such a thing as a mole of charge and, therefore, some sort of atom of electrical charge, some basic unit of electricity. Thomson discovered the electron by using electric and magnetic fields to show that in an electrical discharge in a gas the rays emanating from the cathode are particles with definite mass and charge. Because the cathode-ray particles were the same regardless of what kind of cathode material or residual gas was in the tube, his results supported other evidence that electrons are parts of all atoms. Millikan's precise measurements of the electronic charge confirmed the discrete nature of electricity and established the energy scale of the electron's interaction with an atom. His results yielded precise values of Avogadro's number and of the electron's mass as well as of its charge.

The indivisible "a-tom" had been cut, and two new, very important, questions arose at once. If the electron is in the atom, what does it do there? And if the electron is one part of an atom, what are the other parts? The answers to these questions led to a revolution in physics, but before you can understand this revolution, you need to study properties of light and other forms of electromagnetic radiation. This you do in the next chapter. Before taking up this new topic, however, you might like to see two interesting uses of electric fields to manipulate charged particles.

8.6 USES OF ELECTRIC DEFLECTION

A good idea can have many applications. You have seen how J.J. Thomson used electric deflection to determine that cathode rays are charged particles. Electric deflection of electrons has widespread application as the basis of the cathode ray tube that is the heart of the oscilloscope, arguably the most important instrument in all modern sciences. Electric deflection is also used for low-cost, high-speed, good-quality printing and to hunt for quarks.

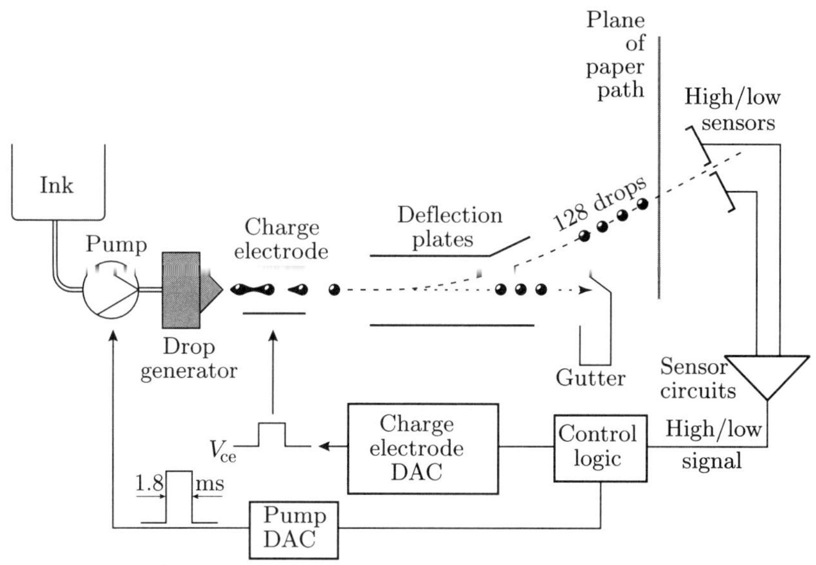

FIGURE 8.5 Inkjet printer showing droplet generator and charger along with deflection plates. ©*1977 International Business Machines Corporation. Reprinted with permission of The IBM Journal of Research and Development, Volume 21, Number 1.*

The Inkjet Printer

The inkjet printer works by electric deflection of objects orders of magnitude greater in size than electrons. In the inkjet printer ink droplets are electrically charged and then steered by electric deflection so that the droplets form characters on paper. The essential features of an inkjet printer are shown in Fig. 8.5. About 10^5 droplets are sprayed onto the paper each second, with about 100 drops forming a single character.[11]

Each droplet has a diameter of 63 μm. Assuming that the droplets have the density of water, this diameter corresponds to a droplet mass of 1.31×10^{-10} kg. A controlled amount of electric charge between 80 and 550 fC can be put on each droplet by varying the voltage on the charge electrode. The droplets pass in a stream at a speed of 18 m/s between two small metal plates 1.3 cm long and 1.6 mm apart. These are the deflection plates.

[11]A quite detailed description of the considerations that went into the design of the inkjet printer developed by IBM for the IBM 46/40 Document Printer is given in "Application of Ink Jet Technology to a Word Processing Output Printer," W.L. Buehner, J.D. Hill, T.H. Williams, and J.W. Woods, *IBM J. Res. Develop.*, 1–9 (Jan. 1977), and in other articles in this issue of the *IBM Journal of Research and Development*.

How Much Can a Droplet Be Deflected?

If the maximum voltage that can be placed across the plates is 3.3 kV, what would be the maximum deflection of the ink drops on a piece of paper 1 cm away?

You could just do this by plug and chug using the formula for electric deflection, but you will acquire more physical insight from a gradual approach. Also, stepwise reasoning is the best way to solve problems in general.

Notice that this printer is a scaled-up version of Thomson's method for deflecting electrons that was described in Section 8.3: A beam of charged particles with momentum p_x passing between a pair of charged plates is deflected through some angle θ because the electric field between the plates imparts to the particles an increment of momentum Δp_y perpendicular to their path. By varying E you can vary Δp_y and control the deflection angle according to the relation

$$\tan\theta = \frac{\Delta p_y}{p_x} = \frac{y}{L},$$

where y is the vertical displacement a distance L from the deflecting plates.

EXERCISES

18. How should the apparatus be designed so it can move the particles to points other than on the y-axis?

How much is Δp_y? It will be the product of the force qE and the time interval Δt during which that force acts. This time interval is just the length of the plates ℓ divided by the velocity of the particles v. Therefore,

$$\begin{aligned}\Delta p_y &= \frac{qE\ell}{v}\\ &= \frac{550\times10^{-15}\times2.1\times10^{6}\times1.6\times10^{-2}}{18}\\ &= 10.1\times10^{-10}\,\mathrm{kg\,m\,s^{-1}}.\end{aligned}$$

Then, to find θ you need to find p_x, the momentum along the x-axis. This is just

$$\begin{aligned}p_x &= mv_x = 1.31\times10^{-10}\times18\\ &= 2.36\times10^{-9}\,\mathrm{kg\,m\,s^{-1}},\end{aligned}$$

from which you see that Δp_y is almost $\frac{1}{2}$ of p_x; in fact,

$$\tan\theta = \frac{\Delta p_y}{p_x} = \frac{10.1 \times 10^{-10}}{2.36 \times 10^{-9}} = 0.428.$$

Over a distance $L = 1$ cm from the deflection plates to the paper, this angle of deflection will produce a displacement y on the paper such that

$$\frac{y}{L} = \tan\theta = 0.428,$$

from which it follows that

$$y = 1 \times 0.428 = 0.428 \text{ cm} = 4.28 \text{ mm}.$$

With the distance to the screen $L = 1$ cm and the length of the deflecting plates $\ell = 1.6$ cm, the IBM inkjet printer differs markedly from Thomson's arrangement in that L is not large compared to ℓ. As a consequence, the amount of displacement that takes place while the charged particle is between the plates is no longer negligible compared to the displacement that occurs as the particle travels to the paper.

It is not very difficult to calculate the vertical displacement that occurs while the particle is between the plates. Between the plates the displacement is just what occurs to a particle moving with a constant acceleration $a = qE/m$. That distance is the old, familiar (?) $y = \frac{1}{2}at^2$, where $t = \ell/v_x$. The value of the acceleration is

$$\begin{aligned} a &= \frac{qE}{m} \\ &= \frac{5.5 \times 10^{-13} \times 2 \times 10^6}{1.31 \times 10^{-10}} \\ &= 8.66 \times 10^3 \text{ m s}^{-2}; \end{aligned}$$

and the value of t is

$$\begin{aligned} t &= \frac{\ell}{v_x} = \frac{1.6 \times 10^{-2}}{18} \\ &= 8.89 \times 10^{-4} \text{ s}. \end{aligned}$$

From these two numbers you can calculate that

$$\begin{aligned} y &= \frac{1}{2}at^2 = 4.33 \times 10^3 \times (8.89 \times 10^{-4})^2 \\ &= 3.42 \times 10^{-3} \text{ m}, \end{aligned}$$

which is 3.42 mm and quite comparable in size to the 4.28 mm of displacement that occurs between the exit point and the paper. The total displacement of a droplet caused by electric deflection is the sum of the deflection while between the plates plus the deflection while traveling from the end of the plates to the paper: $3.42 + 4.28 = 7.7$ mm.

Noticing that this calculation shows that the droplets will be deflected 3.4 mm between two plates that are 1.6 mm apart, you may think that the droplets will hit the deflecting plates. This does not occur for two reasons. First, air resistance reduces the amount of deflection from what you calculate by the arguments given here. Second, as Fig. 8.5 shows, one of the plates is bent upwards to allow the maximally deflected droplets to escape without hitting a plate.

Quark Hunting

Electric deflection of droplets produced and charged like those in an inkjet printer has been used to look for quarks. Quarks are fundamental particles of which most other particles (but not electrons) are thought to be made. The proton is made of three quarks. One very unusual feature of quarks is their electrical charge: It is a *fraction*(!) of the elementary charge e. One kind of quark is thought to have a charge of $+2e/3$ and another to have $-e/3$. A burning question for the past twenty years has been, "Why has no one ever been able to see a single quark with these very unusual charges?"

It has not been for lack of searching. Quarks have been looked for in mine tailings, in seawater, in old stained-glass windows. They have been sought in niobium spheres and tungsten plates. They have been assiduously pursued with high-energy accelerators and experiments of great ingenuity. But no one has ever found one. They seem to occur only in groups inside other particles.

Separating Droplets by Deflection

One search for quarks has used inkjet technology.[12] The idea is to generate a series of droplets with different numbers of charges on them. These are allowed to fall in a vacuum through a pair of electrically deflecting plates. If the number of charges on a droplet is an integer multiple of e, then deflections should occur in discrete amounts. For example, a droplet with a charge of $21e$ will undergo a certain deflection and land a discrete distance away from the landing points of droplets of charge $20e$ or $22e$. The droplets should fall in narrow bands with spaces between them. If a fractionally charged droplet came along, it would land in one of those spaces and so be observable. Figure 8.6 shows the apparatus. Droplets of tetraethylene glycol of mass

[12] "Search for fractional charges using droplet-jet techniques," J. Van Polen, R. T. Hagstrom, and G. Hirsch, *Phys. Rev.* **D36**, 1983–89 (1987).

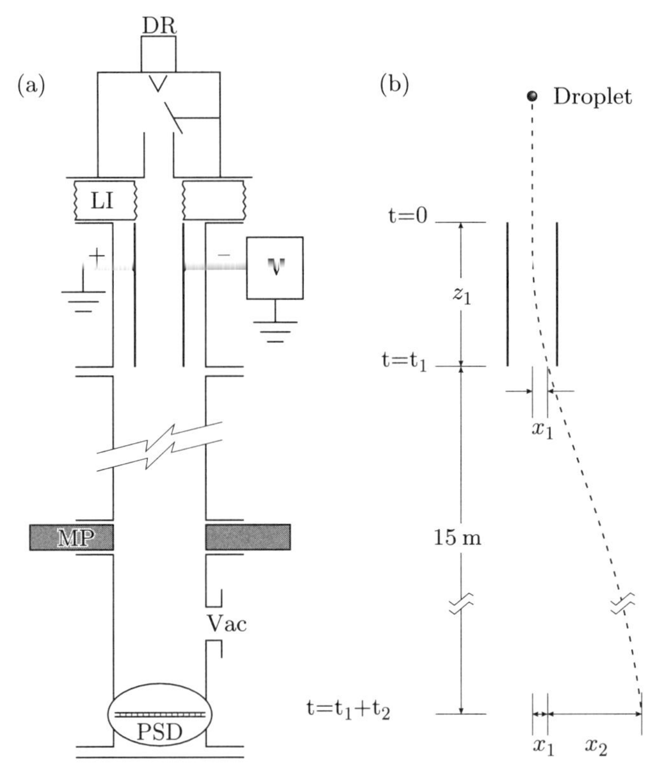

FIGURE 8.6 (a) Schematic diagram of 22 m high evacuated tube inside of which charged droplets fall and are electrically deflected: DR is the droplet source; LI is a vibration isolator; V is the deflection voltage supply; MP is a mounting plate; Vac is the outlet to the vacuum pump; and PSD is the position-sensitive detector that records the fallen drops. (b) Variables and quantities for analyzing the deflection of a droplet; they are described in the text. *Taken with permission from J. Van Polen, R. T. Hagstrom, and G. Hirsch, Phys. Rev. Vol. D36, 1983-89 (1987), ©1987 American Physical Society*

$m = 16$ ng and diameter $2r = 30.8\ \mu$m were injected at the top of a 22 m tall tower. The droplets were traveling at a speed of $v_1 = 8.6$ m/s as they entered the 3.05 m long space between the deflecting plates where they were deflected by the electric field produced by a potential difference of 20 kV between the plates, which were 5 cm apart. After exiting from between the deflection plates, the droplets fell about 15 m to a detector that measured the amount of their sideways displacement.

Does this work? How much will be the separation between droplets when they are detected? Is it enough to be sure that the droplets differ by only one unit of charge? Will there be enough space between droplets that differ by one unit of charge so that you can distinguish any fractionally charged particle if one happens to be present? To answer these questions let's calculate the sideways displacement of a droplet carrying a charge of $q = 20e$ for a deflecting voltage of 20 kV.

Because the tetraethylene glycol droplets are accelerating in the direction of their motion under the force of gravity both as they pass between the deflecting plates and afterwards, it is not adequate to find the angle at which the particles emerge from between the plates as was done in the cases of a beam of cathode rays or a stream of charged droplets of ink. Instead, you must find the sideways displacement x_1 and the sideways velocity v_x at the time t_1 the droplet emerges from between the plates and also calculate the time t_2 that it takes the droplet to fall to the detector 15 m below. The sideways displacement after exiting from between the plates will be $x_2 = v_x t_2$, and the total sideways displacement will be $x_1 + x_2$.

To find the sideways velocity v_x as the droplets exit from between the plates, you need to use the sideways acceleration $a_x = qE/m$ produced by the electric field E during the time t_1 that the droplet is in the field. Then $v_x = a_x t_1$. But how do you find t_1?

To find t_1, use the fact that under constant acceleration the average velocity over some time interval t is $(v_2 + v_1)/2$, where v_1 is the velocity at the start of the interval and v_2 is the velocity at the end of the interval. Then the time t_1 to fall a vertical distance z_1 between the plates is

$$t_1 = \frac{2z_1}{v_2 + v_1},$$

where $v_1 = 8.6$ m/s is the velocity of the droplet as it enters the electric field between the plates and v_2 is its velocity as it exits. Of course, this means you must find v_2, but you can do this using the conservation of energy as was illustrated in Sections 2.8 and 6.4.

EXERCISES

19. Show that a body with an initial velocity $v_1 = 8.6$ m/s will have a velocity of $v_2 = 11.6$ m/s after it falls a distance $z_1 = 3.05$ m near the surface of the Earth.

From these values of v_1 and v_2 you can deduce that

$$t_1 = \frac{2z_1}{v_2 + v_1} = \frac{2 \times 3.05}{8.6 + 11.6},$$
$$t_1 = 0.30 \text{ s}.$$

To find a_x you need the value of the electric field E. You can find E from the voltage $V = 20\,\text{kV}$ across the plates and the spacing $d = 5$ cm between the plates.

EXERCISES

20. Show that $a_x = 0.0800\,\text{m s}^{-2}$, from which it follows that

$$v_x = \frac{qE}{m}t_1 = 0.024\,\text{m/s}.$$

At this point in the calculation you can find the displacement x_1 of the droplet during its passage between the plates. It will be

$$x_1 = \frac{v_{x0} + v_{x1}}{2}t_1 = \frac{0.024}{2}\,0.30 = 0.0036\,\text{m} = 3.6\,\text{mm}.$$

To find the distance x_2 that the drop moves sideways while falling under the acceleration of gravity through the remaining distance $z_2 = 15$ m, it is necessary to find its time of free fall, t_2. The approach is the same as for finding the time of passage between the plates.

EXERCISES

21. Show that $t_2 = 0.93$ s.

At last, knowing the fall time t_2 and the sideways velocity $v_x = 0.024$ m/s, you can calculate the sideways displacement x_2 of the droplet after it has left the plates.

$$\begin{aligned} x_2 &= v_x\, t_2 \\ &= 0.024 \times 0.93 = 0.022\,\text{m} \\ &= 2.2\,\text{cm}. \end{aligned}$$

In the electric deflection of cathode rays the amount of displacement that occurred between the plates was often negligible compared to the amount that occurred afterwards on the trip to the screen. In this search for quarks, the displacement occurring between the plates is not negligible; it is 3.6 mm compared to 22 mm. To find the total deflection you must add the two parts; this gives a total displacement of

$$x_1 + x_2 = 25.6\,\text{mm}.$$

Although it is important to understand the step-by-step argument that has brought you to this answer, a complete algebraic expression is useful.

EXERCISES

22. Show that a droplet of mass m and charge q that has passed through an electric field E in the quark-hunting apparatus will undergo a total displacement of

$$x_1 + x_2 = \frac{1}{2}\frac{qE}{m}t_1^2 + \frac{qE}{m}t_1 t_2. \tag{15}$$

This expression tells you two very important things about the displacement: (1) It is proportional to the charge q on the droplet; (2) it is proportional to the strength of the electric field E across the deflecting plates.

Such an expression has several uses. For example, to check that their apparatus was functioning properly, the experimenters ran a separate set of experiments with 24.2 kV across the plates instead of 20 kV. What would be the displacement produced by a deflecting voltage of 24.2 kV?

Equation 15 shows that you can use a simple proportion to find the answer:

$$x' = \frac{24.2}{20} \cdot 2.56 = 3.1\,\text{cm}.$$

More to the point, you can now answer the basic questions about this experiment: With 20 kV across the plates, by how much will the displacement change if the charge changes by $\pm e$? The answer is

$$\frac{1}{20}2.56 = 0.128\,\text{cm} = 1.28\,\text{mm}.$$

EXERCISES

23. Explain the reasoning by which this answer was obtained.

This result means that a stream of droplets charged with different multiples of e will be spread out into a fan of streams, like fingers of a hand, each stream separated from the next by 1.28 mm. The authors say that the individual streams are only about 0.060 mm wide, so the gaps between the streams were pronounced. They looked with their detector to see whether there were any streams or occasional droplets that fell with a separation $\frac{2}{3}$ or

$\frac{1}{3}$ of 1.28 mm. These might correspond to droplets carrying a single quark as part of its charge. The authors report that they saw no evidence of quarks.

PROBLEMS

1. The official SI equivalent for the English mass unit called the "grain" is 1 grain = 0.0648 grams. Assuming that "mean pressure" is 1 atm, determine from Example 8.1 the value of "mean temperature" used by Faraday.

2. In another experiment, Faraday measured the amount of hydrogen and oxygen gas evolved in electrolysis and also the amount of tin deposited on a tin electrode. He writes, "The negative electrode weighed at first 20 grains; after the experiment it...weighed 23.2 grains....The quantity of oxygen and hydrogen collected...=3.85 cubic inches."

- **a.** Show from the numbers in Example 1 that this corresponds to 0.497 grains of hydrogen and oxygen.
- **b.** Use these data to calculate the atomic weight of tin (Sn) and compare your answer with the currently accepted value.

3. Chapter 6 showed you how to estimate the velocity with which electrons move when they carry a current through a wire. Using the following information, estimate the speed with which ions move between two electrodes in an electrolytic cell.

Assume that the cell is filled with 0.01 N sulfuric acid (pH $\approx$ 2.1, which means about 6×10^{18} H^+ ions in each cm^3) and that the electrodes have surface areas of 0.25 cm^2 and are 4 cm apart.

If the cell is run with a current of 0.2 A, what is the average speed with which ions travel from the anode to the cathode? How does your answer compare with the estimated speed of electrons in a current-carrying wire?

4. Millikan used Stokes's law, which says that a small sphere of radius a falling with velocity v in a homogeneous medium of viscosity $\eta = 18.6 \times 10^{-6}$ Pa s is subject to a retarding force

$$F = 6\pi\eta a v.$$

He observed that a drop of oil of density $\rho = 0.92$ g cm^{-3} would fall 0.522 cm in 13.6 s.

- **a.** What was the radius of the drop?
- **b.** What was its mass?

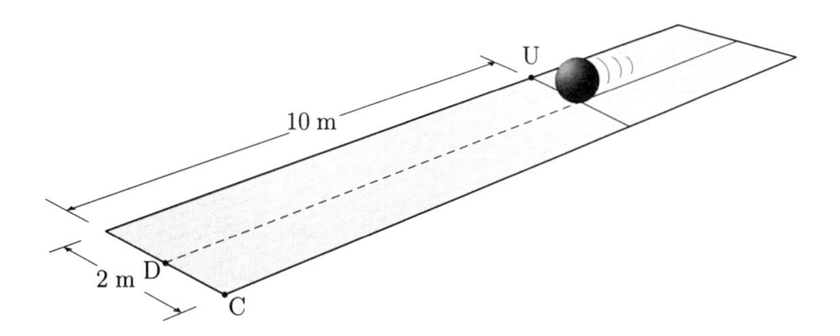

FIGURE 8.7 Ball rolling down a hallway towards D is deflected at point U so as to hit C instead (Problem 6).

5. If a charged oil droplet of density $\rho = 0.92\,\mathrm{g\,cm^{-3}}$ and radius $a = 2\,\mu\mathrm{m}$ is just balanced against gravity's pull by an electric field of 1.9×10^6 V/m, what is the charge on the droplet? Give your answer in multiples of the elementary charge e.

6. A bowling ball of mass $M = 6$ kg is rolling down a hallway at 1.5 m/s. (see Fig. 8.7) If it rolled straight, it would hit an associate dean of students D standing at the end of the hall. You are standing at point U, 10 m from the end of the hall and you intervene heroically by giving the ball a sideways kick as it passes you so that the ball rolls into the corner C instead of hitting D.

a. What must be the angle of deflection θ at point U if the ball is to hit C instead of D?

b. How much sideways momentum Δp_y must you give the ball to produce the desired deflection?

7. What was the first evidence that atoms are electrical in nature?

8. Faraday's work led to the realization that a mole of electric charges is about 96500 C and that this amount of charge releases 1.008 g of hydrogen in electrolysis. From these results find the charge-to-mass ratio of an individual hydrogen ion, H^+, and explain your work.

9. Section 8.3 describes a method for measuring e/m of cathode rays. Suppose that you applied this method and observed a beam of mystery rays emerging from the source and that all that you knew about them was that they were charged particles, each carrying one elementary charge.

Now you accelerate them through a potential difference of $V = 500$ V. Then, as shown in Fig. 8.8, they emerge from the source and enter a uniform

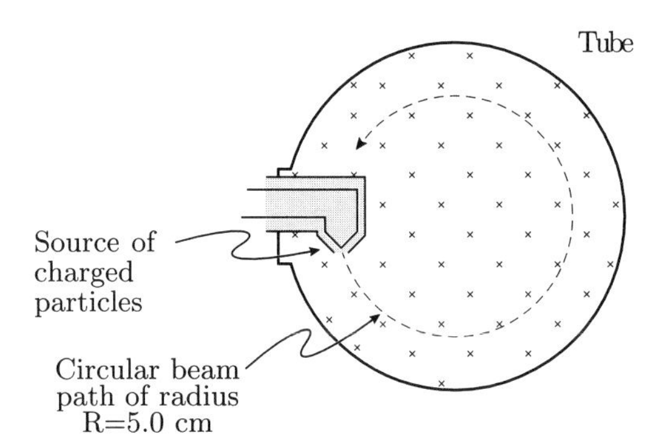

FIGURE 8.8 A beam of charged particles moves in a circle in a magnetic field pointing into the plane of the page (Problem 9).

magnetic field of $B = 65$ mT (pointing into the plane of the page) and bend in a circle of radius 5 cm.

a. Are these particles cathode rays? How do you know?

b. What are some of the important conclusions drawn from Thomson's measurements of the charge-to-mass ratio of cathode rays?

10. In an electrolysis experiment a current of 0.5 amperes flowing for 20 minutes (1200 s) liberates 69.6 mL of gas at the cathode and 23.3 mL at the anode. You may assume that both gases are at STP ($T = 272.3$ K, $P = 1.013 \times 10^5$ Pa).

a. How many moles of gas molecules are produced at the cathode?

b. How many electrons were needed to form one molecule of gas at the cathode?

c. Which of the following chemical reactions is consistent with the information given above?

$$2H_2O \rightarrow 2H_2 + O_2$$
$$2CO_2 \rightarrow 2CO + O_2$$
$$2N_2O_3 \rightarrow 2N_2 + 3O_2$$
$$2NH_3 \rightarrow N_2 + 3H_2$$

11. A droplet 10 μm in radius will fall in air at a steady velocity of 1.1 cm/s because its weight mg (m is its mass, g is the acceleration due to Earth's gravity) is balanced by the viscous force $6\pi\eta av$ (η is the viscosity of air, 18.2 μPa s; a is the radius of the droplet; and v is the velocity of fall).

a. Use these data to find the radius of a droplet that falls with a velocity of 8.59×10^{-4} m/s.

b. When an electric field of 317.5 kV/m is turned on, the droplet in (a) is observed to rise at a steady speed of 2.41×10^{-4} m/s.

c. Show that the upward electrical force on the droplet is 1.28 times the downward gravitational force mg.

d. To within a couple of electrons, how many excess electrons are there on this droplet?

e. What important conclusions do we draw from Millikan's measurements?

12. Faraday's work led to the realization that 96500 C of charge will cause 1 mole of hydrogen or of silver or of chlorine to evolve from electrolysis. Thomson's work led to the measurement of $q/m = 1.76 \times 10^{11}$ C/kg for cathode rays. Millikan measured the elementary charge.

a. What did Faraday's laws of electrolysis imply about atoms and electricity?

b. What did Thomson's work reveal about cathode rays?

c. Combining Faraday's result and Thomson's, find the value of the ratio of the mass of the hydrogen atom to that of the electron.

d. What did Millikan's result reveal about electric charge?

13.a. Explain the significance of the Faraday of charge $Q = 96500$ C?

b. Explain the significance of the discovery that all cathode rays have a definite charge-to-mass ratio of 1.7×10^{11} C/kg?

c. Show by direct calculation how Millikan's experimental measurement of the elementary charge made it possible to determine

i. Avogadro's constant.

ii. The mass of the electron.

iii. The actual masses of atoms.

What fundamental property of electric charge did Millikan's work confirm?

14. In electrolysis 96,500 C of charge releases 1 mole of hydrogen atoms.

a. From this information show how to determine the charge-to-mass ratio of the proton. Clearly state any assumptions you make.

b. Suppose in measuring e/m of the electron you observed that electrons accelerated through a voltage acquire velocity of $v = 0.03\,c$ and bend in the circle of radius $r = 3$ cm in a magnetic field of $B = 2$ mT. What would you determine e/m to be? (Your answer should be different from the accepted value.)

c. Explain why results like your answers to (a) and (b) implied to J.J. Thomson the existence of a new particle. What was it and in what ways is it important?

15. Use a spreadsheet least squares routine to derive a best value for e from Millikan's data in Table 8.2. Do this in the following way: Plot a graph of $1/t_{\prime F} - 1/t_F$ vs n' (columns 4 and 5 in the table). Take the slope of this graph and multiply it by the factor (which you must determine from the information Millikan provides) that converts this slope into e. Explain why this calculation gives e. (Millikan corrected his results for inaccuracies in Stokes's law that occur when droplets are very small.)

16. The diagram in Fig. 8.9 shows a pair of parallel plates 60 mm long and separated by 12 mm to which a power supply (not shown) is connected. An electron beam enters the region between the plates from the left, as shown.

a. Assuming that the electric field between the plates is uniform, find the magnitude of the force exerted on any electron by the field given that the potential difference between the plates is 48 V.

b. If the speed of the electrons is 5×10^7 m/s, find the electron's change of momentum in the y direction that results from passing between the plates.

c. Suppose that the beam of electrons is replaced by a beam of negative hydrogen ions of the same speed (a hydrogen ion, H^-, has the same charge as an electron and a mass equal to 1836 electron masses). Which, if any, of your answers to (a) and (b) above would have to be changed? Why?

17. Give brief, compact answers to each of the following. Use specific facts, equations, or diagrams to support your answers.

a. Explain why we think that there is such a particle as the "electron."

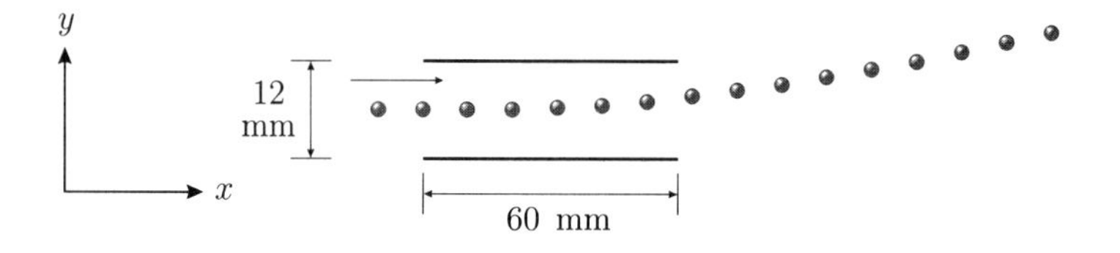

FIGURE 8.9 Electron beam passing between parallel plates (Problem 16).

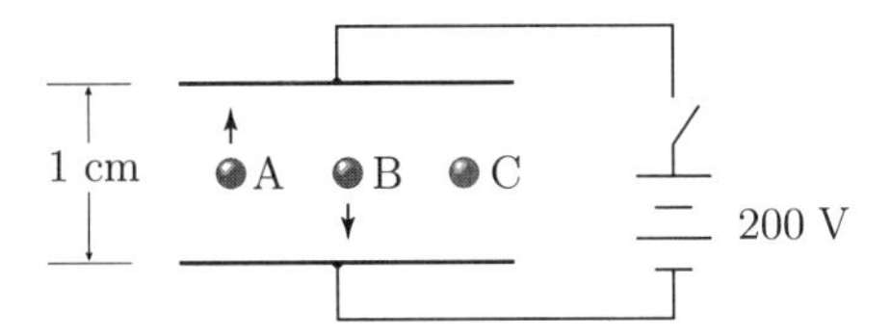

FIGURE 8.10 Charged glass microspheres between two plates to which a voltage can be applied (Problem 18). Arrows show direction of motion when the switch is closed.

b. Explain how Millikan's measured value of e can be used to find both Avogadro's number and the mass of the electron.

c. What is the evidence for thinking that atoms are electrical in nature?

18. Olive Oyal, a promising young physicist, has decided to recreate Millikan's famous oil-drop experiment using modern materials. Rather than employing oil, Olive chooses precision glass microspheres of diameter 1.0 μm and density 2.5 g cm^{-3}. Figure 8.10 illustrates her experimental apparatus and three spheres A, B, and C. Olive observes that
(1) before the switch is closed, all three spheres fall with the same speed v_f, and
(2) after the switch is closed, sphere A rises with speed v_f, sphere B *falls* with speed v_f, and sphere C comes to rest.

a. Based on the information given above, specify whether the charge on each of the three drops is +, −, 0, or indeterminate.

b. Prove that the mass of each sphere is equal to 1.31 pg.

c. Now calculate the charge on each of the three spheres. From these results what might Olive conclude is the basic, elementary, unit of charge?

19. The magnitude of the charge on an electron is 1.6×10^{-19} C; it is negatively charged.

a. Could Millikan have done his experiment with the air pumped out of the volume of space where he made an electric field, i.e., between the plates? Why?

b. Find the mass of a droplet of oil of density 0.92 g cm^{-3} if the radius of a droplet is 3 μm.

c. An electric potential difference of 5000 V is put across two large conducting plates separated by 2 cm. What is the value of the electric field in the space between the plates?

d. If an electric field of 300 kV/m exerts a force of 4.8×10^{-14} N on an electrically charged drop of oil, how much is the electric charge of the oil drop?

20. We measure electrical charge in units of coulombs; a hundred years ago it was common to measure electrical charge in units called esu. The conversion is 3×10^9 esu = 1 C.

a. What is the charge of the electron measured in esu?

b. What are cathode rays?

c. Why was Thomson the first to see cathode rays be deflected by an electric field?

d. Who was J.J. Thomson? Give a few biographical facts about the man.

21. Thomson measured e/m to have a value of 2.3×10^{17} esu/g. Convert this to C/kg and compare the result with the currently accepted value.

22. After Thomson measured e/m for cathode rays, he noted two significant properties of the rays. What were they and why were they important?

23. A charged oil droplet is suspended motionless between two parallel plates ($d = 0.01$ m) that are held at a potential difference V, as shown in Fig. 8.11. Periodically, the charge on the droplet changes, as in Millikan's original experiment. Each time the charge changes, V is adjusted so that the droplet remains motionless. Here is a table of recorded values of the voltage V:

i. 350.0 volts.

ii. 408.3 volts.

iii. 490.0 volts.

iv. 612.5 volts.

a. From the data above, determine the charge on the droplet for case (i) above. What assumptions do you need to make? (Hint: the ratio of voltages = ?)

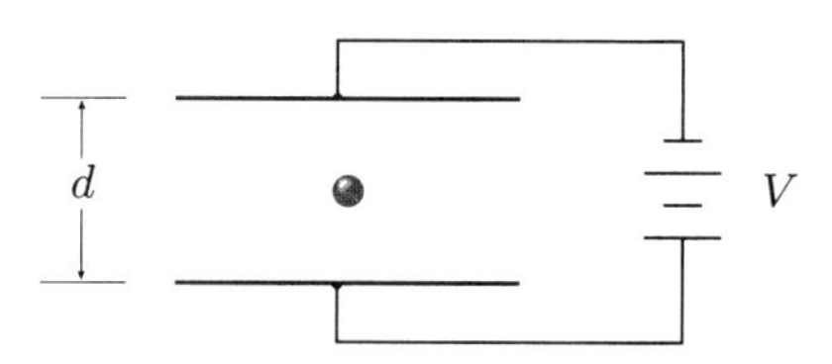

FIGURE 8.11 An oil drop suspended in an electric field. (Problem 23).

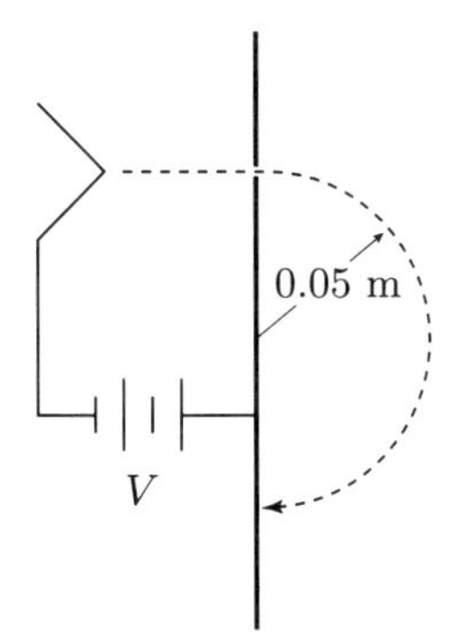

FIGURE 8.12 Accelerated charged particles enter a uniform magnetic field. (Problem 25).

b. Find the magnitude and direction of the electric field in case (i) above.

c. What is the mass of the drop?

24. From Faraday's experiment you find that during the electrolysis of water, 0.01 g of H_2 gas is liberated by the passage of 965 C of charge. Find the charge-to-mass ratio of a proton from these data.

25. Particles each with mass m and charge q accelerate through a potential difference of $V = 100$ volts and enter a region of uniform magnetic field $\vec{B}$, where their paths are circular with radius $R = 0.05$ m, as shown in Fig. 8.12. The magnitude of the field is $B = 6.75 \times 10^{-4}$ T.

a. Indicate the direction of $\vec{B}$ on a drawing.

b. Derive the following relationship between the radius R and q, m, V, and B:

$$R^2 = \frac{m}{q}\frac{2V}{B^2}.$$

c. Find the value of the charge-to-mass ratio (q/m) of these particles and identify them. Justify your answer!

26. In a laboratory electrolysis experiment, 40 mL of N_2 gas and 60 mL of O_2 gas are liberated at the cathode and anode, respectively. The room pressure and temperature are 730 Torr and 25°C.

a. Which of the following overall reactions has taken place?

$$2N_2O \rightarrow 2N_2 + O_2$$
$$2NO_2 \rightarrow N_2 + 2O_2$$
$$2N_2O_3 \rightarrow 2N_2 + 3O_2$$
$$2N_2O_5 \rightarrow 2N_2 + 5O_2$$

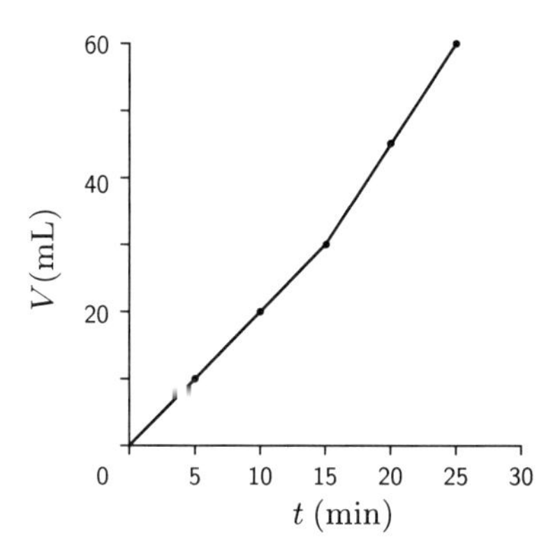

FIGURE 8.13 Rate of electrolytic evolution of N_2 (Problem 26).

b. Figure 8.13 shows a plot of the volume of N_2 evolved at the cathode vs. the time in minutes. Which of the following statements most likely explains the "kink" in the graph?

i. The current was increased after 15 minutes.

ii. A gas leak developed above the cathode.

iii. A gas leak developed above the anode.

iv. Data points were collected more frequently after the first 15 minutes.

c. At the temperature and pressure given above, what volume would 1 mole of N_2 gas occupy?

d. What is the total mass of N_2 gas that has evolved after 25 minutes?

27. In a modern version of Millikan's famous experiment, a charged radioactive sphere of mass 1 ng (= ? kg) is suspended motionless between two parallel conducting plates spaced by 1.0 mm.

The apparatus (plates and sphere) is in vacuum. Initially, a potential difference of 15.3 kV is required to counteract the gravitational force.

a. Determine the magnitude of the following forces:

i. Gravitational force.

ii. Electric force.

iii. Viscous force.

b. What is the charge on the sphere?

c. By radioactive decay, the sphere periodically emits a charged particle, and the voltage required to balance the gravitational force changes suddenly. You observe the following:

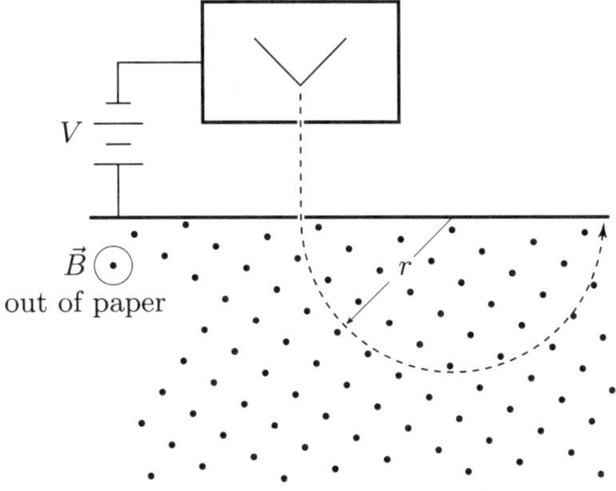

FIGURE 8.14 Charged particles accelerate through a voltage V and then travel in a region of uniform magnetic field $\vec{B}$ (Problem 28).

i. The sphere is balanced by a voltage of 15.3 kV for a short time, whereupon....

ii. the sphere begins to fall, and the voltage must be increased to 30.6 kV in order once again to levitate the sphere, and then....

iii. the sphere falls and is unaffected by the magnitude or sign of the potential difference applied to the plates.

Explain these observations and determine the charge of the particle emitted in each radioactive decay.

28. In a lab experiment, an unknown species of charged particle is produced and accelerated through a potential difference of 200 V. The particles then enter a region of constant magnetic field $B = 1.0\,\text{mT}$, where they trace out a circular orbit of radius $r = 4.8\,\text{cm}$. The apparatus, particle trajectory, and magnetic field are as shown in Fig. 8.14.

a. Starting from the expressions for the magnetic force, the centripetal force, and the kinetic energy vs. accelerating voltage, prove that

$$\frac{m}{q} = \frac{B^2 r^2}{2\,V}.$$

b. From the information given, identify the particle.

c. Suppose the above particles were replaced by "negative muons," which have the same charge as an electron and a mass 207 times larger than an electron mass. What magnetic field $\vec{B}$ (direction as well as magnitude) is required if the muons (accelerated through the same potential difference) are to follow the same circular path as the original particles?

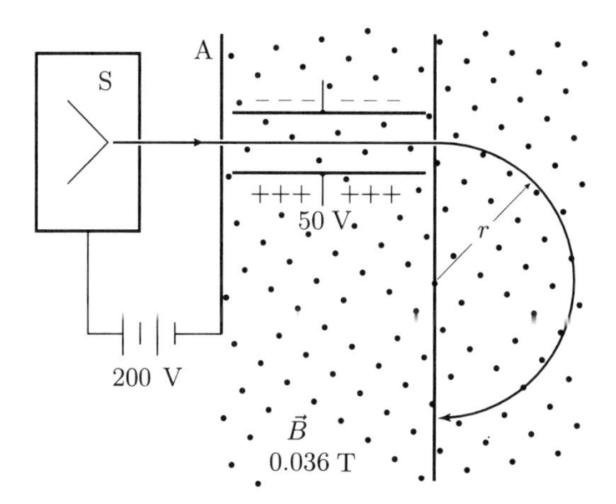

FIGURE 8.15 Experimental setup for Problem 29. Note that the region to the right of the anode A is filled with a magnetic field $B = 0.036$ T pointing out of the paper.

d. Explain the significance of the discovery that all cathode rays have a definite mass-to-charge ratio of 0.57×10^{-11} kg/C.

29. As shown in Fig. 8.15, charged particles are produced in a source S and accelerated through a potential difference $V_{\text{acc}} = 200$ V. The particles enter a region of constant magnetic field $B = 0.036$ T (directed out of the paper) and pass undeflected through two charged parallel plates spaced by 1.0 cm. The voltage across the plates is 50 V.

a. Study the figure carefully. Are the particles positively or negatively charged? How do you know?

b. Prove that the particle velocity $v = 1.39 \times 10^5$ m/s.

c. Find the mass-to-charge ratio of the particles, and from this identify the particle, choosing from the following list:

i. Electron.

ii. Alpha particle.

iii. Proton.

iv. None of the above.

30. From what you know about J.J. Thomson's discovery of the electron:

a. Explain how one can use a combination of electric and magnetic fields to determine the ratio of charge to mass of a particle when you don't know either.

b. Tell what there is about such experiments that leads you to conclude that cathode rays are particles?

c. Describe the ways in which Thomson's experiments suggest that atoms contain electrons, i.e., that electrons are parts of atoms?

SPREADSHEET EXERCISE: THE MILLIKAN OIL DROP EXPERIMENT

Millikan's oil drop experiment had two important outcomes: it found the value of the charge on the electron, and, equally important, it provided convincing evidence that all electrical charge is some integer multiple of an elementary charge e. This computer exercise illustrates how to use a computer spreadsheet to analyze Millikan's data and show both of these outcomes. The instructions given are for the QuattroPro computer spreadsheet software, but with small changes they will work for Excel. The data to be analyzed are shown in Table 8.3. The original table is in Millikan's paper "The isolation of an ion," *Phys. Rev.* **32**, 356–358 (1911); in the version given here the values of the charges and viscosity have been converted to SI units.

THEORY

Millikan measured the terminal velocity v_1 of a charged droplet of mass m falling in air in a gravitational field g; he then applied an electric field F to the droplet and measured its terminal velocity as it rose against gravity under the effect of the field. He showed that for this situation the electric charge q on the droplet is given by

$$q = \frac{mg}{F}\left(\frac{v_1 + v_2}{v_1}\right). \tag{16}$$

His experiment was designed to work with the same oil droplet for hours at a time. In column G of Table 8.3 the values of t_1, the time the drop falls between the crosshairs when the electric field is turned off, are essentially constant because they are for the same droplet. Using radiation from a radioactive source, Millikan changed the charge on the droplet from time to time. Changes in charge changed the time t_2 that it took the droplet to rise in the applied electric field. These changes show up clearly in the variations of the times recorded in column F of Table 8.3. It is useful to note that the change in charge, Δq, can be found from Eq. 16:

$$\Delta q = \frac{mg}{Fv_1}\left(v_2' - v_2\right) = \frac{mgt_1}{F}\left(\frac{1}{t_2'} - \frac{1}{t_2}\right). \tag{17}$$

The droplet's mass m is also found from the time of fall t_1. This is done by connecting the mass of the droplet to its size and density and then finding its size from its terminal rate of fall using Stokes's law. The mass m of a spherical droplet is

$$m = \frac{4}{3}\pi a^3 \sigma, \tag{18}$$

where σ is the density of oil and we neglect Millikan's correction for buoyancy due to ρ, the density of air. The drop's radius a is obtained from Stokes's law for the terminal velocity of a small sphere falling in a viscous medium,

$$a = \sqrt{\frac{9\eta v_1}{2g\sigma}}, \tag{19}$$

where η is the viscosity of air. If you replace a in Eq. 18 with Eq. 19, you get

$$m = \frac{4}{3}\pi \left(\frac{9\eta v_1}{2g}\right)^{\frac{3}{2}} \left(\frac{1}{\sigma}\right)^{\frac{1}{2}}. \tag{20}$$

The measurement of t_1 divided into the distance between the cross hairs d_c determines the value of v_1, which substituted into Eq. 20 gives the value of m.

The values of the constants needed for Eq. 20 and Eq. 17 are:

$$\begin{aligned}
g &= 9.8\,\text{m/s}^2,\\
\sigma &= 896\,\text{kg/m}^3,\\
\rho &= 1.28\,\text{kg/m}^3,\\
\eta &= 1.836 \times 10^{-5}\,\text{N s/m}^2,\\
d_p &= 0.016\,\text{m},\\
d_c &= 0.0101\,\text{m},\\
V &= 7800\,\text{V}.
\end{aligned}$$

The following procedure has two goals: First, it will enable you to show that Δq is always an integer multiple of some basic quantity, and, second, it will enable you to find the value of that quantity. The step-by-step instructions given below are for QuattroPro with Windows.

PROCEDURE

a. In a QuattroPro spreadsheet, type in a header that contains your name and the title and date of the exercise. Then, in two columns type in the name and value of the parameters listed above. (It is usual in computers to enter a number like 1.836×10^{-5} by typing 1.836E-5; the computer uses"E-5" instead of 10^{-5}.) In cell A11 enter 23.4, the average value of t1 in seconds.

b. In Cells A18 and A19 type "t2" and "(sec)". This will be a header for the column of numbers that you will type underneath. On each row of column A, starting from row 20, type the values of t_2. Take them from column F in Millikan's data in Table 8.3, and use only the values that appear below the words "Change forced with radium." This will be 45 values in all. If you get ambitious, you can type in all 78 values. From the times t_2 you could use d_c to calculate the velocities v_2 as Millikan did, but since the velocities are all proportional to $1/t_2$, it is simpler to calculate just the reciprocal times by setting up column B to contain $1/t_2$. To do this, type the header: "^t_2" in B18 and "(^1/sec)" in cell B19 (using "^" will center the labels in their cells). Then, in cell B20 type "1/A20." Once you type this in you can copy this cell to all the other cells underneath so that you obtain $1/t_2$ for all the data entries. (In QuattroPro you copy by selecting the cell you want to copy by clicking on it with the mouse; a little black box outlines the cell. Then go up to the tool bar and click on the copy icon—probably the second icon from the left; then use the mouse to move the cursor back to cell A21. There click and drag until the cells A21 to A64 are outlined in black; release the mouse button; move the cursor up to the paste button—the third icon from the left—and click on it. Now column B should contain all the values of $1/t_2$.)

c. You now have enough data in your spreadsheet to see whether there is anything to this idea of charges being integer multiples of some smallest quantity of charge.
Do this by looking at the differences between the values of $1/t_2$. You can do this by plotting your data in a bar graph. To do this, click on the graph icon, fifth from the left in the tool bar, and change the cursor into a little bar graph. Now, to put a graph at some convenient place on your spreadsheet, say, cell C2, move the cursor there and click and drag to make a square area with a dashed line around it. With the cursor inside that area click the right button of your mouse. A menu will appear. Click on the word "Type"; then click on the upper right-hand icon of the little array of pictures that appears—this selects "bar-graph type"; click on OK. Now right-click on the graph again, and this time select "series." Click on series and enter "B20..B64" in the third line labeled "1st series," and click OK. This should produce a bar graph, and if you look at it carefully, you may see that the differences between the heights of the bars are pretty much multiples of some smallest difference. Indeed, the differences are equal.

Not convinced? Use the spreadsheet to make the evidence clearer. Sort the data in ascending order and see whether the successive changes do not all have the same step size, as they must if they are changing by the smallest, elementary, amount. To do this, copy the values of $1/t_2$ to another column and free them from their cell addresses by using the Block Values commands. Click on "Block" in the menu line; then click on "Values." Enter B20..B64 in the "from" line and D20..D64 in the "to" line; click OK. The numbers in column D should be the same as those in column B.

Now sort the numbers in column D. Click on "Data" and then on "Sort." Enter D20..D64 in the "Block" line and D20 in the first row of "Sort key." Click OK, and the numbers in column D should now appear sorted in ascending order. To see how convincing they are as evidence for the existence of an elementary charge, make a bar graph of them. The resulting staircase with steps of equal height is very strong evidence of the "quantization" of charge.

d. The next step is to find the value of the step in charge Δq. You can do this from the sorted values. One possibility is to average all the values that are similar and then subtract adjacent values. You might try the @AVG function. For example, writing @AVG(D20..D29) in a cell will give you the average of the first 10 values of $1/t_2$. Find the averages of the next groups of similar values, and then calculate their differences. The average of these differences, call it $\langle \Delta q \rangle$, is proportional to the value of the elementary charge.

Now, to find the value of the elementary charge, multiply $\langle \Delta q \rangle$ by the constants specified by Eq. 17.

A. To find the electric field, here called F, divide the voltage of 7800 V by the distance between the plates, $d_p = 0.016$ m. Millikan's data show that the voltage varied slowly as he did his experiment, but the variation is not significant, and an average value of 7800 V will do very well.

B. You can calculate m by entering the formula of Eq. 20 in a convenient cell. To type in an exponent use the ^ symbol. For example, to raise the contents of cell F17 to the power 3/2, you type "F17^(3/2)."

When you have all the correct constants together, multiply them by the average of the differences. You should get the same value for the elementary charge that Millikan reports at the end of Table 8.3. This is slightly higher than the value accepted today because of a small error in Millikan's value for the viscosity of air.

Hand in a printout of your data sheet, graphs, and the numerical value of Δq.

TABLE 8.3 Millikan's Oil-Drop Data in Modern Units

Distance between crosshairs = 1.010 cm
Distance between plates = 1.600 cm
Temperature = 24.6°C
Density of oil at 25°C = 0.8960 g cm^{-3}
Density of air at 25.2°C = 18.36 μPa·s

	G (sec)	F (sec)	n	e_n (10^{-19} C)	e_1 (10^{-19} C)
	22.8	29.0	7	11.50	1.642
	22.0	21.8	8	13.16	1.645
	22.3	17.2			
$G = 22.28$	22.4	--	9	14.82	1.647
$V = 7950$	22.0	17.3			
	22.0	17.3			
	22.7	14.2	10	16.48	1.648
	22.9	21.5	8	13.16	
	22.4	11.0	12	19.72	1.644
	22.8	17.4	9	14.82	
	22.8	14.3	10	16.48	
$V = 720$	22.8	12.2	11	17.99	1.635
$G = 22.80$	23.0	12.3			
	22.8	14.2			
	--	--	10	16.48	1.648
$F = 14.17$	22.8	14.0			
	22.8	17.0			
$F = 17.13$	--	17.2	9	14.82	1.647
	22.9	17.2			
$F = 10.73$	22.8	10.9			
	22.8	10.9	12	19.72	1.644
$V = 7900$	22.8	10.6			
$G = 22.82$	22.8	12.2	11	17.99	1.635
$F = 6.7$	22.8	8.7	14	22.90	1.636
	22.7	6.8	17	27.76	1.633
	22.9	6.6			
	22.8	7.2			
$F = 7.25$	--	7.2			
	--	7.3			
	--	7.2	16	26.13	1.634
	23.0	7.4			
$F = 8.65$	--	7.3			
	--	7.2			

TABLE 8.3 Millikan's Oil-Drop Data *(cont.)*

	G (sec)	F (sec)	n	e_n (10^{-19} C)	e_1 (10^{-19} C)
	22.8	8.6	14	22.90	1.636
	23.1	8.7			
	23.2	9.8	13	21.24	1.635
$F = 10.63$		9.8			
	23.5	10.7	12	19.72	1.644
	23.4	10.6			
	23.2	9.6			
	23.0	9.6			
$V = 7820$	23.0	9.6			
$G = 23.14$	23.2	9.5			
$F = 9.57$	23.0	9.6	13	21.24	1.635
	--	9.4			
	22.9	9.6			
	--	9.6			
$F = 8.65$	22.9	9.6			
	--	10.6	12	19.72	1.644
	--	8.7	14	22.90	1.636
	23.4	8.6			
$F = 12.25$	23.0	12.3			
	23.3	12.2			
	--	12.1	11	17.99	1.635
	23.3	12.4			
			Change forced with radium		
	23.4	72.4			
	22.9	72.4			
$F = 72.10$	23.2	72.2	5	8.207	1.641
	23.5	71.8			
	23.0	71.7			
	23.0	39.2	6		
$V = 7800$	23.2	39.2			
$G = 23.22$	--	27.4	7	11.50	
	--	20.7	8	13.14	1.642
	--	26.9	7	11.50	1.642
	--	27.2			
	23.3	39.5			
$F = 39.20$	23.3	39.2	6	9.881	1.647
	23.4	39.0			
	23.3	39.1			

TABLE 8.3 Millikan's Oil-Drop Data *(cont.)*

	G (sec)	F (sec)	n	e_n (10^{-19} C)	e_1 (10^{-19} C)
	23.2	71.8	5	8.207	1.641
	23.4	382.5	4		
	23.2	374.0			
	23.4	71.0	5	8.207	1.641
$V = 7760$	23.8	70.6			
$G = 23.43$	23.4	38.5	6		
	23.1	39.2			
	23.5	70.3			
	23.4	70.5			
	23.6	71.2	5	8.207	1.641
	23.4	71.4			
	23.6	71.0			
	23.4	71.4			
	23.5	380.6			
	23.4	384.6			
$F = 379.6$	23.2	380.0			
	23.4	375.4	4	6.559	1.640
	23.6	380.4			
	23.3	374.0			
	23.4	383.6			
	--	39.2			
$F = 39.18$	23.5	39.2	6	9.881	1.647
$V = 7730$	23.5	39.0			
$G = 23.46$	23.4	39.6			
	--	70.8			
$F = 70.65$	--	70.4	5	8.207	1.641
	--	70.6			
	23.6	378.0	4	6.559	0
	Saw it, here, at end of 305 sec, pick up two negatives				
	23.6	39.4	6	9.881	1.647
	23.6	70.8	5	8.207	1.641
					Mean of all e_1s = 1.640

Differences	8.207 − 6.559	=	1.65
	9.881 − 8.207	=	1.67
	11.50 − 9.881	=	1.62
	13.14 − 11.50	=	1.64
	Mean dif.	=	1.64

C H A P T E R

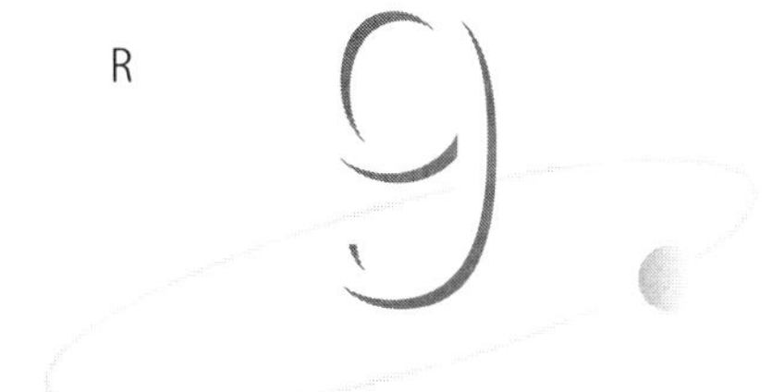

Waves and Light

9.1 INTRODUCTION

The preceding chapters have illustrated the power of the atomic hypothesis. From the observations of Dalton and Gay-Lussac we were able to deduce the chemical composition of molecules and determine ratios of atomic masses. In the kinetic theory we modeled gas atoms and molecules as featureless hard spheres, which allowed us to interpret physical quantities such as temperature and pressure in terms of the more fundamental concepts of kinetic energy and momentum. Using values of the mean free path determined from measurements of the viscosity of gases, we could estimate Avogadro's number and, consequently, the diameter and mass of single atoms. Later, the experiments of Faraday, Thomson, and Millikan proved conclusively that atoms have internal structure, i.e., they are themselves composed of smaller, more fundamental particles. We have identified one of these particles—the electron—and have found that it is removable, replaceable, and interchangeable. What other particles are contained in atoms? How are they assembled, and what holds them together?

Some answers were found by analyzing the light emitted and absorbed by atoms; other information came from probing atoms with electromagnetic radiation not visible to the human eye. The tools and techniques for such analysis and such probing could only be developed as physicists learned more about light and other forms of electromagnetic radiation. Because the wave properties of light are central to understanding how it can be used to reveal the inner secrets of the atom, you should now spend some time understanding waves and the wave behavior of light.

9.2 THE NATURE OF WAVES

After a brief description of the general nature of waves, we will look closely at a special class of periodic waves, those with a pure sinusoidal waveform, the so-called "harmonic," or "sine," waves. Such waves represent simple physical situations: A sinusoidal sound wave is a pure tone; a sinusoidal light wave is a pure color.

Harmonic waves also have the virtue of being mathematically fairly simple. A few parameters are enough to completely describe a harmonic wave, and it is a major goal of this section to define and describe these: "amplitude," "wavelength," "frequency," "velocity," and "phase."

The manner in which waves combine when two or more come together at the same place is a fundamental property of waves, and it plays a major role in exploring matter in general and atoms in particular. Therefore, a large portion of this chapter is devoted to studying the combination, or "interference," of waves.

A Traveling Disturbance

What is a wave? A wave is a traveling disturbance without any transport of matter. For example, when you snap a jump rope, a pattern of deformation passes from one end of the rope to the other, but the parts of the rope stay put. Or again, consider a long line of upright dominoes. You can start a wave of falling dominoes by striking the end domino so that it topples against its neighbor, which then falls and strikes its neighbor, and so on. The disturbance—falling dominoes—propagates from one end of the line to the other, yet no domino moves far from its initial position.

The same behavior holds for other kinds of waves. Imagine tossing a pebble into a quiet pond. The resulting circular ripples—surface water waves—travel outward for several meters before disappearing. The individual water molecules, however, do not move more than about 1 cm. In fact, they return to their original positions after the waves die out. The sound waves reaching your ear during a physics lecture are a further example of this property. They have traveled a distance of 10 m or so, yet the individual air molecules have not moved more than a few microns in response to the passing wave. Consider the alternative: If molecules actually traveled from speaker to listener, then you not only would hear the lecture, but you would also smell the lecturer's most recent meal. "This lecture stinks!" would take on a new meaning.

Light waves are somewhat more abstract than the examples of the previous paragraph. For one thing, they can travel in a vacuum: They do not need a "medium" such as air or rope or dominoes or a water surface. For another, the disturbance is in the form of a mixture of changing electric and magnetic fields. Nevertheless, it turns out that the description of wave properties that works well for sound waves in air or deformations traveling through solids works equally well for light.

Velocity, Wavelength, and Frequency

All waves travel with a finite velocity. This is certainly as true for light as it is for each of the other examples discussed above. The succession of observations and experiments over the last three centuries that have led to the determination of the speed of light ($\approx 3 \times 10^8\,\mathrm{m\,s^{-1}}$) makes a fascinating story.

EXERCISES

1. Anyone who has witnessed a fireworks display has direct evidence that light travels much faster than sound. It takes about 5 seconds for sound to travel 1 mile (1.6 km). How long does it take light to travel the same distance? If you hear an explosion three seconds after you see its flash, how far away did it occur? Write an equation relating distance to the time delay. Amaze your friends next July 4!

Consider again the example of circular water ripples. Figure 9.1 contains two cross-sectional views of the water surface, one at time t_1 and the other at a later time t_2. Each graph shows the height of the surface as a function of the radial distance r from the center, i.e., each is a plot of the waveform. The wave velocity is given by:

$$v = \frac{r_2 - r_1}{t_2 - t_1}. \tag{1}$$

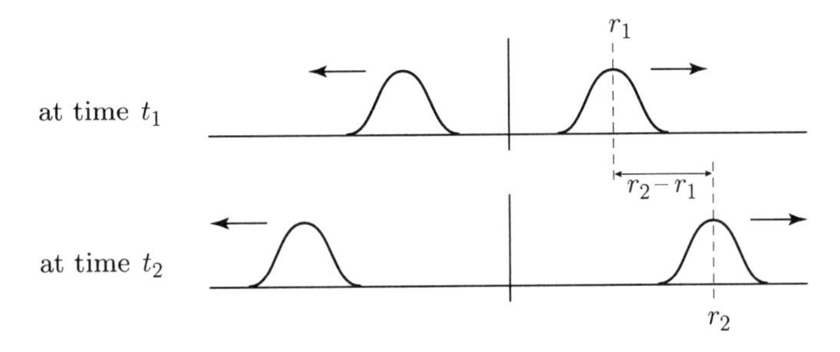

FIGURE 9.1 Views at two different times of water waves traveling across a surface.

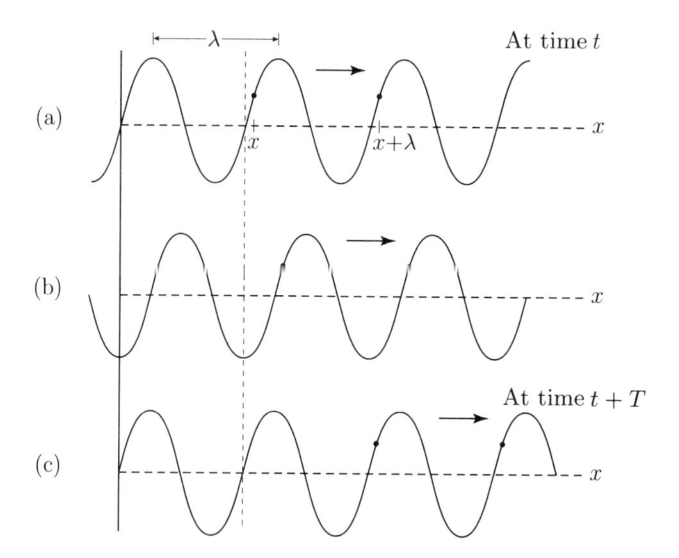

FIGURE 9.2 Views at three different times of a sinusoidal wave with period T traveling to the right: (a) at time t; (b) at time $t + \frac{T}{4}$; (c) at time $t + T$.

Now study Fig. 9.2, which illustrates a different waveform called, for obvious reasons, a sinusoid or sine wave. The graphs might represent the surface of water, the shape of a stretched string, the pressure (or density) fluctuations associated with a sound wave, or the variation in electric field in a light wave. The sinusoidal waveform is a particularly useful one for us to study, and much of the following discussion will be devoted to it.

The sine wave in Fig. 9.2 is an example of a wave that is periodic in space. It is periodic because the pattern of the wave repeats regularly. As Fig. 9.2 shows, the form of the wave at any arbitrary point x is the same at distances $\pm\lambda$, $\pm 2\lambda$, $\pm 3\lambda$, $\pm 4, \ldots$ from x. The shortest repeat distance λ is called the "wavelength."

Because the waveform is traveling, it is also periodic in time. An observer stationed at x_0 watching the waveform pass by will see the basic pattern repeat itself. The number of repeats in a unit time (e.g., the number of crests that pass in one second) is called the "frequency" f of the wave, and is measured in number of occurrences per second. The SI units are s^{-1} and are called "hertz" (Hz).

Humans can hear sound waves with frequencies between 20 Hz and 20 kHz. Dogs can hear up to 35 kHz. Medical sonograms are taken with sound waves with frequencies above 1 MHz. The range of electromagnetic wave frequencies that has been studied is enormous, and the uses are numerous, and they vary depending upon the magnitude of the frequency. For example, the AM radio broadcast band uses frequencies of electromagnetic radiation between 540 and 1500 kHz. The FM band lies between 88 and

TABLE 9.1 Some Different Kinds of Electromagnetic Waves

Commonly Used Name	Frequency		Wavelength	
AM radio waves	540–1500	kHz	–	
FM radio waves	88–108	MHz	–	
TV channels 2–6	54–88	MHz	–	
TV channels 7–13	174–216	MHz	–	
Microwaves	–		30 –0.03	cm
Infrared	–		300–0.7	μm
Visible	–		700–400	nm
Ultraviolet	–		400–30	nm
X-rays	–		30 –0.03	nm
Gamma rays	–		< .03	nm

108 MHz. Your microwave oven cooks food with 2450 MHz electromagnetic radiation. Visible light has frequencies between 430 THz and 750 THz (that's nearly 10^{15} Hz). Other examples of electromagnetic waves are given in Table 9.1.

Since the distance between successive crests is λ, and the number of crests passing per second is f, then the wave velocity must be given by

$$v = \lambda f. \tag{2}$$

The situation is exactly analogous to a passing railroad train. If the length of each car is L meters, and N cars pass per second, then the velocity of the train is NL meters/sec.

The time between arrival of successive crests is called the period T of the wave. If f crests pass per second, then the time between crests must be $1/f$, i.e., the period is

$$T = \frac{1}{f}. \tag{3}$$

EXERCISES

2. "Concert A" is a pure sinusoidal wave having a frequency of 440 Hz. The speed of sound in air at room temperature is about 340 m/s. How far apart are the pressure maxima in the sound wave? The pressure minima? What is the period of the wave?

3. Complete Table 9.1 by calculating the missing entry (frequency or wavelength) in each row. What is the wavelength of your favorite AM radio station? your favorite FM station?

Amplitude

If you were asked what the amplitude of a wave is, you could probably give a reasonable answer based on your intuition. It certainly is a measure of the magnitude of the disturbance, but for a sine wave like that shown in Fig. 9.2, amplitude usually refers to the *maximum* displacement from the undisturbed state. Thus in Fig. 9.2 the amplitude is the maximum height of the sine curve above the x-axis.

Let's represent the amplitude by A. Then you can write an equation for the wave in Fig. 9.2(a) as

$$y = A \sin\left(2\pi \frac{x}{\lambda}\right). \qquad (4)$$

Notice carefully what is going on in Eq. 4. The argument of a sine function has to be an angle, so we make sure that it is, and, furthermore, we express it in *radians*. Because this periodic function is to repeat every wavelength—that's the definition of a wavelength—the sine's argument must increase by 2π at one wavelength, by 4π at $x = 2\lambda$, by 6π at $x = 3\lambda$, and so on. That is exactly what Eq. 4 accomplishes.

EXERCISES

4. Suppose Eq. 4 describes a wave on a long string. Let $A = 5$ cm; $\lambda = 10$ cm. What would be the displacement of the string at $x = 2.5$ cm? $x = 5$ cm? $x = 7.5$ cm? $x = 10$ cm? Graph the displacement of the string vs. its length.

5. In the previous exercise, what would be the amplitude of the wave at the given values of x?

If you answered 5 cm for each x, pat yourself on the back. Strictly speaking, "amplitude" means the value of A, and this does not change with x. On the other hand, you should be aware that many physicists are sloppy in their terminology and would say "amplitude" when they really mean "displacement." There is not much you can do about that except be alert to the context.

EXERCISES

6. Suppose Eq. 4 describes a wave traveling down a gas-filled pipe. Let $A = 5$ Pa; $\lambda = 10$ cm. What would be the extra pressure in the pipe at $x = 2.5$ cm? At $x = 7.5$ cm?

7. Suppose we are talking about a light wave traveling down a thin transparent fiber. Let $A = 5$ V/m; let $\lambda = 10\,\mu$m. What would be the electric field in the fiber at $x = 2.5\,\mu$m? At $x = 5\,\mu$m? At $x = 7.5\,\mu$m?

The three preceding exercises show you how Eq. 4 retains the same form while representing different kinds of waves.

Phase

The argument of the sine function in Eq. 4 is called its "phase." Referring to Fig. 9.2(a), you can see that at $x = 0$ the phase of the wave is 0. At $x = \lambda/8$ the phase is $\pi/4$ radians; at $x = \lambda/4$ the phase is $\pi/2$ radians.

EXERCISES

8. Show that the phase difference, $\Delta\phi$, between two points x_1 and x_2 is

$$\Delta\phi = 2\pi\frac{x_2 - x_1}{\lambda}. \tag{5}$$

Notice that the three sinusoids of Fig. 9.2(a), (b), and (c) are identical except for their phases. The curve of Fig. 9.2(b) will have the same equation as (a) except that it is shifted to the right. Do you recall that a function $f(x)$ is shifted to the right a distance h just by writing it as $f(x - h)$? So you can write the equation of Fig. 9.2(b) simply by shifting Eq. 4 by a constant amount of phase. From the figure you can see that this amount is $\pi/2$ radians, and therefore, the equation for Fig. 9.2(b) is just

$$y = A\sin\left(2\pi\frac{x}{\lambda} - \frac{\pi}{2}\right).$$

There is always more than one way to write the phase. For one thing, there is no way to tell whether the curve of Fig. 9.2(b) was shifted to the right by $\pi/2$ or shifted to the left by $3\pi/2$ radians. It would have been just as correct to write

$$y = A\sin\left(2\pi\frac{x}{\lambda} + \frac{3\pi}{2}\right).$$

For another, changing the phase by any multiple of 2π does not change the value of the function. There is also the fact that $\sin(\theta - \pi/2)$ is the same as $-\cos(\theta)$, so that Fig. 9.2(b) could also have been written

$$y = -A\cos\left(2\pi\frac{x}{\lambda}\right).$$

EXERCISES

9. Can you think of some other equivalent forms of Eq. 4?

To represent a moving sinusoidal waveform, all you need to do is add to the phase of Eq. 4 a term $-2\pi t/T$. This has the effect of shifting the wave to the right at a steady rate.

Putting all these pieces together gives the most general form for a one-dimensional traveling harmonic wave:

$$y = A\sin\left(2\pi\frac{x}{\lambda} - 2\pi\frac{t}{T} + \phi\right), \tag{6}$$

which is described using the following notation and terminology:

Properties of Periodic Waves

y	displacement
A	amplitude
x	position
λ	wavelength
t	time
T	period
$f = \frac{1}{T}$	frequency
ϕ	phase constant
$2\pi\frac{x}{\lambda} - 2\pi\frac{t}{T} + \phi$	phase

Transverse and Longitudinal Waves

Waves are also characterized by the direction of the displacement of the wave relative to its direction of propagation. In Fig. 9.3(a), the wave is traveling on a taut string, and each element of the string moves up and down vertically as the wave moves in the positive x-direction. When the displacement associated with a wave is perpendicular to the direction of propagation as in Fig. 9.3(a), the wave is said to be a "transverse" wave. If the displacement is parallel to the direction of propagation as in Fig. 9.3(b), the wave is said to be a "longitudinal" wave. Sound waves in air are longitudinal waves; light waves in a vacuum are transverse waves.

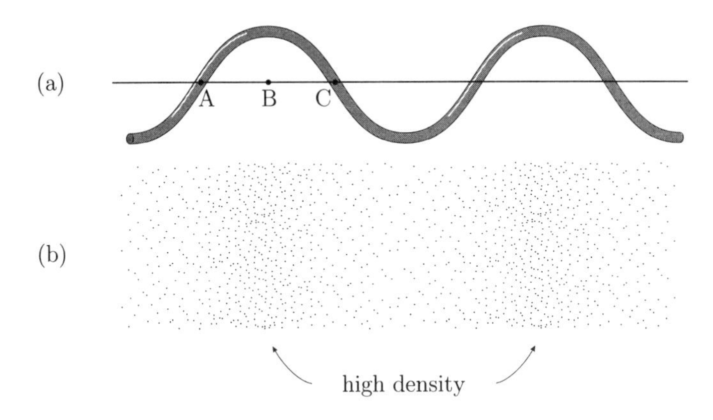

FIGURE 9.3 (a) Transverse wave on a string. (b) Longitudinal sound wave in a gas.

EXERCISES

10. Imagine that Fig. 9.3(a) is a snapshot of the wave traveling to the right along the string. At the exact moment that the photo is taken, what is the direction of motion of the string at point A? At point B? At point C? How would your answers change if the wave were traveling to the left?

Intensity

Although waves are not themselves the transport of matter, they do carry momentum and energy from one point in space to another. It is by means of such transported momentum and energy that waves have effects. The warmth of sunshine tells you that light waves carry energy. The pain in your ears tells you that rock music transports momentum.

The specific definition of "intensity" of a wave in three dimensions is the amount of energy carried across a unit area in a unit time. Thus the SI units of intensity are $\mathrm{J\,m^{-2}\,s^{-1}}$ or equivalently $\mathrm{W\,m^{-2}}$.

EXAMPLES

1. The intensity of sunlight on the upper atmosphere of Earth is about $1.39\,\mathrm{kW\,m^{-2}}$. How much solar energy does Earth receive in one second?

The Earth with its radius of about $R = 6400\,\mathrm{km}$ offers to the Sun a circular target with an area of $\pi R^2 = \pi(6.4 \times 10^6)^2 = 1.3 \times 10^{14}\,\mathrm{m}^2$. Therefore, the Earth will receive

$$1.39 \times 10^3 \times 1.3 \times 10^{14} = 1.8 \times 10^{17}\,\mathrm{J}$$

in one second.

For the purposes of this book it is enough for you to know that *the energy carried by any wave is proportional to the square of its amplitude*. This should seem plausible to you if you know that the energy stored in a stretched spring is proportional to the square of the maximum amount of its stretch. Sound waves and waves along strings all stretch and compress the matter through which they travel just the way a spring can be stretched and compressed. It is not obvious that this rule should work for water waves and light waves, but it does

9.3 INTERFERENCE OF WAVES

Imagine the following peculiar situation. After a long day's drive, you have just checked into the Bates Motel and are about to enjoy a warm, relaxing shower. When you enter the shower stall, you notice that it is equipped with two independent water nozzles, each of which operates normally and delivers a fine spray distributed uniformly over the entire stall. Imagine your surprise when, turning on both shower heads together, you notice that there are positions under them where you do not get wet at all, and other positions where you get four times as wet as with a single shower head! This does not really happen, of course, but it would if the nozzles emitted waves instead of particles of water.

Here is a more plausible example to illustrate the point. Suppose you and a dozen friends lug a CD player and two stereo speakers out onto a flat, grassy, open field. You set up the speakers a distance d apart, and then you and your friends form a line at a perpendicular distance D from the speakers as shown in Fig. 9.4. Now play a rather dull recording consisting of a single pure note, i.e., a sinusoidal wave, sustained for the entire duration of the tape. If only one speaker is plugged in, then you all will hear the note, with the central observer O_0 recording the highest intensity. When both speakers are activated, some of you will hear nothing at all, while O_0 will report an intensity four times greater than for the single-speaker case! This phenomenon is called "interference."

Interference is a unique and defining property of waves. If any kind of radiation exhibits interference, you know it is a wave. Thus, if the water from the shower actually behaved the way described above, you would know immediately that the water spray was wavelike. This idea that interference tells you when something is a wave is important later in this book. Interference is also very important for the study of atomic properties.

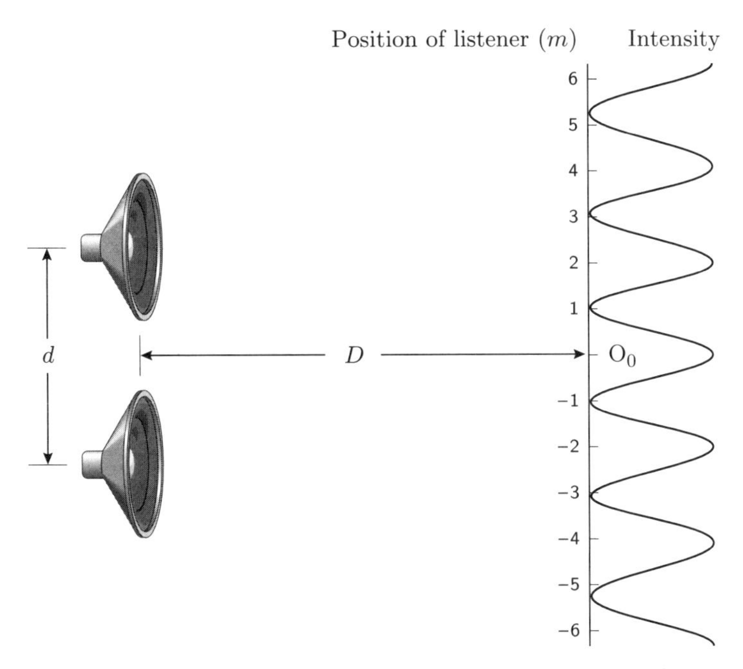

FIGURE 9.4 Acoustic interference experiment between two stereo speakers. Each tic mark signifies an attentive student listener.

Interference in One Dimension

Interference is a consequence of the property of superposition, a somewhat fancy word that means that when two one-dimensional waves occupy the same points in space, their displacements just add together algebraically. To understand this principle consider a one-dimensional case in which two pulsed waves with the same shape approach each other on a long stretched string, as shown in Fig. 9.5. To simplify the discussion, suppose that each pulse has a trapezoidal shape. What happens as the two pulses come together? As shown in Fig. 9.5, the string's shape is exactly what you would calculate if you imagined that each pulse passed undistorted through the other and their heights simply added together at each point along the string. This simple additivity occurs for all small-amplitude waves and is called the "principle of superposition." It means that to find the waveform due to a combination of waves, you just add, point by point, the displacements of each wave.

EXERCISES

11. To see that superposition can be quite interesting, consider what happens in Fig. 9.5 when the wave coming from the right has negative displacement, i.e., if the right-hand trapezoid is flipped over with respect to the axis.

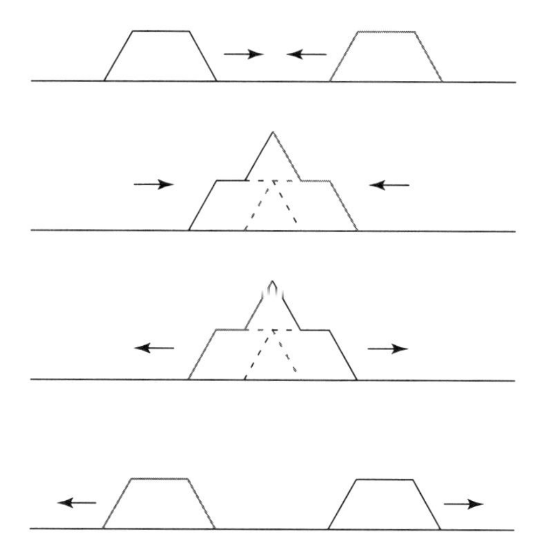

FIGURE 9.5 Two waves approaching on a string and combining. At each moment the resulting wave shape is the point-by-point sum of the two waveforms.

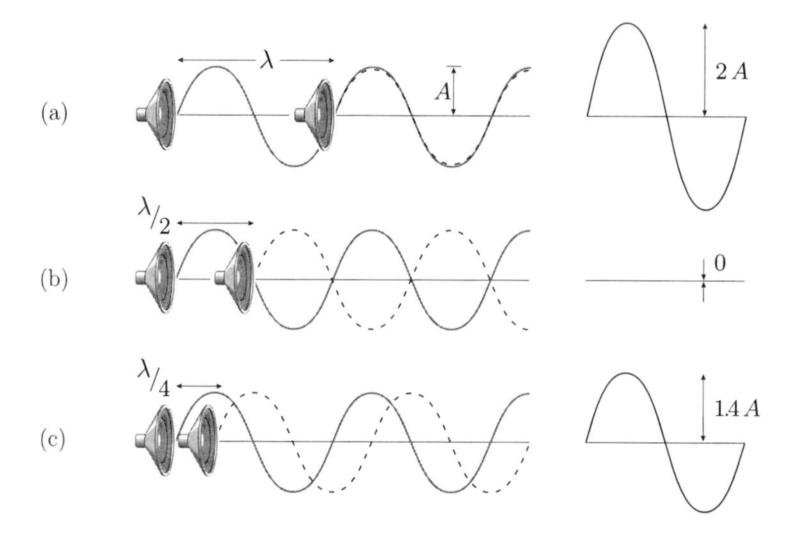

FIGURE 9.6 Constructive and destructive interference of sine waves. The output of the left speaker is the solid curve; the output of the right speaker is the dashed curve. One cycle of the sums of these two curves is shown at the extreme right.

Now look at Fig. 9.6. Two stereo speakers, one behind the other, emit sine waves of the same frequency f and amplitude A in the same direction toward an observer. The speakers are displaced from each other by a distance L along the line from speaker to observer. When both speakers are turned on, the listener will hear a sine wave of frequency f having an amplitude that depends critically on the speaker separation L. As the figure shows, when $L = 0, \lambda, 2\lambda, \ldots$, the total amplitude equals $2A$, i.e., the waves add together to form a larger wave. This is called "constructive interference." If $L = \lambda/2, 3\lambda/2, \ldots$ the waves cancel exactly, and the listener hears nothing, even though both speakers are emitting sound! This is called "destructive interference." For other speaker separations, the resulting amplitude will take on values between 0 and $2A$. Since the sound-wave intensity is proportional to the energy carried by the wave, and since the energy is proportional to the square of the amplitude, the intensity for constructive interference is four times as large as it was for a single speaker.

EXERCISES

12. Each speaker in Fig. 9.6 is emitting a 1 kHz sound wave with an amplitude A. Calculate three speaker separations L that lead to destructive interference, i.e., $A_{tot} = 0$.

So far we've looked at special cases where identical waves add (they are in phase) or where they exactly cancel (they are exactly out of phase).

In the first case shown in Fig. 9.6(a) the two waves line up exactly and add with no cancellation because the waves have the same phase at every instant in time. In the second case the waves cancel because the first wave is $A \sin \theta$ and the second wave is $-A \sin \theta$. The two waves differ by a phase constant. You can see that this is so because $-A \sin \theta = A \sin(\theta - \pi)$. We say that these waves are π radians or 180°out of phase.

In the third case, Fig. 9.6(c), the addition is more complicated because the two waves add at some points and cancel at others. Again we can characterize the way the waves combine by their difference in phase: If the solid-line waveform is $A \sin \theta$ and the dashed-line waveform is $A \sin(\theta - \frac{\pi}{2})$, these two waves are $\pi/2$ radians, or 90°, out of phase. Figure 9.7(a) shows how a shift of one wave by a quarter of a wavelength corresponds to a phase shift of 90°.

What happens in general? A more general case would be a wave $A \sin \theta$ adding to a wave $A \sin(\theta + \Delta\theta)$. We say that these two waves are out of phase by $\Delta\theta$. To be able to see clearly the results of combining $A \sin \theta + A \sin(\theta + \Delta\theta)$, we can use one of those long-forgotten trigonometric identities:

$$A \sin \theta + A \sin \theta' = 2A \sin\left(\frac{\theta + \theta'}{2}\right) \cos\left(\frac{\theta - \theta'}{2}\right).$$

Then if you let $\theta' = \theta + \Delta\theta$, you see that

$$A \sin \theta + A \sin(\theta + \Delta\theta) = 2A \cos\left(\frac{\Delta\theta}{2}\right) \sin\left(\theta + \frac{\Delta\theta}{2}\right). \tag{7}$$

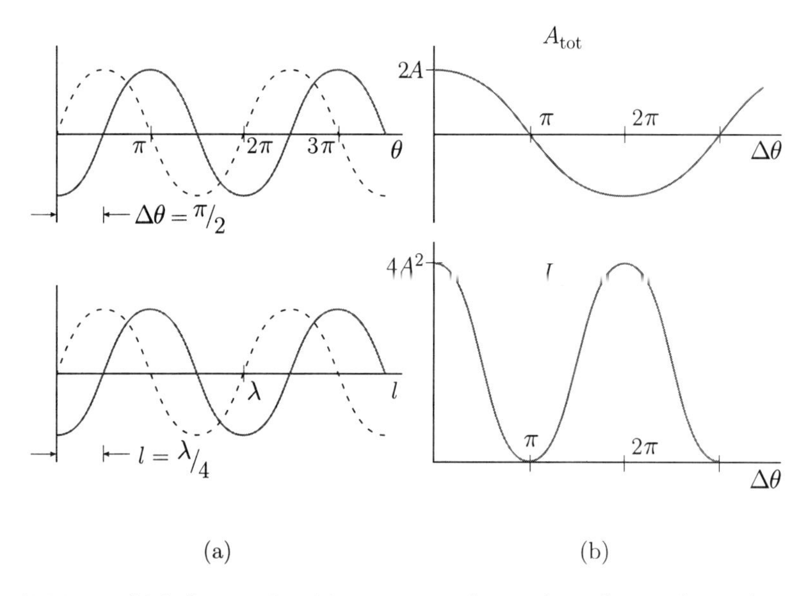

FIGURE 9.7 (a) A change of position corresponds to a phase change; (b) resultant amplitude and intensity vs. phase shift.

Equation 7 provides a good way to think about the combined waves. Because the right hand side of Eq. 7 has a factor of $\sin\left(\theta + \frac{\Delta\theta}{2}\right)$, the result must be a wave with the same shape as the two waves being combined, but shifted in phase by $\Delta\theta/2$ from either. The shape is the same, but the amplitude is different and is $2A\cos\left(\frac{\Delta\theta}{2}\right)$. Figure 9.7(b) shows how this amplitude factor and its square vary as the phase difference $\Delta\theta$ goes from 0 to 3π radians. The behavior of the square of the amplitude of the summed waves is important because it is proportional to the energy carried by a wave. For this reason the curve in the lower part of Fig. 9.7(b) represents how the loudness of sound or brightness of light will vary as the phase between two combining waves is changed.

Notice that Fig. 9.7(b) and Eq. 7 give the amplitudes of the special cases in Fig. 9.6(a) and (b). For case (a) there is no phase difference, i.e., $\Delta\theta = 0$ and $\cos 0 = 1$, so the amplitude is $2A$ and the resulting combined wave is $2A\sin\theta$. For a phase difference of $\Delta\theta = \pi$, the amplitude is $2A\cos\frac{\pi}{2} = 0$, so the combined wave is zero everywhere, as it should be for total destructive interference.

EXERCISES

13. What will Eq. 7 predict for case (c) in Fig. 9.6 where $\Delta\theta = -\frac{\pi}{2}$?

Let's express the resulting amplitude in terms of L, the separation between the speakers. A spacing of L corresponds to a difference in θ, or a phase shift $\Delta\theta$, in radians given by the proportion

$$\frac{L}{\lambda} = \frac{\Delta\theta}{2\pi}. \tag{8}$$

Therefore, the amplitude of the new sinusoid is

$$A_{\text{tot}} = 2A\cos\left(\frac{\Delta\theta}{2}\right) = 2A\cos\left(\frac{\pi L}{\lambda}\right).$$

Note that the absolute value of A_{tot} is a maximum of $2A$ when $\pi L/\lambda = 0, \pi, 2\pi, \ldots$, i.e., when $L = n\lambda$ (where n is any integer), exactly as expected. The intensity I of a wave is proportional to the square of its amplitude, so that

$$I \propto 4A^2\cos^2\left(\frac{\pi L}{\lambda}\right). \tag{9}$$

EXERCISES

14. Show that Eq. 9 is consistent with our expectations for destructive and constructive interference, as discussed above.

15. Suppose that for the two speakers one behind the other as shown in Fig. 9.6, total destructive interference first occurs at a separation of $L = 15$ cm, and then at 45 cm, 75 cm, ...
(a) What is the frequency of these sinusoidal waves?
(b) If the amplitude of each wave is A, calculate the ratio of the intensity recorded by the listener at $L = 10$ cm to the intensity she would hear if one of the speakers was turned off.

Now let's extend the discussion beyond the one-dimensional case. Recall the previous example of the two stereo speakers in an open field. As shown in Fig. 9.8, an observer at position $y \neq 0$ is not equidistant from the two speakers, i.e., $r_1 \neq r_2$. If $d \ll D$, so that r_1 and r_2 are very nearly parallel, then $\angle\text{ACE} \approx 90^\circ$, and to a good approximation the path difference $\overline{\text{CE}}$ is $r_2 - r_1 = d\sin\theta$, where $\theta = \tan^{-1}(y/D)$. As in the previous example, this path difference gives rise to a phase difference between the waves emitted by the two speakers. In this two-dimensional case there will be constructive interference with $A_{\text{tot}} = 2A$ when the path difference $\overline{\text{CE}} = n\lambda$. For this situation the angles satisfy the following relations:

$$d\sin\theta = n\lambda, \tag{10}$$

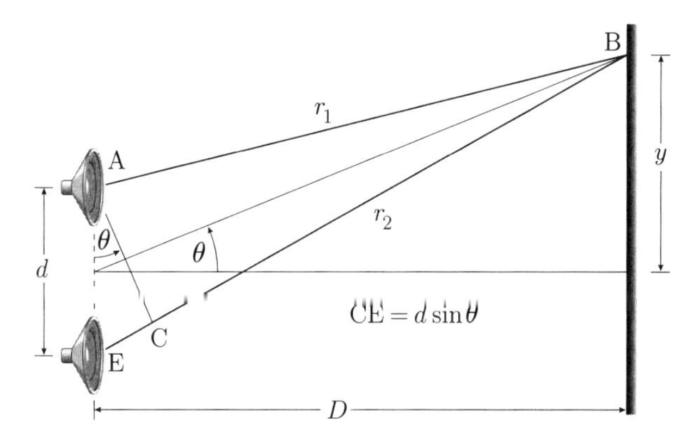

FIGURE 9.8 Geometry for the two-speaker interference experiment.

and there will be destructive interference with $A_{\text{tot}} = 0$ when

$$d \sin\theta = \left(n + \frac{1}{2}\right)\lambda, \tag{11}$$

where $n = 0, 1, 2, 3, \ldots$.

Once again, at certain locations observers would hear nothing, while at other places observers would record an intensity four times that coming from a single speaker. Using the previous results, you could plot the sound intensity (proportional to the square of the amplitude) along the line y. According to Eq. 9, the plot would vary as the square of a cosine curve and look like the graph in Fig. 9.9. This plot, the pattern of intensity vs. position, is an example of what is called an "interference pattern." The large spatial variations in intensity are the unmistakeable signature of interference, which tells you that a wave is present. The shape and size of the interference pattern contain a great deal of information about the wave and its sources.

EXERCISES

16. Two speakers, such as those shown in Fig. 9.8, are separated by a distance $d = 1$ m and emit sound waves of frequency 1 kHz. Calculate the smallest three angles for (a) constructive and (b) destructive interference. If $D = 30$ m, where on the line y should observers stand in order to observe constructive interference?

17. If the frequency in the previous exercise is slowly increased to 2 kHz, should the observers move away from or toward the center ($y = 0$) to follow the intensity peaks?

Light Is a Wave.

What does all this have to do with light? In 1801 Thomas Young demonstrated that light forms interference patterns, and thereby proved conclusively that *light is a wave*. Young's experimental apparatus was the exact analogue of the sound-wave experiment described above. He directed a strong light source (sunlight) through a tiny aperture and then onto a card containing two closely spaced small holes. Light passing through the holes illuminated a distant screen and formed unmistakable intensity maxima and minima—called bright and dark fringes. The effect is small because the wavelength of visible light is very much smaller than the dimensions of ordinary objects.

The wavelength of visible light tells you its color. White light is a mixture of all colors of the rainbow. If white light passes through a glass prism, it spreads out into all its component colors—red, orange, yellow, green, blue violet (roygbv). You will often need to know roughly what wavelengths of light correspond to which colors. A list is given in Table 9.2. Notice the basic behavior: *The longest visible wavelengths are red; the shortest visible wavelengths are blue or violet.*

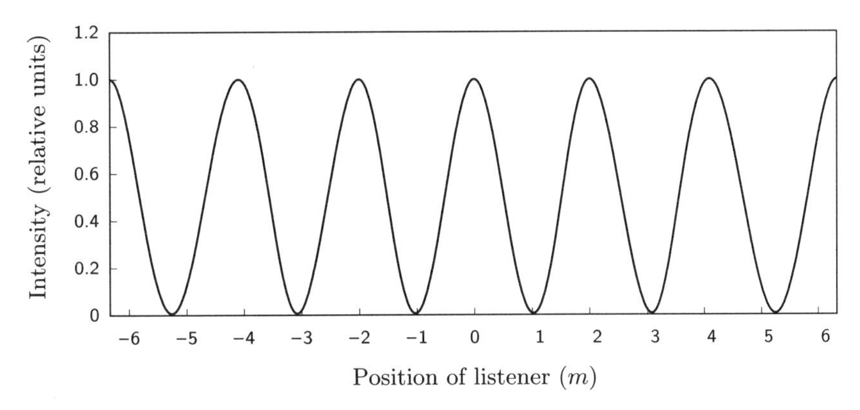

FIGURE 9.9 A listener moving along the y-line in Fig. 9.8 will hear the intensity from the two speakers vary as shown above.

TABLE 9.2 Rough Correspondence of Wavelengths and Colors of Visible Light

Color	red	orange	yellow	green	blue	violet
Wavelength (nm)	660	620	580	520	440	380

9.4 INTERFERENCE OF LIGHT FROM SLITS

Atoms emit and absorb light; different atoms emit and absorb different mixtures of different wavelengths. By measuring these wavelengths and their intensities to great accuracy, it is possible to deduce the internal structure and behavior of atoms and molecules. An important tool for making such measurements, called the "diffraction grating," uses interference to sort out component wavelengths with remarkable precision. The diffraction grating's importance as a way to look into the interior of atoms would be reason enough for studying its principles of operation, but these principles also turn out to be essential for understanding basic ideas of quantum theory. Because slits are the basic elements of a diffraction grating, we start our exploration of these principles by examining the interference of light from two narrow slits spaced close together.

Double-Slit Interference

Let us examine Young's famous experiment in greater detail. As shown in Fig. 9.10(a), a monochromatic light source is focused on a small slit S_0 in an otherwise opaque screen. Light passing through S_0 illuminates two identical slits S_1 and S_2 situated equidistant from S_0. Light passing through the two

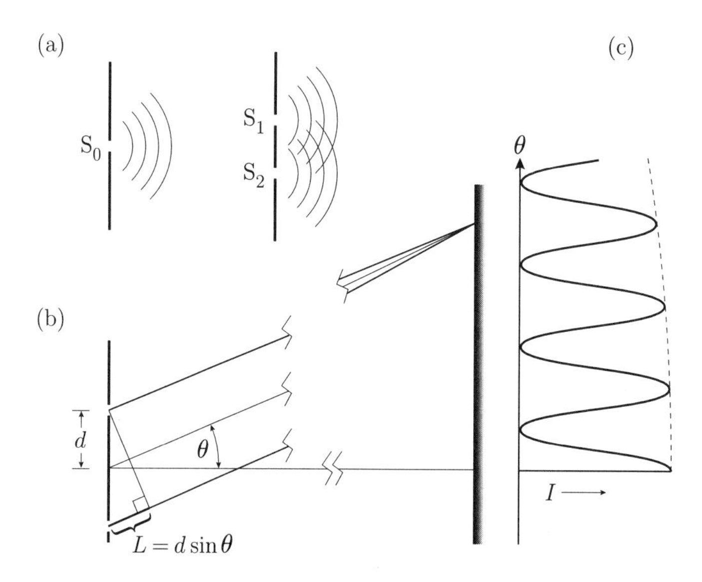

FIGURE 9.10 Geometry for a double-slit interference experiment. (a) Two slits serve as in-phase sources of light; (b) variation of the path difference L with angle θ; (c) intensity of interference pattern on a screen.

slits forms an interference pattern, which can be seen on a screen at the right. Notice the sets of concentric circular arcs spreading from left to right. These arcs, reminiscent of the crests of surface water waves, are lines of equal phase, and the phase difference between successive arcs is 2π radians. The electric field E (and the magnetic field B) has the same value everywhere along such an arc. The spacing between arcs is λ, and they propagate at the speed of light $c = 3 \times 10^8$ m/s. Because S_1 and S_2 are equidistant from S_0, an arc of constant phase reaches both slits simultaneously, and the phases of the light at S_1 and S_2 are the same, or as we say, "the slits are illuminated in phase." In effect the slits become identical, in-phase light sources, as suggested by the ripple patterns of Fig. 9.10(a). Now, the geometry of Fig. 9.10(b) is exactly the same as for the two stereo speakers, and so the same equations (Eqs. 10 and 11) hold here for constructive and destructive interference. Therefore, the maxima in the double-slit interference pattern occur when the path difference

$$d \sin\theta = n\lambda,$$

where n is any integer, and the intensity pattern on the screen will be

$$I = I_0 \cos^2\left(\pi \frac{d \sin\theta}{\lambda}\right),$$

the same as Eq. 9, where L has been replaced with $d \sin\theta$. Since visible light has such a small wavelength (500 nm as compared to $\approx$ 30 cm for sound), we need a small value of d and a large value of D in order to see a clear interference pattern. In many cases, θ is much smaller than 1 radian, and then the approximation $\sin\theta \approx \tan\theta \approx \theta$ (radians) is valid.

EXERCISES

18. Two slits spaced 0.5 mm apart are illuminated by monochromatic light of wavelength 650 nm.
(a) What color is this light?
(b) Using the notation of Fig. 9.10 write down the angles θ at which the first three bright fringes appear to one side of the central fringe. Express these angles in radians.
(c) Is the small-angle approximation valid for this exercise?
(d) If the interference pattern is projected onto a screen a distance $D = 2$ m from the slits, what is the distance between the fringes?

19. An electrical discharge in hydrogen gas emits a mixture of red, green, blue, and violet light. The source illuminates a pair of slits, causing an interference pattern to appear on a distant screen. Make a sketch showing

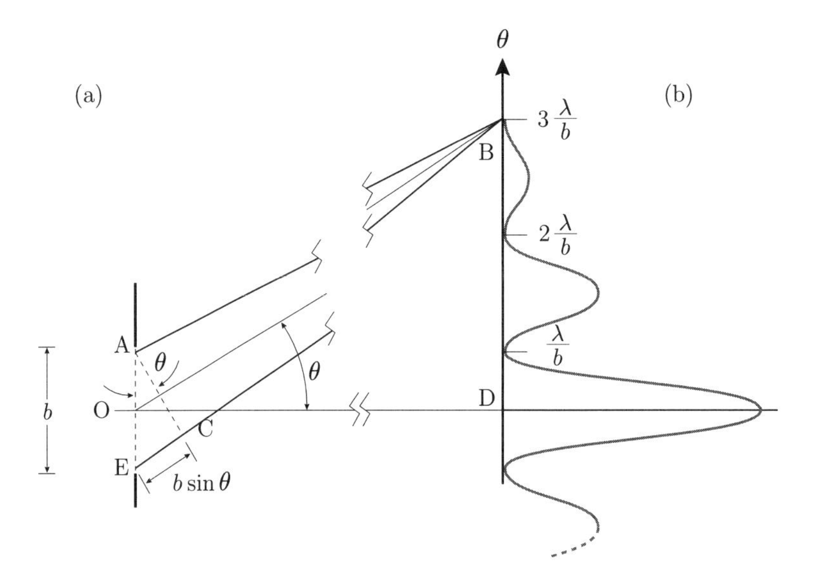

FIGURE 9.11 Single-slit diffraction: (a) geometry; (b) interference pattern.

the relative position of the first few bright fringes of each color. What color is the center of the pattern ($n = 0$)?

Single-Slit Diffraction

Notice, in Fig. 9.10, that the interference pattern is drawn as a cosine-squared curve with an amplitude that falls off as θ moves away from 0. The falloff occurs because the individual slits do not radiate light uniformly in all forward directions. To understand this, consider a single slit of width b, as shown in Fig. 9.11(a). To do the analysis, imagine the entire width of the slit to consist of a very large number of closely spaced identical light sources. Now consider light incident on the screen at the right at an angle θ to the line $\overline{\text{OD}}$ normal to the screen. If $b \ll \overline{\text{OD}}$, then the rays of light from the slit are all essentially parallel, and $\angle$ ACE $\approx 90^\circ$, so that there is a path difference $\overline{\text{CE}} = b \sin\theta$ between light emanating from the top and bottom edges of the slit. The $\angle$ODB $= 90^\circ$ also, so the path difference between light originating at the bottom ($y = 0$) and middle ($y = \frac{b}{2}$) of the slit is $\frac{b}{2}\sin\theta$. If this distance is equal to $\lambda/2$, then the sources at $y = 0$ and $y = \frac{b}{2}$ interfere destructively, *as do all other pairs of sources* at $y = \delta y$ and $y = b/2 + \delta y$, where the value of δy specifies which pair of sources you are considering. In other words, each source interferes destructively with another, so that the slit radiates no light in the direction θ given by $b \sin\theta = \lambda$. The same reasoning applies if $b \sin\theta = 2\lambda$, since pairs of sources at $y = \delta y$ and $y = \frac{b}{4} + \delta y$ interfere destructively.

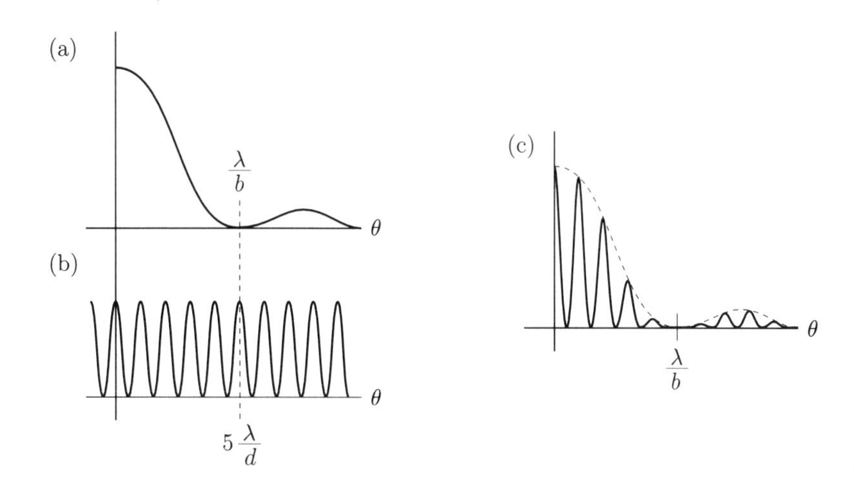

FIGURE 9.12 (a) Single-slit diffraction pattern; (b) double-slit pattern for $d = 5b$; (c) resultant interference pattern.

Thus, *no light is radiated by the slit in directions given by* $b \sin\theta = n\lambda$. A plot of the intensity vs. θ, which is called the "single-slit diffraction pattern," is shown in Fig. 9.11(b). The most important features of the plot are the central maximum at $\theta = 0$ and the *minima* given by

$$b \sin\theta = n\lambda, \tag{12}$$

where $n = 1, 2, 3, \ldots$.

Combined Double- and Single-Slit Patterns

Returning to the double-slit interference pattern, you can now see that the result of Eq. 9 must be modified to include the pattern of radiation from each slit, i.e., the single-slit diffraction pattern. Figure 9.12 illustrates the resulting intensity distribution for the special case $d = 5b$. Notice how the peak intensities of a double-slit pattern decrease for large θ; they are brought down by the effects of diffraction occurring at each slit.

Equation 12 can also be used to understand why the wave nature of light is not immediately obvious. For slit widths b that are large relative to the wavelength of light λ, the secondary maxima of the single-slit pattern occur at such small values of θ that they are difficult to see next to the central maximum. Now that you know what to look for, however, put the first two fingers of your hand together as for a two-finger salute, and make a small adjustable slit between them. Hold this slit up close to your eyes and look through it at a lamp or a window or the sky. Make the slit small enough and you will see black striations. These are hand-made interference fringes.

EXERCISES

20. A single slit of width $b = 0.2$ mm is illuminated with 633 nm light from a helium–neon laser.
(a) Calculate the angular positions of the first 3 minima.
(b) Is the small-angle approximation valid for these angles?
(c) If the interference pattern is projected on a screen 2 m away, how far apart are the dark fringes?

21. Sketch, as in Fig. 9.12, the interference pattern for $d = 3b$.

Multislit Interference Patterns

As the number of slits producing the interference is increased from 2 to many (N), the pattern changes and becomes more and more useful for analyzing the wavelength composition of light. Figure 9.13 illustrates what happens to the interference patterns as N takes on the values of 2, 3, 4, and 10. The assumption here is that $b \ll d$, so that the first single-slit diffraction minimum is located at large values of $\theta \approx \lambda/d$. Note that there are large-intensity maxima at the same θ values as for the double-slit case, but for N slits there are $N - 2$ smaller maxima between the large ones. The smaller ones are called "secondary" maxima; the large ones are called "principal"

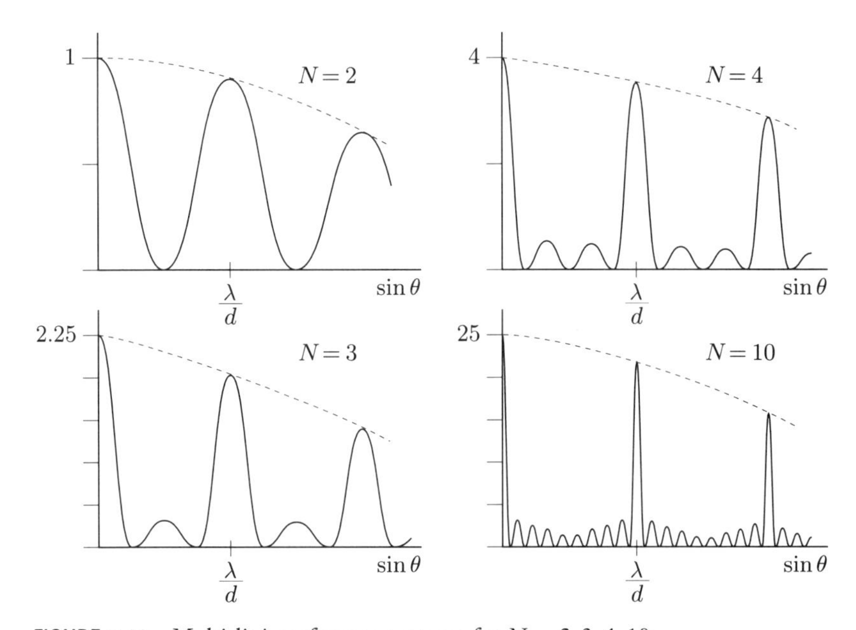

FIGURE 9.13 Multislit interference patterns for $N = 2, 3, 4, 10$.

maxima. As N increases, the principal maxima become narrower and more sharply defined. They also get brighter because of the larger amount of light collected by the N slits.

Such N-slit interference devices are called "diffraction gratings."[1] As N is made larger and d is made smaller, the principal maxima become sharper and spread out more, and the secondary maxima become smaller and negligible. The principal maxima occur at the angles such that

$$d \sin\theta = n\lambda, \tag{10}$$

just as for the double-slit case, but now they are extremely sharp and well defined. If θ and d are well known, then the wavelength λ can be determined with six-digit precision.

Equally important, a good diffraction grating will separate light consisting of two wavelengths that are nearly the same into two distinct principal maxima that can be located and compared with high precision. Without proof we just tell you that the ability to separate and distinguish between two wavelengths improves with increasing N. The smallest difference in wavelength $\delta\lambda$ that can be separated is $\delta\lambda = \lambda/N$. The ratio $\delta\lambda/\lambda = 1/N$ is called the "resolution" of the grating.

EXAMPLES

2. Modern technology can produce as many as 30,000 slits per centimeter on a glass plate. When a 1 cm wide piece of such a closely ruled grating is used to examine green light, $\lambda = 550$ nm, the grating can distinguishably separate wavelengths that differ by as little as $550/3 \times 10^4 = 0.018$ nm.

Diffraction gratings can work by reflection as well as by transmission. You probably know that compact disks spread out reflected incident light into bands of colors. The reason is that the grooves on a compact disk are 1.6 μm wide and make the surface of the disk a nice diffraction grating.

EXERCISES

22. If you look at the light reflected from a compact disk, what angular separation would you expect to see between the red light ($\lambda = 625$ nm) that corresponds to $n = 1$ and that which corresponds to $n = 2$ in Eq. 10?

[1]The word "grating" seems to have been chosen because the closely spaced slits reminded someone of spaced slits in iron grillwork of a fireplace grating.

Spectra, Spectrometers, Spectroscopy

The collection of different wavelengths emitted by a source is called its electromagnetic "spectrum." The rainbow produced when water droplets spread out the visible wavelengths present in sunlight is the *spectrum* of sunlight. Because there is present some amount of almost every wavelength, the Sun's spectrum is said to be "continuous." Many sources, especially hot gases, emit light that is not continuous in its spectrum but is instead a mixture of quite distinct separate wavelengths. Because a spectrum is often produced by defining a beam of light from the source by narrow slits and then spreading out the beam with a prism or diffraction grating, the observer sees the separate colors spread out into lines of light that are images of the slits. These images are called "spectral lines." Spectra composed of spectral lines are called "line spectra." The study of line and continuous spectra is called "spectroscopy," and those who perform such studies are called "spectroscopists." The study of the spectra of atoms is called "atomic spectroscopy."

Precision optical instruments for spreading out light into its component wavelengths and then measuring their values λ are called "spectrometers." The principal components of a spectrometer that uses a diffraction grating are shown in Fig. 9.14. Modern optical spectrometers can measure wavelengths to a precision of 1 part in 10^5 and differences in wavelengths $\lambda_1 - \lambda_2$ to a precision of 1 part in 10^7. Powerful instruments such as these have been

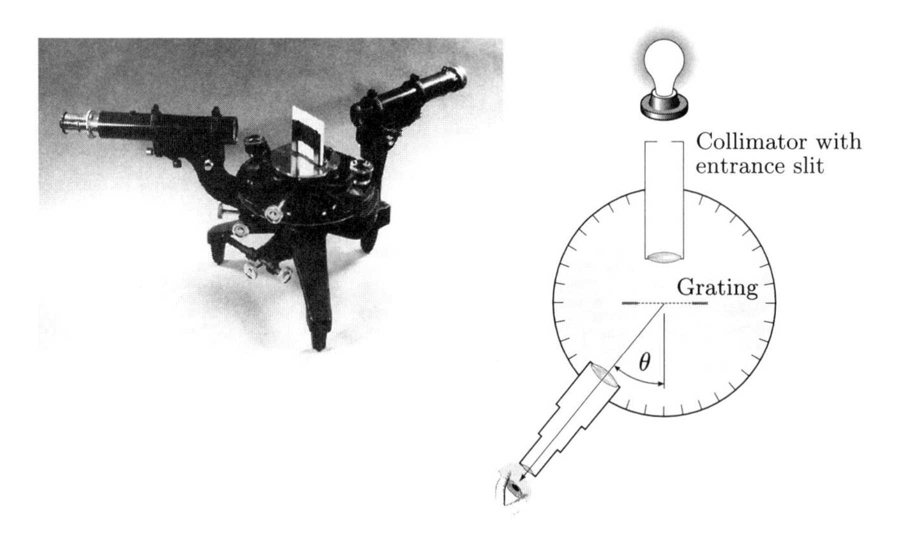

FIGURE 9.14 Components of a grating spectrometer.

essential for exploring the structure of atoms and developing and testing the quantum theory.

9.5 ATOMIC SPECTROSCOPY

Helium, the lightest of the inert gases and the second-most-abundant element in the universe was first discovered on the Sun in 1868. It was not found on Earth until 27 years later. Preposterous? Not at all. The discovery of helium illustrates the power of spectroscopic techniques.

Atomic spectroscopy reveals a most important experimental result: Isolated (i.e., gaseous) atoms gain and lose energy by absorbing and emitting light at certain precisely defined wavelengths (or frequencies); each atomic species has its own unique set of wavelengths—often referred to as "spectral lines." The collection of these lines is called the "spectrum" of the atom. An atom's spectrum is its fingerprint. If the light emitted by a hot, glowing gas contains the spectral lines of, say, oxygen, then oxygen must be present in the gas. Spectral lines of several elements are listed in Table 9.3.

Helium was discovered when its spectral lines were seen in the Sun. Physicists affixed a spectrometer to a telescope to study the spectrum of the hot gases of the solar corona when it was visible for several minutes during the total eclipse of 1868. The measured wavelengths were then compared to spectra obtained on laboratory specimens of the known elements. The spectral lines of hydrogen and sodium were easily identified, but the corona also emitted a bright yellow line with a wavelength that did not match up with light from any of the known elements. Physicists concluded that they had discovered a new element, and they named it helium from *helios*, the Greek word for the Sun. Twenty-seven years later, when the same yellow line was observed in the spectrum of gas released from a sample of uranium ore, researchers knew that they had discovered helium on Earth.

A common way to produce atomic spectra in the laboratory is a gas discharge tube like the one shown in Fig. 9.15. In this device electrons are accelerated through a low-pressure gas in a glass container, or "tube." The electrons collide with the gas atoms, transferring some of their kinetic energy to the atoms, which then shed their excess energy by radiating light at the wavelengths of their characteristic emission spectrum.

EXERCISES

23. Light from a gas-filled discharge lamp is analyzed by a spectrometer using a diffraction grating with 10,000 slits per cm. Bright lines are recorded

TABLE 9.3 Wavelengths (in nm) of Some Representative Atomic Spectral Lines—Only the Strongest Emission Lines Are Shown

Hydrogen – H	Helium – He	Neon – Ne
656.28	667.82	626.6
486.13	587.56	621.7
434.05	501.57	610.21
410.17	447.15	585.25

Mercury – Hg	Sodium – Na	Argon – Ar
614.95	588.995	706.72
579.07	589.592	696.54
576.96	568.82	487.99
546.07	498.28	476.49
435.83		442.60
404.66		434.81

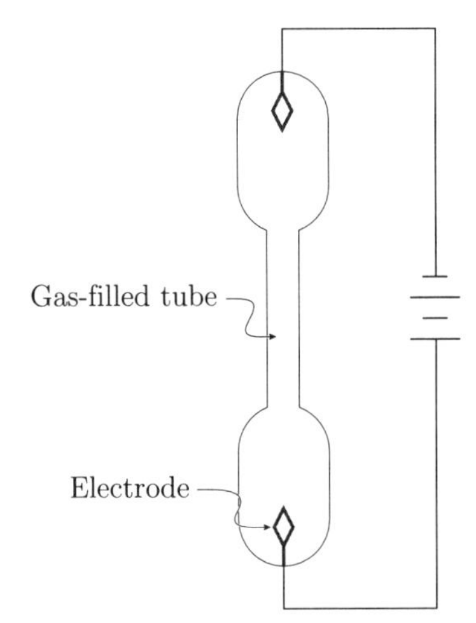

FIGURE 9.15 Discharge tube for producing visible spectral lines.

at the angles $\theta = 26°, 29°, 41°, 62°$, and $75°$. Using Table 9.3, identify the gas and specify the color of each line.

The spectroscopy of visible spectral lines has played a central role in the development of modern physics. When in the early twentieth century physicists sought to explain and understand the bewildering complexity of the spectra of even the simplest atoms, they concluded that some of Newton's basic ideas about motion do not apply to atomic-sized systems and that a more fundamental description of the behavior of matter, now called "quantum theory" or "quantum mechanics," had to be developed. Spectroscopy has been a major tool for carefully testing the predictions of quantum theory and also for stimulating further theoretical advances.

In astronomy all of our information is conveyed by electromagnetic waves, especially visible light. Using spectroscopy, astronomers can measure the temperature and composition of the stars, their rotation rates, and also their motion relative to Earth.

9.6 PROBING MATTER WITH LIGHT

We can also use interference patterns to infer the geometrical structure of the objects that produce them. This makes it possible to learn about the structure of bits of matter far too small to see. To see how to reverse the reasoning and extract information about the number, shape, and size of slits responsible for the interference patterns for configurations of N slits when neither N nor d is known, consider the pattern shown in Fig. 9.16(a). It arises from a double-slit assembly with $d/b = 2.5$. If you know that $\lambda = 550$ nm (yellow), then you can determine d and b without ever examining the slits directly.

The ability to deduce things about a structure from the interference pattern it produces is important, because the structure causing the interference may be out of reach, as are stars, or perhaps too small to be measured with ordinary instruments, as are atoms.[2] Moreover, we are not restricted to collections of slits. Although the mathematics may become involved, we can reconstruct rather complicated structures from the interference patterns they generate. Indeed, this is the way holography and x-ray analysis (to be discussed in Chapter 13) work.

[2]The American physicist A.A. Michelson had a remarkable talent for devising instruments and techniques for precision measurements. For a fascinating and readable account of his exploits see A.A. Michelson, *Light and Its Uses*, University of Chicago Press, 1902.

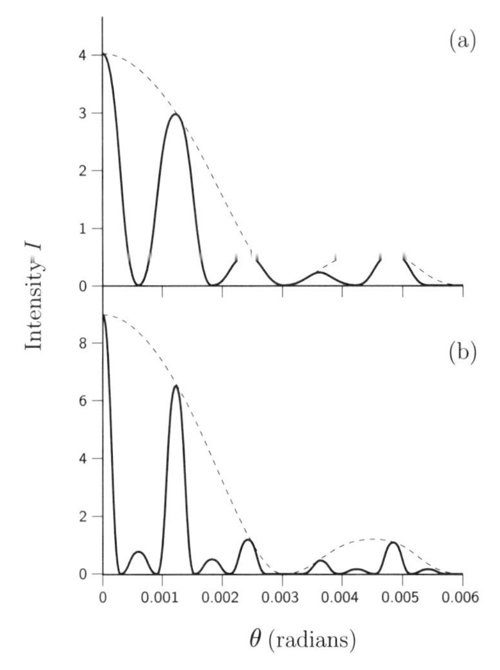

FIGURE 9.16 Interference patterns from two "unknown" slit structures. The dashed-line envelope of the fringes is shown to help you do Exercise 24.

EXAMPLES

3. Consider Fig. 9.16 (a). Why do we say that this pattern is produced by two slits with $d/b = 2.5$?

The clue that it is produced by $N = 2$ slits is that the principal maxima have no secondary maxima between them. Notice also that the principal maxima are nearly equally spaced; indeed, the minima are exactly equally spaced. Notice also that the first diffraction minimum lies on top of the third interference minimum, where $d \sin\theta = 2.5\lambda$ and $b \sin\theta = \lambda$. Since these two minima occur at the same angle, you can divide the first equation by the second, and $\sin\theta$ and λ cancel to leave $d/b = 2.5$.

EXERCISES

24. Given $\lambda = 550\,\text{nm}$, find d and b from Fig. 9.16(a). Describe the slit structure (number, width, and separation) giving rise to the interference pattern in Fig. 9.16(b).

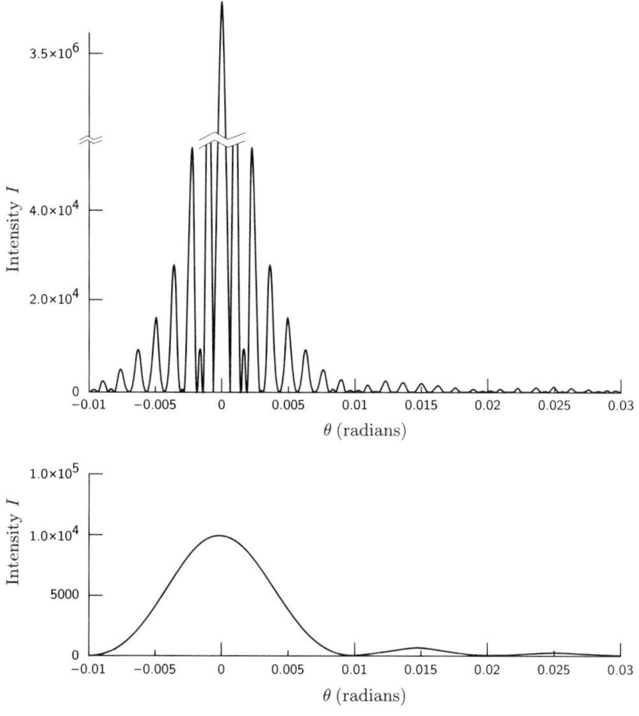

FIGURE 9.17 The lower curve shows the pattern of intensity from a single slit of width 100λ. The upper curve is the pattern of intensity of laser light diffracted around an opaque strip of width 100λ. The laser beam is about 20 times the width of the strip.

Figure 9.17(a) shows the pattern of intensity of light that will appear on a screen after it has been diffracted through a single slit 100 wavelengths wide. Figure 9.17(b) shows the intensity pattern of light diffracted by an opaque strip 100 wavelengths wide when it is illuminated by a laser beam 15 times wider than the strip. Notice that the envelope of the intensity pattern in (b) is very similar to the pattern in (a). In particular, the diffraction minima occur at the same angles in the two cases.

Because of these similarities, it is possible to infer the width of a solid object from its diffraction pattern, just as you can deduce the width of a transparent slit from its diffraction pattern. Figure 9.17 shows that for the simple case of a single slit or strip you can use exactly the same technique: Measure the angles at which the minima occur and then use Eq. 12 to find b.

The similarities between the patterns are not coincidences. The fact that in the right sort of experimental arrangement an obstacle that blocks light forms the same pattern of interference minima as an opening of the same size is related to "Babinet's principle." We mention the principle here only to emphasize that interference can be interpreted as arising from spaces between matter (*e.g.*, slits), or, equivalently, from the matter itself (e.g., the screen material surrounding the slits).[3]

Regardless of the detailed shape of the object (slits, obstacles, lines, circles, whatever), there are fundamental limits on the information available from interference patterns. Consider once again a single-slit diffraction pattern such as that shown in Fig 9.17(a). In order to determine the size of the object, we must be able to locate the positions of the intensity minima or maxima. For minima to be observable they must occur for $\theta < 90^\circ$. Because the positions of the minima are given by $b \sin\theta = n\lambda$ and because $\sin\theta \leq 1$, you must have $\lambda \leq b$ if you are to measure b.

Now consider a double-slit pattern such as that of Fig. 9.12(b). The same reasoning as in the previous paragraph shows that in order to measure d you must have $\lambda \leq d$. In other words, if you are to distinguish structural features of size x or resolve two features separated by a distance x, then you must use radiation with wavelength $\lambda \leq x$. *The tinier the objects you study, the shorter must be the wavelengths you use.* For example, the spacing between atoms in a solid is about 0.2 nm. To measure this spacing you need radiation of comparable wavelength. Electromagnetic radiation of these wavelengths exists but is not visible to the eye; this kind of "light" is called "x-rays."

EXERCISES

25. Suppose the interference pattern of Fig. 9.17(b) was created by a thin wire illuminated by light from a helium–neon laser at a wavelength of 633 nm. Determine the diameter of the wire.

26. Describe the interference pattern that would arise if Isaac Newton had placed a strand of his hair in the beam of a helium–neon laser. If the pattern were projected onto a screen 2 m from the hair, how far from the center of the pattern would the first three minima fall?

9.7 SUMMARY

A wave is a propagating disturbance. It travels with a finite speed, about $340\ \mathrm{m\,s^{-1}}$ for sound waves in air and $3 \times 10^8\ \mathrm{m\,s^{-1}}$ for all electromagnetic waves in vacuum.

[3]The notable differences between the two intensity patterns in Fig. 9.17 arise largely because of the finite size of the illuminating laser beam, but there are other possible sources of differences that become significant when the width of the structure gets close to the size of the wavelength. For example, see R.G. Greenler, J.W. Hable, and P.O. Slane, "Diffraction around a fine wire: How good is the single-slit approximation?" *Am. J. Phys.* **58**, 330–331 (1990).

Sine waves are the building blocks of all waveforms. A sine wave is characterized by its amplitude A, wavelength λ, frequency f (or, equivalently, its period T), and its phase. Displacement y of a sine wave traveling to the right is described by the equation

$$y = A \sin\left(2\pi\frac{x}{\lambda} - 2\pi\frac{t}{T} + \phi\right),$$

where $(2\pi\frac{x}{\lambda} - 2\pi\frac{t}{T} + \phi)$ is the phase, and ϕ is the phase constant.

If the displacement is along the line of travel of the wave, it is a longitudinal wave; if the displacement is perpendicular to the direction of propagation of the wave, it is a transverse wave.

Both the energy carried by a sine wave and its intensity are proportional to the square of the wave's amplitude.

The velocity of propagation of a sinusoidal wave is given by

$$v = \lambda f.$$

Waves interfere with one another. Any phenomenon exhibiting interference must have a wave nature. Since light forms interference patterns, then light must be a wave.

The two most basic interference patterns are single-slit diffraction and double-slit interference. The *minima* of the single-slit diffraction pattern occur at angles θ given by the expression

$$b \sin\theta = n\lambda,$$

where b is the width of the slit and n is any integer $\neq 0$.

The *maxima* of the double-slit interference pattern occur at angles θ given by the expression

$$d \sin\theta = m\,\lambda,$$

where d is the separation between the slits and m is any integer.

An array of N equally spaced slits ($N \geq 2$) has principal maxima satisfying the double-slit equation above. For $N \gg 2$, the array is called a diffraction grating. The integer m is called the "order" of the corresponding intensity maximum.

Diffraction gratings are used to measure accurately and precisely the wavelengths of light emitted from or absorbed by atoms. Each atomic species has a unique spectrum that identifies the atom and provides clues to its internal structure. Visible-light, or optical, spectroscopy has played an extremely important role in the development of modern physics and especially atomic physics.

If λ is known, an interference pattern can be used to study the object that generates it. Interference can be used to determine the sizes and structures of objects over an enormous range, from smaller than atoms to larger than stars. To distinguish structural features of size x with waves requires waves with wavelength λ of the order of x or smaller.

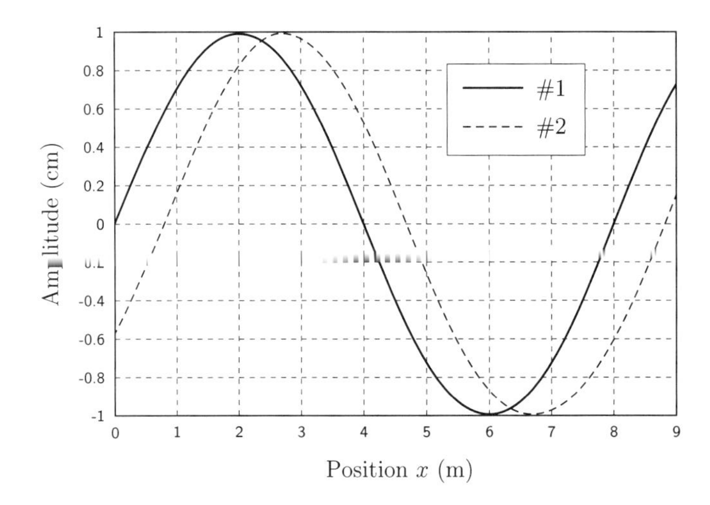

FIGURE 9.18 Two waves at time $t = 0$ (Problem 1).

Light waves are remarkable in that their speed c in a vacuum is constant regardless of how fast an observer is moving toward or away from their source. In Newtonian physics this is impossible, and physics had to be completely restructured to take this property of light into account. The next two chapters will discuss this restructuring, Einstein's special theory of relativity, and some of its surprising consequences.

PROBLEMS

1. The diagram in Fig. 9.18 shows two waves labeled #1 and #2 as they appear at time $t = 0$.

a. Write an equation for wave #1 in terms of its wavelength at $t = 0$.

b. Write an equation for wave #2 at $t = 0$. Be particularly sure to get the phase constant correct.

2. In North America electricity is sent over wires in waves that have a frequency of 60 Hz. If these waves travel over wires at the speed of light, 3×10^8 m/s, what is their wavelength? Compare this to the wavelength of visible light.

3. A TV picture is generated by an electron beam that makes 512 parallel passes across the screen 30 times each second. In the days before cable TV, "ghosts," or secondary images, would appear due to reflections of the signal from nearby buildings. (See Fig. 9.19.) If a ghost appears 1 inch to the right

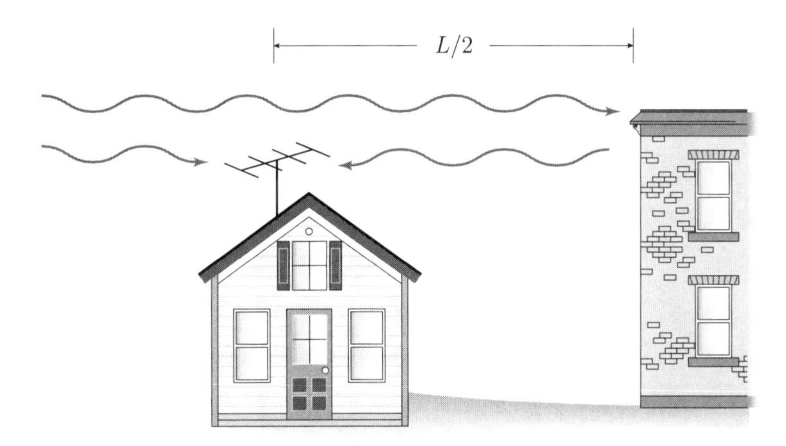

FIGURE 9.19 How reflections produce ghost TV signals (Problem 3).

of the main image on the screen, and the screen is 10 inches wide, what is the extra distance L that the reflected wave travels to reach the TV antenna?

4. Using a traveling microscope a student finds that there are 50 slits per cm in a diffraction grating. She then takes light from a laser of unknown frequency and shines it through the grating onto a wall 6 m distant from the slits. On the wall she finds that the displacement of the first maximum from the center is 1.5 cm.

What is the wavelength of the laser light?

5. **a.** Consider the diagram in Fig. 9.20 showing two waves. If these are snapshots taken at t = 0 s, what is the phase difference between the two waves?

b. What is the wavelength of the above waves?

c. If snapshots taken four seconds apart show that the waves have moved 16 cm during that interval of time, what is the frequency of the waves?

6. Figure 9.21 shows two small loudspeakers in open air, A and B, emitting sinusoidal sound waves of equal amplitude and of frequency 840 Hz. An observer listens to the sound from the speakers while moving along the line OO′. At point M, directly opposite the midpoint of the line AB, she is aware that the sound is the loudest; as she continues toward O′, she notices that the intensity has dropped to zero at point Y. She measures the distances AY and BY to be 3.35 m and 3.15 m, respectively.

a. From this information, find

i. the wavelength of the sound and

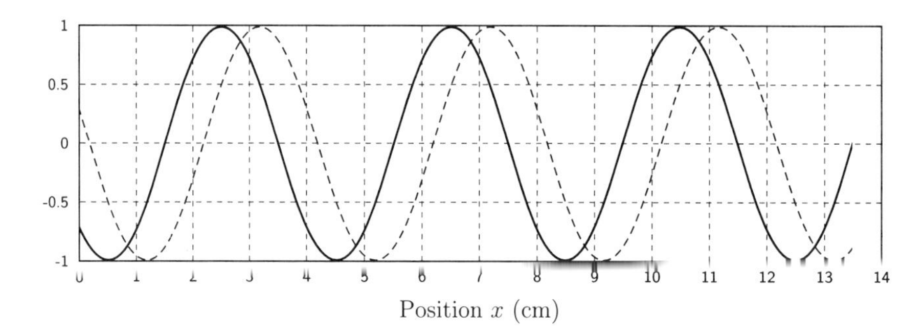

FIGURE 9.20 Two waves identical except for phase (Problem 5).

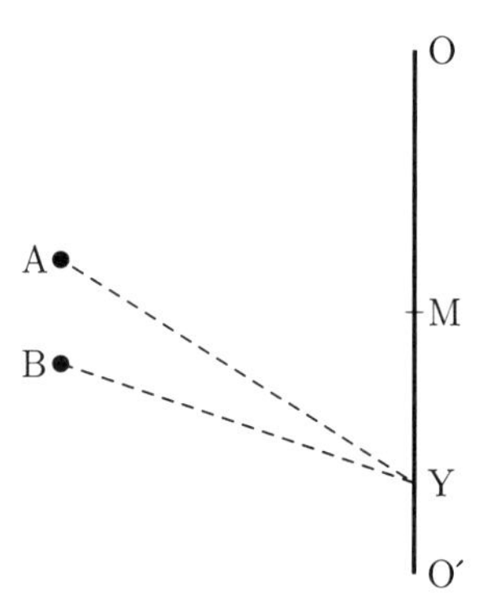

FIGURE 9.21 Speakers and listener for Problem 6.

ii. the speed of sound in air.

b. Suppose the observer returns to point Y and stays there as the frequency of the sound is slowly raised. At what frequency will she hear the sound intensity go through a maximum? Can this happen at other frequencies? Explain clearly.

7. A laser emitting light of wavelength 600 nm illuminates a long, thin wire 20 cm from a screen and parallel to it. The illumination produces an interference pattern as shown in Fig. 9.22.

a. At what angle does the first minimum of the intensity pattern occur? The second minimum?

b. What is the diameter of the wire?

c. If the wire used above were replaced by one having twice the diameter, what would be the new positions on the screen of the first and second intensity minima?

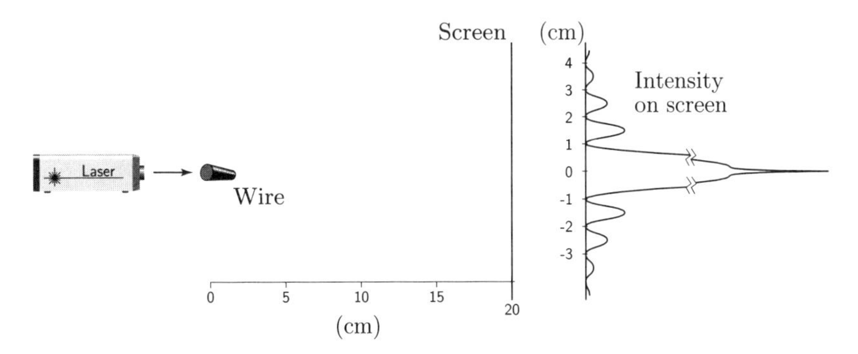

FIGURE 9.22 Intensity pattern produced when a laser illuminates a thin wire (Problem 7).

d. How thin could a wire become and still produce at least one intensity minimum on the screen? What does this imply about using light to measure small objects?

8. Two identical small loudspeakers emit sound waves of wavelength λ, which are in phase. The loudspeakers are separated by a distance of 0.9 m, as shown in Fig. 9.23.

a. If $\lambda = 30\,\text{cm}$, will the interference at point P be constructive or destructive? Explain.

b. What will the frequency of the waves be?

c. If the frequency is halved, what kind of interference occurs at point P? Explain.

9. Suppose that you wish to measure the exact width of a human hair using light. You also know that the diameter of the hair is roughly 0.1 mm. Using the diffraction pattern as a measuring technique, would it be better to use far-infrared radiation ($\lambda = 1000\,\mu\text{m}$) or visible light ($\lambda = 600\,\text{nm}$)? Explain.

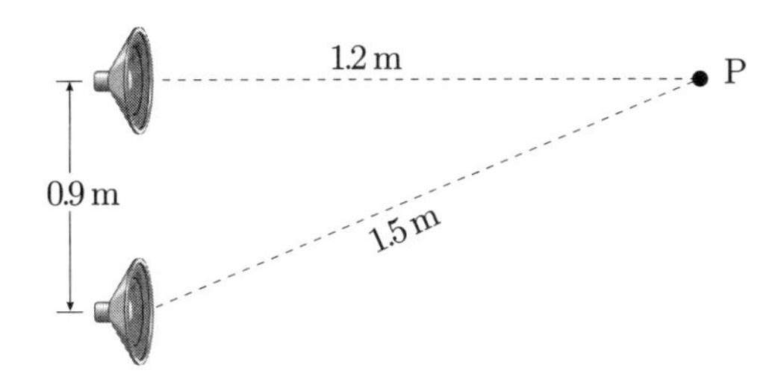

FIGURE 9.23 Two speakers emitting the same wavelengths in phase with each other—(Problem 8).

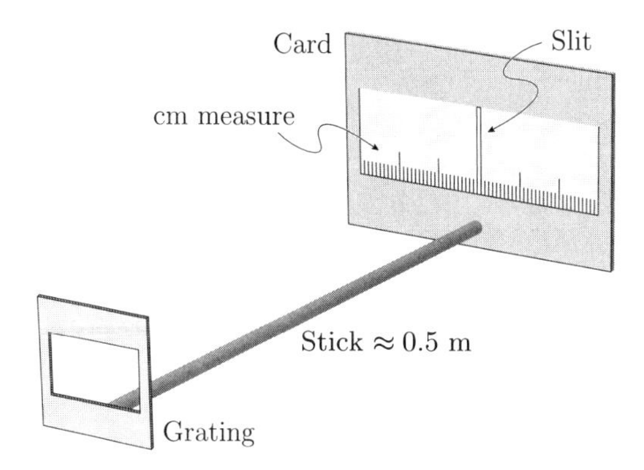

FIGURE 9.24 Basic components of a diffraction spectrometer (Problem 10).

10. It is quite possible to build a useful spectrometer from simple components. Figure 9.24 shows such an instrument. Construct your own spectrometer using an inexpensive replica grating (your instructor should be able to get you one) and Fig. 9.24 as a guide. Use your spectrometer to measure the spectrum emitted by the streetlamps on your campus or in your town. Identify the gas emitting the light. Table 9.3 may be of help.

11. The bright colors given off by aerial fireworks are due to metallic powders mixed with the explosive. The rapid combustion of the explosive heats the metal, causing it to emit radiation at wavelengths included in its emission spectrum. What metals are responsible for the various colors? Search the literature for answers and write a paragraph or two describing the physics of fireworks. Hint: A 1991 issue of *Scientific American* carried a full article devoted to fireworks displays.

SPREADSHEET EXERCISE: ADDING WAVES–INTERFERENCE

In this exercise you will use QuattroPro for Windows or some other spreadsheet to generate two sine waveforms that differ in phase but that have the same wavelength λ and unit amplitudes A. Adding the waves together will show you the effect of a phase difference $\phi_2 - \phi_1$.

You will show the effect of interference by making a couple of graphs of the waveforms and their sum. You will also show that the formula for the sum of two waves

$$B \sin\left(\frac{2\pi x}{\lambda} + \frac{(\phi_1 + \phi_2)}{2}\right),$$

where

$$B = \sqrt{A_1^2 + A_2^2 + 2A_1 A_2 \cos(\phi_2 - \phi_1)},$$

gives the same result as adding the waves together point by point. The value of B is correct for the addition of any two waves; the phase constant $(\phi_1+\phi_2)/2$ is correct only if $A_1 = A_2$.

When you write such formulas in Quattro you may want to save some typing by setting $\phi_1 = 0$ and letting ϕ_2 be the phase difference, but be sure you have the correct sign for ϕ_2.

INSTRUCTIONS

The following step-by-step instructions are for use with QuattroPro for Windows. They are supplied for two purposes. First, they are to help you through what may be unfamiliar uses of Quattro and to show you how to fix up the graphs to be useful. Second, and more important, the instructions show you how to set up a spreadsheet with some parameters that you can vary. If you follow these instructions, the spreadsheet will be set up so that you can change the phase difference by typing in a single number. Such an arrangement is very useful for doing a succession of calculations with different parameters. In this case you will be able to vary the phase difference and show how adding the waves can give constructive, destructive, or partial interference depending on the difference in phase between the two waves.

In what follows, the material on the left side of the arrows is what you should type; the address to the right of the arrow tells you in what cell of the spreadsheet to type it.

Begin by setting up some headings and putting in some initial values for the wavelength and the phases of the waves.

```
^Phase        →A5
^Radians      →B5
^Degrees      →C5
"phi1         →A6
0             →B6
"phi2         →A7
@pi * 1/3     →B7
@degrees(B7)→C7
4             →B8
^x            →A10
```

Then fill up 101 cells with successive values of x. The following instructions do this from 0 to 10 in steps of 0.1:

/BF and enter A11..A111 in the Blocks row,
0 in the Start row,
0.1 in the Step row,
10 in the Stop row.

Now put in some more headings.
^Wave1 →B10
^Wave2 →C10
^Sum →D10
"Theory→E10

Now load some formulas, one for wave1 and another for wave2:
@sin(2*@PI*$A11/$B$8 + $B$6) → B11,
@sin(2*@PI*$A11/$B$8 + $B$7) → C11.

Then put the sum of these two waves into D11
+$B11 + $C11 → D11.

Then copy these formulas into the rest of the cells of the block by typing
/BC
B11..D11 in the From row,
B11..D111 in the To row.

You should now have the two waves in the B and C columns and their sum in the D column. Notice that if you want to change the phase difference, all you need to do is change the entry in cell B7. Try it.

Notice also that this program is set up to have you enter the phase angle in radians into cell B7. When you do that, it automatically calculates the phase angle in degrees in cell C7. Perhaps you would like to do it the other way around.

MAKING A NICE GRAPH

The following instructions will help you to set up the graph with a nice grid and uncluttered lines.
First make your graph by clicking on the Graph Tool icon, which is fourth from the right. The cursor will then turn into a little graph. Move it to the place where you want your graph to be. Click and drag to make a rectangular region for your graph.
Now choose the graph type; you want it to be an *xy* plot:
Put the cursor in your graphing area and click the right button of your mouse

Click on Type in the Graph Properties menu that appears.
Click on the *xy* graph icon, which is the second one in the second row from the top.
Click OK.

Now to tell the graph what data to plot:
Put the cursor in the graph area and right-click the mouse.
Select Series from the Graph Properties menu.
Enter A11..A111 in the *x*-values row.
Enter B11..B111 in the 1st row.
Click OK, and you should have a nice graph of the wave.

Plot the wave in column C in series 2 and the sum of B and C in series 3.

To be quantitatively useful a graph should have a grid. To put one on your graph use the Graph Editor. To get to the Graph Editor type /GE; then select Graph1 and then click OK.

A second box will appear with your graph in it. To change anything use the mouse to put the cursor on the part of the graph you wish to change and then right-click the mouse. For example, to put vertical grid lines on your graph move the cursor to point at the *x*-axis. The words "x1axis" should appear at the lower left-hand corner of the screen. Then right-click, and a menu should appear with the top line reading "x-axis properties."
Click on X-Axis Properties.
Click on Major Grid Style.
Click on Line Style.
Click on one of the choices, say the 2nd from left in the top row.
Click OK.

Do the same for the *y*-axis.

Change the line style of at least one of the three curves on your graph so that it is visually distinct from the others.

TASKS

a. Put some useful titles and identifying labels on your graphs.

b. Print two graphs, one for a phase difference of $\pi/3$ and another for a phase difference of $0.9\,\pi$.

c. Add to the spreadsheet a column that calculates the sum of the waves according to the formula for adding waves of the same amplitude and wavelength. (If you need to take a square root, use the spreadsheet function @SQRT( ).) Print one page of your spreadsheet, the first 35 lines or so, showing that direct addition of the waves and the addition formula give the same result.

d. Show the formula that you used above the column containing the evaluations of the formula.

Hand in the three printouts.

CHAPTER 10

Time and Length at High Speeds

10.1 INTRODUCTION

Electrons and light are two of the most important tools we have for learning about the structure of atoms. In order to use these tools correctly, you need to know how energy and momentum are transferred by objects moving with speeds approaching the speed of light. This means that you need to be able to use some of the ideas of Einstein's special theory of relativity.

The fundamental ideas of the special theory of relativity have to do with the nature of space and time. At high speeds it becomes apparent that these two concepts are interconnected in ways not suspected until Einstein proposed his theory. These connections mean that for objects moving with speeds approaching the speed of light, energy and momentum are related in ways quite different than experience with everyday speeds suggests. Although energy and momentum are the tools we need for studying atoms, the ideas of space and time that come from the special theory of relativity are so interesting in themselves that you should learn a little about them first.

It is a fundamental axiom of the special theory of relativity that the speed of light c is constant for all observers regardless of their relative motion. Equally fundamental is the idea that the laws of physics are the same in all frames of reference moving at constant velocity. This is called the "principle of relativity." In the following sections, after a look at some early experimental evidence for the constancy of c for all observers moving or at rest, we will see how these facts of nature lead us to expect moving clocks to run slow and moving lengths to become shorter. These curious predictions of the special theory of relativity have a surprising implication: *Moving and stationary observers will disagree as to whether two events occurring at different places in space occur at the same time.*

But your main goal should be to get a working understanding of what it means to say $E = mc^2$. You especially need to know how to describe kinetic energy at high speeds and how kinetic energy is connected to momentum. Therefore, the next chapter will deal with these topics and the related fact that mass depends on velocity.

The ideas presented in this chapter mean that Newton's description of matter and motion and time and space is wrong. In other words, most of the physics you have been taught so far is wrong! Yet Newton's physics works very well for the world in which we live, because at low speeds it is the best approximation to the special theory of relativity in a precise mathematical sense; Newtonian physics is the limiting case of relativity for speeds v much smaller than c. Thus Newton's physics is an excellent description of the world in which you live, and you have not wasted your time learning it.

This idea of one formulation of physics as a limiting case of a more general theory is important. You will see this sort of thing happen often as you study more physics. To fully appreciate how one formulation can be obtained as the limiting case of another, you need a mathematical tool that permits you to create simple approximate equations from complicated exact ones. This tool is so important and so useful that we are going to introduce it to you before going any further.

10.2 APPROXIMATING A FUNCTION

In physics we like theories that give us formulas. A formula is a functional relation between some independent variables and a dependent variable. You have run into lots of these: $y = \frac{1}{2}g\,t^2$, $K = \frac{1}{2}m\,v^2$, $2\,a\,y = v^2 - v_0^2$, and so on. The algebraic equation is a particularly convenient representation of a simple function, but there are many other representations. For example, your math and physics teachers are always nagging you to use graphs. In fact, with graphs you can represent many more different, arbitrary functions than with formulas.

Figure 10.1 shows the graph of a smooth, single-valued function $y(t)$ vs. t. Perhaps it describes a mass falling and rising in some weird way that no formula can describe. But formulas are quite convenient, so let's look for a way to *approximate* the behavior shown in the graph with a simple formula.

Straight-Line Approximations

We can do this by looking at such a small part of the curve that the small part is smooth and simple. Looking at the graph in Fig. 10.1, you can see that if you take a small enough part of $y(t)$ centered on some time t_0, the

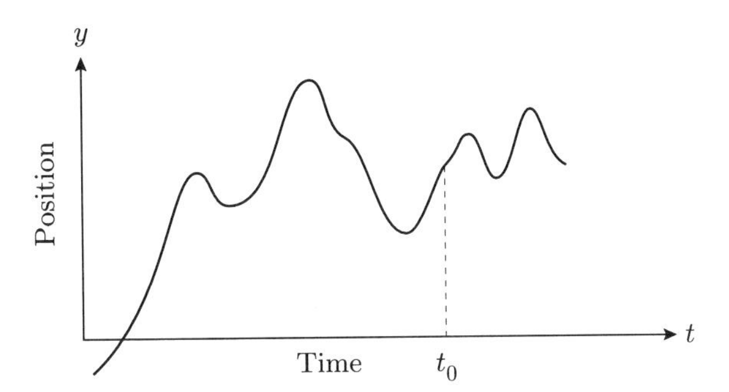

FIGURE 10.1 A graph of position as a function of time.

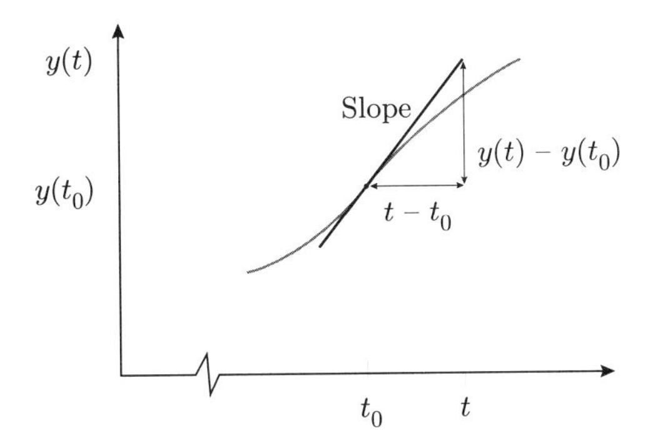

FIGURE 10.2 The best straight-line approximation to a curve near some point $(y(t_0), t_0)$ on the curve is the straight line tangent to the curve at that point.

function will look like a straight line and can be well approximated by a linear function, i.e., the formula for a straight line.

What do you need to know in order to write down the equation of the straight line that best approximates the curve in the graph around t_0? It seems that the best straight-line approximation would be a line tangent to the curve and passing through the point $y(t_0)$. Figure 10.2 shows that if you know the slope m of the curve at $(t_0, y(t_0))$, you can write the equation of the straight line from the definition of the slope,

$$\frac{y(t) - y(t_0)}{t - t_0} = m,$$

and you can rewrite this as

$$y = y(t_0) + m\,(t - t_0). \tag{1}$$

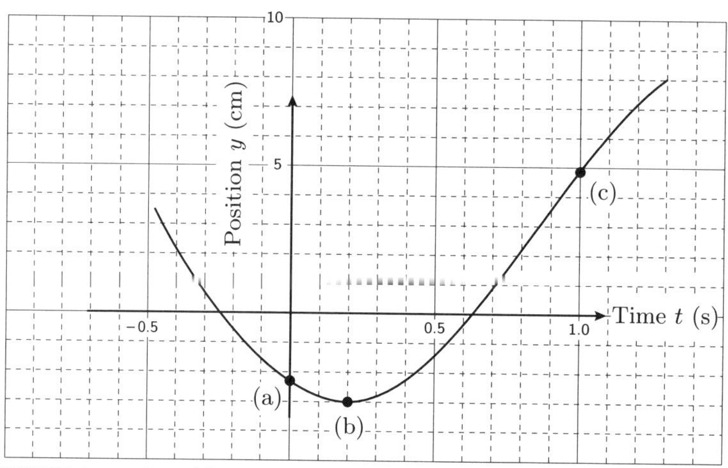

FIGURE 10.3 An arbitrary curve.

EXERCISES

1. Consider the curve in Fig. 10.3. Find equations that give the best straight-line fit to the curve at points (a), (b), and (c).

Because Eq. 1 applies to any curve that has a slope, it also applies to curves for which formulas—perhaps quite complicated ones—exist. If there is a formula, it is easier to find the approximate straight line if you make use of the fact that the slope m of $y(t)$ at the point t_0 is the same thing as the derivative $y'(t)$ evaluated at that point, that is, $m = y'(t_0)$. For then,

$$y(t) \approx y(t_0) + y'(t_0)\,(t - t_0), \tag{2}$$

and all you need to do to find the best straight-line approximation to your function is to evaluate the function y at t_0 and evaluate the first derivative of the function y' at t_0 and substitute into Eq. 2.

EXAMPLES

1. Suppose the function is $y = 5/t^2$. What would be the best straight-line approximation of this function at $t = 5$?

The derivative of $5/t^2$ is $-10/t^3$. The value of this derivative at $t = 5$ is -0.08. The value of the function at this point is $y(5) = 0.2$. Therefore, the equation of the straight-line approximation is

$$y = 0.2 - 0.08\ (t - 5).$$

TABLE 10.1 Comparison of Values of the Exact and Approximate Versions of the Function $5/t^2$.

Time t	Exact	Approx	Percent difference
4	0.3125	0.2800	−10.4
4.4	0.2583	0.2480	−4.0
4.6	0.2363	0.2320	−1.8
4.8	0.2170	0.2160	−0.5
5.	0.2000	0.2000	0
5.2	0.1849	0.1840	−0.5
5.4	0.1715	0.1680	−2.0
5.6	0.1594	0.1520	−4.7

Let's evaluate $y(t)$ and its approximation at some points in the vicinity of $t = 5$. Table 10.1 shows how this kind of approximation works. For any value of t between 4.8 and 5.2 the straight line gives the same answer as the exact formula to within 0.5%. If that is precise enough, you can use the approximate formula.

EXERCISES

2. Use a spreadsheet to evaluate the function of Table 10.1 and its approximation. Then have the spreadsheet plot the two functions on the same graph. Choose your ranges of values and scale your plots to show clearly that the approximation works well around $t = 5$ and not so well for values more distant from 5.

3. Find the linear function that best approximates $5/t^2$ in the vicinity of $t = 10$. Plot both the exact function and the approximate one for values of t around $t = 10$.

Binomial Expansions

Functions of the form $y = (1 + x)^n$ are common, and they deserve particular attention. Because the argument of the function has two terms, 1 and x, it is called a "binomial function." We often want to approximate a binomial function around $x = 0$, so let's look at linear approximations for this case only.

To apply Eq. 2, first evaluate $y(0)$. For binomial functions of this form and for any value of the exponent n, $y(0) = 1$. Next evaluate $y'(0)$. The derivative of the binomial function is

$$\frac{dy}{dx} = f'(x) = n(1 + x)^{n-1},$$

and when you evaluate this at $x = 0$ you get $f'(0) = n$ for any value of n. This means that *around the origin* any binomial function of the form $y = (1 + x)^n$ can be approximated by the straight-line equation

$$y = 1 + nx \tag{3}$$

for any value, integer or fractional, of n. The advantage of remembering Eq. 3 is that you then never have to do any differentiation. All you need to do is look at the exponent of the binomial. Another advantage is that it is easy to show that the approximation is quite good as long as x is much smaller than 1. If $x \leq 0.1$, the approximation will be accurate enough for most of our purposes. Equation 3 is an example of the first step of what is called the "binomial expansion."

EXAMPLES

2. Find a linear approximation to the function $y = \sqrt{1 + x}$ at the origin. To use Eq. 3, recognize that $y = \sqrt{1 + x}$ is the same thing as $y = (1 + x)^{1/2}$ so that $n = \frac{1}{2}$, and the expansion is

$$y = 1 + \frac{x}{2}.$$

EXERCISES

4. Find the linear equation that best approximates the function

$$y = \frac{1}{\sqrt{1 + x}}$$

near the origin.

To use Eq. 3 instead of Eq. 2 you may need to manipulate a given binomial into the form $(1+x)^n$. For example, suppose the binomial were $y = \sqrt{4 + 8x}$. If you realize that you can factor 4 from under the radical, then you can rewrite it as $y = 2\sqrt{1+2x}$. Next imagine the substitution $z = 2x$, so that you get $y = 2\sqrt{1+z}$. Then apply Eq. 3 to get

$$y = 2\left(1 + \frac{1}{2}z\right) = 2 + 2x.$$

With practice you can do a lot of this in your head.

Fairly often, you may want to find an approximate function that is linear but not in a linear variable. To see what this double talk means, look at the following.

EXAMPLES

3. Suppose you have a function

$$K(p) = \sqrt{m^2 + p^2} - m,$$

where m is constant and p is the independent variable. Find an approximate formula for K when p is near 0. There is no variable that is linear in this equation, but think of p^2 as a single entity; think of it as the "x" in Eq. 3.

Then factor m^2 out from under the radical to get

$$K(p) = m\left(\sqrt{1 + \frac{p^2}{m^2}} - 1\right) \approx m\left(1 + \frac{1}{2}\frac{p^2}{m^2} - 1\right) = \frac{p^2}{2m}.$$

Not only does the above example show you an important trick for finding an approximate formula, it also illustrates one of those cases where the approximate formula is a lot simpler than the exact one.

Amaze Your Friends!

You can use the binomial expansion to do mental arithmetic. Suppose you wish to know the square root of 1.03. You now know that this is the same as $(1+0.03)^{1/2}$, which has a binomial expansion of $1 + 0.03/2 = 1.015$. When you compare this to the 1.0149 you get with your calculator, you see that the approximation is very good.

EXERCISES

5. What is the cube root of 1.05? Do it in your head.

6. What is the square root of 0.96? Do it in your head.

7. Use the binomial expansion to find the square root of 4.8. Do it in your head.

8. Evaluate the cube root of 0.00103. Do it in your head.

The Small-Angle Approximation

EXERCISES

9. Draw a graph of $y = \sin\theta$ for $-\pi/4 < \theta < \pi/4$. From the graph obtain the equation of the straight line that best approximates the graph in the vicinity of $\theta = 0$. Do your work in radians.

You have been using the answer to the above problem ever since you first took physics, but now by applying the ideas of the previous sections you can understand where the "small-angle approximation" comes from. Let's work out an algebraic answer to Exercise 10.9.

EXAMPLES

4. Equation 2 says that the best straight-line approximation around a point x_0 is determined by the value of the function at the point around which you are making the approximation, $y(x_0)$, and the slope of the function at that point, $y'(x_0)$. To apply Eq. 2 to $\sin\theta$ in the vicinity of $\theta_0 = 0$ you make the following changes,

$$\begin{aligned} y(x_0) &\longrightarrow y(\theta_0) = \sin 0 = 0, \\ y'(x_0) &\longrightarrow y'(\theta_0) = \cos 0 = 1, \\ x - x_0 &\longrightarrow \theta - \theta_0 = \theta, \end{aligned}$$

and find that the best linear approximation to $\sin\theta$ around $\theta = 0$ is

$$\sin\theta \approx \theta,$$

which is just the small-angle approximation.

EXERCISES

10. Find the best straight-line approximation to the function $y = \tan\theta$ around the point $\theta = 0$.

11. Find the best straight-line approximation to the function $y = a\,e^x$ around the point $x = 0$.

12. Find the best straight-line approximation to the function $y = \ln(1+x)$ around the point $x = 0$.

In discussing and listening to physics you will often need to know the linear approximations to $\sin\theta$, $\tan\theta$, e^x, $\ln(1 + x)$, and $(1 + x)^n$, so just learn them. In this and the next chapter you will see how the approximation $(1 + x)^n = 1 + n\,x + \cdots$ shows that at low speeds Newtonian mechanics is the best approximation to the exactly correct special theory of relativity.

10.3 FRAME OF REFERENCE

The idea of a "frame of reference" is central to any discussion of motion. When we speak of an object's frame of reference, we mean the object and the collection of all things that are at rest relative to the object. Thus, assuming that you are not doing anything very weird as you read this, your frame of reference right now probably includes yourself, your table, chair, lamp, room and its surrounding buildings and landscape.

Right away it should be apparent that different objects can have different frames of reference. For example, if you are reading this while riding on a bus, then you, the vehicle, its seats, windows, aisle, and driver make up a reference frame. But someone standing on the roadside as you drive past clearly has a different reference frame, as do people in cars moving past your bus in any direction.

In important ways the appearance of the world depends on your choice of reference frame. If you are riding in a bus, the trees and houses move past you. If you are standing under a tree at the side of the road, you see the bus move by. If you are driving along in a car, you might see the bus approach from one direction and pass you while the tree approaches and passes you from the other direction. If you think about it, you can see that the velocities you will measure depend on what frame of reference you choose.

Velocity Depends on Reference Frame

For everyday events we so often choose the Earth as the frame of reference that we usually forget that a choice was made. If someone told you that the bus is going 60 mph, you probably would not ask what reference frame was being used. And you might chuckle if someone said that a telephone pole leaped at his car and dented the bumper. The statement is wry and ironic, but in the reference frame of the driver of the car it is an accurate description of what happened.

Clearly, the description of motion depends on what frame of reference is used. Imagine that you are standing out on the highway as a bus drives by going east at 60 mph and overtakes a rusty pickup truck going east at 30 mph. There are three different reference frames involved here.

EXERCISES

13. What reference frame has been used to describe this situation?

14. Describe the velocities using the reference frame of the bus. In the reference frame of the bus, you are passing by at 60 mph toward the west, and the pickup is traveling westward at 30 mph.

15. What do the motions look like in the reference frame of the pickup truck?

16. Simplicio says that quantity of motion, mv, is an intrinsic property of a body like its color. What do you tell him? Give some numerical examples to illustrate your argument.

Does Physics Depend on Reference Frame?

A major historical achievement of physics has been the ability to describe motions of bodies as the result of forces. But if these motions depend on the reference frame in which they are described, does this mean that the laws of physics are different in different reference frames?

We need to be careful about what we mean here. Clearly, the numerical values of physical quantities depend on the reference frame. If you take a 6 kg bowling ball at rest and set it rolling at 3 m/s down a bowling alley, it goes from having a kinetic energy of 0 J to having a kinetic energy of 27 J in your frame of reference. But if someone is traveling along a conveyor belt at 1 m/s beside the alley, then in her frame of reference the initial kinetic energy is first 3 J, and then 12 J. In the first case the kinetic energy changes by 27 J; in the second case the change is 9 J, quite different numbers.

But a law of physics is not the same thing as the number it predicts in given circumstances. In this case it might be relevant that a constant force F applied over some distance s changes the kinetic energy of a mass m by an amount equal to Fs. If these quantities are different in different frames of reference but differ in just the right way to give the numbers that are correct for the different reference frames, then the law would be the same in different reference frames even if it gave different numbers.

Suppose the bowling ball was thrown in 0.5 s with a constant force of 36 N, so that the force was applied over a distance of 0.75 m in the reference frame of the bowling alley. Then in that frame of reference the change in kinetic energy would be $36 \times 0.75 = 27$ J. Now suppose that in the reference frame of the conveyor belt, the force and the time of throw are the same. Because the belt is moving away from the thrower at 1 m/s, the force is applied over a distance shorter by $0.5 \text{ s} \times 1 \text{ m/s} = 0.5$ m. Therefore, in this frame of reference the force is applied over a distance of only 0.25 m and so produces a change in kinetic energy of $36 \times 0.25 = 9$ J. Thus, the law of physics $Fs = \Delta$K.E., the change in kinetic energy, is the same in the two different frames of reference although it gives numerical results that are different but correct in each reference frame.

This idea that laws of physics are the same in different reference frames has led physicists to propound "the principle of relativity." The principle is most easily understood if we use only frames of reference that are in uniform, straight-line motion relative to one another. The car, the bus, and you the observer discussed above are a good approximation of three such frames. (They are not exactly the right choice, because they are all rotating around Earth's axis while going around the Sun in an ellipse.) Then experience leads us to say that the laws of physics are the same in all such frames of reference. This is the "principle of relativity."

You can see right away that if the laws of physics are identically the same in all uniformly moving frames of reference, then there is no experimental basis for thinking that any one frame is the "right" or "special" one. You choose a particular reference frame for describing some set of motions not because the physics will be correct in one and wrong in another, but because the chosen frame is convenient—maybe it makes calculations easier; maybe it makes interrelationships more evident.

The principle of relativity means that for physicists in enclosed laboratories moving uniformly relative to each other there is no experiment that can be done that will tell who is moving and who is not. *There is no special frame of reference*, no place in the universe that is absolutely at rest relative to everything else. Motion is a property that depends only on your choice of reference frame.

How Motion Described in One Frame is Described in Another

Given a description of motion in one reference frame, we often need to find its description in another. You have already transformed one description to another when you thought about the bus, the car, and an observer (you). In the observer's frame of reference, let's call it the S frame, the bus was traveling

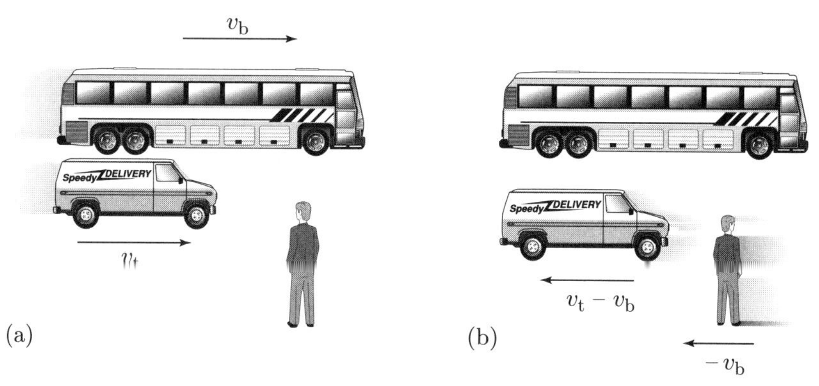

FIGURE 10.4 (a) Motion of bus and truck seen from observer's frame of reference S; (b) motion of observer and truck seen from bus's frame of reference S′.

with a speed of $v_b = 60$ mph, the truck with $v_t = 30$ mph, and the observer was at rest, $v_o = 0$ mph. However, in the bus's frame of reference, let's call it S′, the velocities were $v'_b = 0$ mph, $v'_t = -30$ mph, and $v'_o = -60$ mph. You probably made the calculations without much thought about what you were doing. Now let's codify what you did.

First, notice that relative to the S frame, from which you are transforming, the S′ frame, to which you are transforming, has a velocity $V = 60$ mph. (We're using + for eastward and – for westward motion.) Then, to transform the velocities of objects in the S frame to their velocities in the S′ frame, you do the following calculations:

$$v'_b = v_b - V,$$
$$v'_t = v_t - V,$$
$$v'_o = v_o - V.$$

The general rule to go from the S frame to the S′ frame is to subtract the velocity of S′ relative to S from the velocity of the object in S, i.e.,

$$v' = v - V. \tag{4}$$

Equation 4 is called a "transformation," because it transforms the velocity from the S frame to the S′ frame.

EXERCISES

17. Use Eq. 4 to find the velocities of the bus, truck, and observer in the frame of reference of the truck.

The most astonishing feature of Eq. 4 is that it is **WRONG**. It is an excellent approximation as long as all the velocities are small compared to the speed of light $c = 3 \times 10^8$ m/s, but *it is never exact.*

10.4 THE CONSTANCY OF c

The incorrectness of Eq. 4 follows from a most surprising fact: *The speed of light is the same in all reference frames regardless of their relative motion.*

You can see at once that this fact contradicts Eq. 4. Imagine a parked car that turns on its headlights. The light will come past you at a speed c. Now imagine that the car is moving toward you at 120 mph and it turns on its lights. Equation 4 predicts that the speed of the light going past will be greater by 120 mph. However, no change in the speed is observed experimentally, and so Eq. 4 must be wrong.

Albert Einstein has written that when he was twelve years old he wondered what light would be like for someone moving beside it at the speed of light. From early in the 1800s, when interference phenomena showed that light is a wave, physicists thought that light must be a disturbance in some medium and that light passed from one point to another like sound waves through a solid. As they learned more about electromagnetism, physicists deduced more properties of the medium and gave it a name. The hypothetical medium that carried light waves was called the "ether." If there were an ether and it carried light waves, then Eq. 4 would be correct. For a number of reasons, particularly for logical consistency, Einstein concluded there was no ether and that c would be the same for all observers independent of their relative motion.

In this conclusion he was supported by the evidence of a remarkable experiment performed by the American physicists Albert A. Michelson and Edward W. Morley.

The Michelson–Morley Experiment

Although physicists thought that there must be some medium to support the propagation of electromagnetic waves, there was no experimental evidence for the ether. Michelson set out to measure its presence using the Earth's motion around the Sun to provide a relatively large velocity, 30 km/s. He wanted the largest possible velocity relative to the velocity of light in order to have detectable consequences. If there were an ether through which light traveled like a boat sailing on water, then a beam of light traveling in the direction of the motion of Earth through the ether would take less time to complete a round trip than a second beam traveling the same distance but at right angles to Earth's motion.

Notice that even using the high speed of Earth's motion in space, the expected effect was likely to be small. The Earth's speed is only $3 \times 10^4/3 \times 10^8 = 10^{-4}$ of the speed of light, and, as we shall shortly see, the effect he could hope to measure was proportional to the square of this number, i.e., to 10^{-8}. He needed measurement techniques of great sensitivity to have any hope of obtaining a significant result.

In his work with interference of light waves, Michelson developed a device that he named the "interferometer," and he used it to measure lengths to an accuracy of *a few tens of nanometers*. The interferometer split a light beam into two beams and sent them on round trips along two paths at right angles to each other and recombined them at the end of their trips. The two paths at right angles were exactly the right arrangement for testing the existence of the ether. Moreover, because knowledge of a length to within 10 nm corresponds to knowing the travel time for light to within $10^{-8}/3 \times 10^8 = 3 \times 10^{-17}$ s, he realized that an interferometer could measure the very small difference in travel times along the two paths that was to be expected for light traveling in the ether.

Figure 10.5 shows a diagram of the interferometer devised by Michelson. Its essential features are the two legs of equal length L at right angles to each other, the partially silvered mirror (G_1) that splits the incoming beam of light between the two legs, and the fully reflecting mirrors (M_1 and M_2) at the ends of the legs. Light from a source to the left that is very nearly a pure single wavelength is made into a nearly parallel beam and passed into the interferometer. Two beams are formed when the first mirror reflects half of the light into one arm and transmits half of the light into the other. The

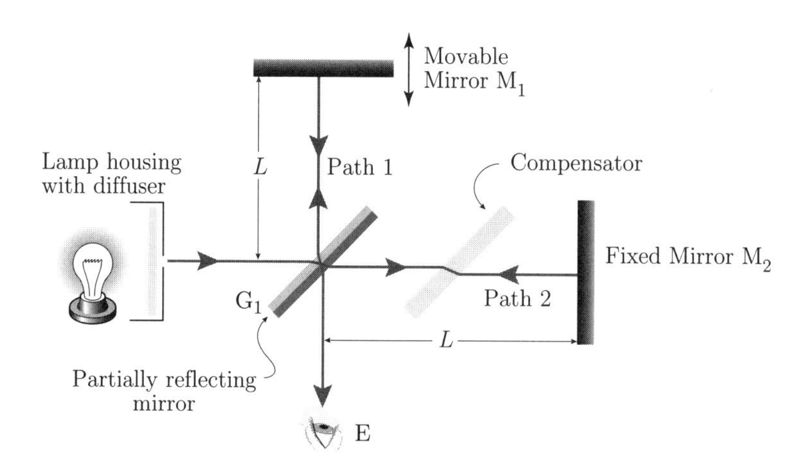

FIGURE 10.5 Schematic diagram of a Michelson interferometer.

two beams then go along their respective arms separately and are reflected by mirrors M_1 and M_2. When the beams arrive back at G_1, half of each is reflected and half is transmitted. Parts of the two beams recombine and travel toward the observer E.

In an ideal case the two beams would travel exactly the same distance and arrive at G_1 exactly parallel. Under these circumstances the two outgoing beams would be π radians out of phase over the whole field of view, and a viewer would see a uniform darkness.[1] The ideal situation is hard to achieve, and usually phase difference varies a number of times from constructive interference to destructive interference across the field of view. This shows up as a pattern of dark and light rings or bands called "fringes," some examples of which are shown in Fig. 10.6.

Now suppose the return of one beam were delayed by a small fraction of a period of the wave. Then the delayed beam would be in phase with the other at slightly different places, and the positions of cancellations would also shift slightly. The whole fringe pattern would thus shift as a result of the delay and would shift in a continuous fashion as the delay increased. When the delay is exactly one period, the original phase relations would again hold, and the pattern would have shifted by exactly one fringe. Figure 10.6 shows fringe patterns that occur for different arrangements of the interferometer mirrors.

EXERCISES

18. What is the smallest difference in the path lengths of light along the two arms of the interferometer that will cause the fringes to shift by exactly one fringe?

According to Newtonian physics, Earth's motion through the ether should produce detectable results in an interferometer. To see why, suppose that the interferometer's arm #1 is in the direction of Earth's motion. Then, in the outward path, the ether, in which light is traveling at speed c, is flowing toward Earth at speed v. As a result, according to Eq. 4, the speed of light relative to the interferometer should be $c - v$. Coming back, the light is moving with the stream, and its velocity relative to the interferometer should

[1] You might think that the two waves should be exactly *in* phase, but a light wave traveling through air undergoes a phase change of about π radians when it reflects from a glass (or metal) surface; but undergoes very little phase change when it is traveling through glass and reflects from a glass (or metal) surface. In the Michelson interferometer one wave makes two reflections at air–glass interfaces, and the other wave makes one. As a result, the two waves arrive essentially π radians out of phase when the two arms of the interferometer are exactly equal.

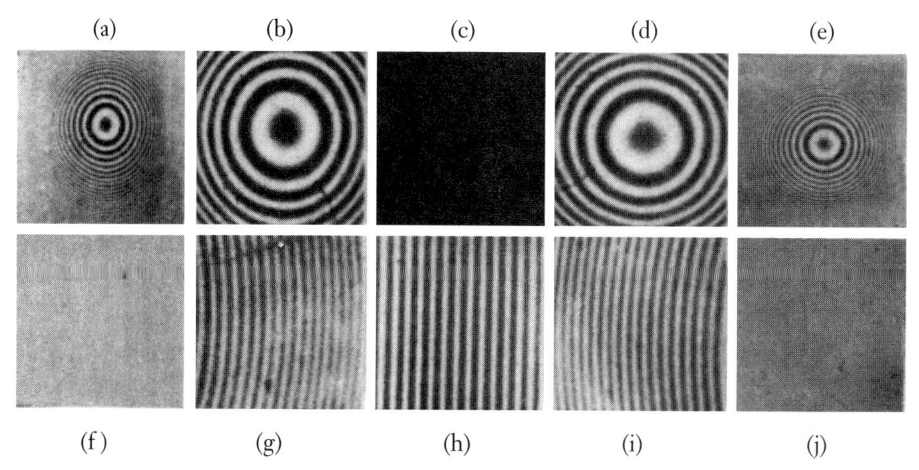

FIGURE 10.6 The white represents fringes; the dark is the space between them. (a) Fringes when the distance to M_1 is somewhat less than the distance to M_2; (b) when the distances are nearly the same; (c) when the distances are exactly the same; (d) when the distance to M_1 is a bit more than the distance to M_2; (e) when the distance to M_1 is somewhat more than the distance to M_2; (f)–(j) are for the same mirror separations respectively as (a)–(e), except that now the mirrors are not parallel but slightly tilted relative to one another. *Taken with permission of the McGraw-Hill Companies from F. A. Jenkins and H. E. White,* Fundamentals of Optics *2nd edition, McGraw-Hill, 1950.*

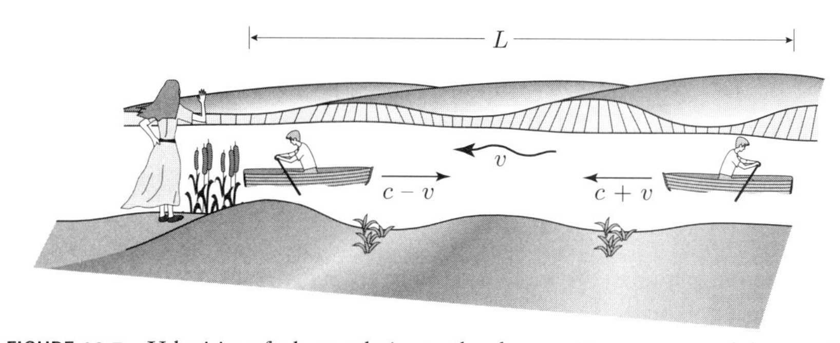

FIGURE 10.7 Velocities of a boat relative to the shore as it moves up and down a river.

be $c + v$. These results are identical to the problem of rowing up and down a river, where the boat is the light, the river is the ether, and the river banks are the interferometer.

Suppose as in Fig. 10.7 a rower rows with a velocity c with respect to the water. If the water is flowing at a velocity v, the boat will have a velocity $(c - v)$ with respect to the shore when rowing up stream, and a velocity $(c + v)$ when rowing downstream. This is just an application of Eq. 4.

The time t_1 to travel a distance downstream is $L/(c + v)$. The time t_2 to travel that distance upstream is $L/(c - v)$. Combining the two times to get the total time t gives

$$
\begin{aligned}
t = t_1 + t_2 &= \frac{L}{c+v} + \frac{L}{c-v}, \\
&= L\left(\frac{c+v\,+\,c-v}{(c+v)(c-v)}\right) \\
&= \frac{2Lc}{c^2-v^2} = \frac{2L}{c}\left(\frac{c^2}{c^2-v^2}\right) \\
&= \frac{2L}{c}\left(\frac{1}{1-\frac{v^2}{c^2}}\right).
\end{aligned}
\tag{5}
$$

The last step of algebraic manipulation has been performed to get a form more convenient for later comparisons.

Now consider motion across the path of the ether stream. This process works out to be the same as another boat rowing across a river and back, as in Fig. 10.8. We suppose that the river is a distance L wide. But now, for the rower to reach a point directly across from the starting point and to return to it, she must point the boat upstream, both going and coming back. Then the path taken relative to the shore will be a straight line back and forth, but relative to the water the path will form an isosceles triangle with an altitude of L and base vt^*, where t^* is the roundtrip travel time. To find this time t^* you can use the facts that the time to cross the river equals the time to come back and that the speed relative to the water is c. Then from the Pythagorean theorem the distance traveled at c (i.e., relative to the water) is

$$
ct^* = 2\sqrt{L^2 + \left(v\frac{t^*}{2}\right)^2},
$$

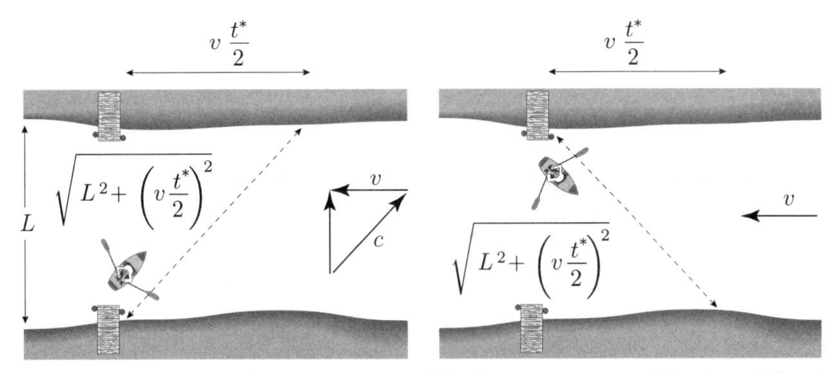

FIGURE 10.8 Diagrams for motion over and back across a stream. The dotted lines indicate the path of the boat as seen by an observer moving along with the water.

which you can solve to find t^*. (For clarity, angles in the diagram have been exaggerated. With $v \ll c$, the deviation from perpendicular to the shore would be very small.)

EXERCISES

19. Show that

$$t^* = \frac{2L}{c}\frac{1}{\sqrt{1-(\frac{v}{c})^2}}. \tag{6}$$

Comparing these results for the two types of trips shows a difference in the return times. Two boats starting out together would *not* get back at the same time. If light travels at a speed c in some ether, the same would be true for light beams traveling through an ether stream.

To predict what would happen in an interferometer, we need to calculate by how much the fringes would shift because of the difference in travel times along the two paths to mirrors M_1 and M_2. Taking c to be the velocity of light in the ether and v the velocity of Earth through the ether, you can show that the difference in travel times Δt will be

$$\Delta t = t - t^* = \frac{2L}{c}\left\{\frac{1}{1-\frac{v^2}{c^2}} - \frac{1}{\sqrt{1-\frac{v^2}{c^2}}}\right\}.$$

In this form it is hard to see what the result is telling you about Δt. Now the binomial approximation becomes useful to obtain an approximate equation for the time delay:

$$\begin{aligned}\Delta t &= \frac{2L}{c}\left\{\left[1-\left(\frac{v}{c}\right)^2\right]^{-1} - \left[1-\left(\frac{v}{c}\right)^2\right]^{-1/2}\right\}\\ &\approx \frac{2L}{c}\left\{1+\left(\frac{v}{c}\right)^2 - 1 - \frac{1}{2}\left(\frac{v}{c}\right)^2 + \cdots\right\}\\ &\approx \frac{2L}{c}\frac{1}{2}\left(\frac{v}{c}\right)^2.\end{aligned} \tag{7}$$

The quantity $2L/c$ is the roundtrip time for light to travel through the interferometer when there is *no* motion relative to the ether. In an early version of Michelson's apparatus, L was 1 m and v was the orbital speed of Earth,

30×10^3 m/s. Thus

$$\Delta t = \frac{2}{3 \times 10^8} \frac{1}{2} \left(\frac{3 \times 10^4}{3 \times 10^8} \right)^2 = 0.33 \times 10^{-16} \text{ s.}$$

Michelson and Morley could hope to measure such a small time interval because the period of oscillation of a light wave is so short. The period T of light of wavelength $\lambda = 5 \times 10^{-7}$ m (green light) is

$$T = \frac{\lambda}{c} = 1.7 \times 10^{-15} \text{ s.}$$

Therefore, $\Delta t/T = 0.02$, or 1/50 of a period. As a result, the interferometer's pattern of fringes would be shifted by 1/50 of a fringe by the effect of the ether.

Of course, there was no way to observe where the fringes would have been if they had not been shifted by motion relative to the ether, so there was no way to observe a shift with a single measurement. To produce an observable effect Michelson measured the fringe pattern and then rotated the entire apparatus through 90°. This reversed the roles of the two arms and should have caused the fringe pattern to shift by $2\Delta t/T$, i.e., by about 1/25 of a fringe.

While this calculation did not predict a very impressive effect, it was sufficient to make a trial worthwhile. Michelson had optics good enough to detect a shift of about 0.01 fringe, so he knew he could reliably detect a shift of $1/25 = 0.04$ fringe.

EXERCISES

20. Suppose by an improved design Michelson could lengthen the arms of the interferometer from 1 m to 10 m. What then would be the predicted shift in fringes when the interferometer was rotated through 90°?

Michelson's Results

Michelson's first trials did not produce any detectable fringe shift, even though he was convinced that he should have been able to see the expected shift. Therefore, he and Morley built a larger interferometer, shown in Fig. 10.9, with mirrors arranged so that the path of the light in each leg could be lengthened by reflecting back and forth several times before the beams combined. To permit easy rotation without mechanical distortion, they mounted the interferometer on a large stone base and floated it in mercury. With the path length increased to about 11 m, the total fringe shift

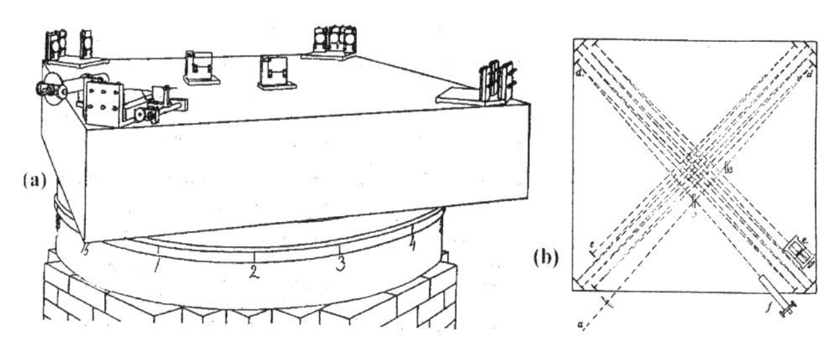

FIGURE 10.9 (a) The interferometer used by Michelson and Morley in 1886. The apparatus sits on a stone base floating in mercury. This makes it easy to rotate the interferometer through 90°. (b) A diagram of how the pathlength L was lengthened by multiple reflections. *Taken from A.A. Michelson and E.W. Morely,* Am. J. Sci. *Vol. 34 No. 203, 333–345 (1887) as reproduced in* Selected Papers of Great American Physicists, *S.R. Weart, editor, published by the American Institute of Physics, ©1976 The American Physical Society*

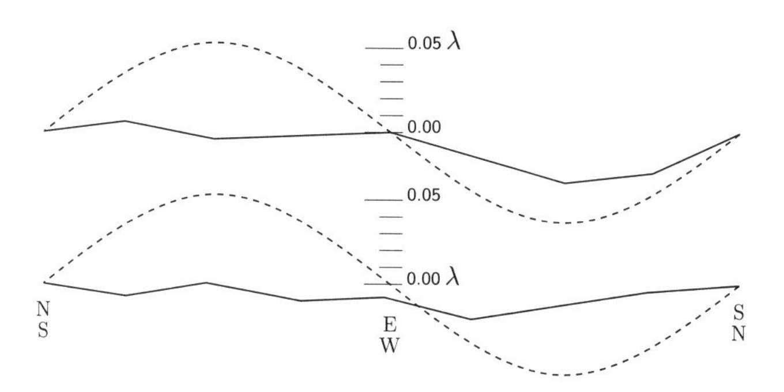

FIGURE 10.10 The solid lines are the observed fringe shifts vs. compass heading of the interferometer. The dotted curves are $\frac{1}{8}$ of the expected fringe shifts. *Taken from A.A. Michelson and E.W. Morley,* Am. J. Sci. *Vol. 34 No. 203, 333–345 (1887) as reproduced in* Selected Papers of Great American Physicists, *S.R. Weart, editor, published by the American Institute of Physics, ©1976 The American Physical Society*

should have been 0.4 of a fringe. The experiment was carried out in 1886. Figure 10.10 shows their published results; there still was no fringe shift!

"Now what?" he probably said to himself. Well, just possibly the motion of the Sun through the Galaxy was effectively canceling Earth's motion, producing zero relative velocity. If so, then waiting half a year until Earth was on the other side of the Sun would double the effect. But no luck there either. Many other variations were tried, including checks to be sure that the arms did not change length by thermal expansion. In all these cases the result of the experiment was still a "null." That is, nothing was observed. The null result could not have been an accidental failure of the experiment. The only conclusion was that motion through the ether could not be observed.

Was there no relative motion? Was it present but for some reason unobservable? Einstein's theory explained the result. There are no special reference frames, he said. Ether, if it existed, would be the only reference frame in which the speed of light was c, and so it would be quite different from all other frames of reference. Therefore, there is no ether. His theory asserts, consistent with the observations of Michelson and Morley, that the speed of light is the same for all observers regardless of their relative motion. Michelson and Morley observed no fringe shift because with c the same in all frames of reference, the speed of light was the same along each arm of the interferometer, and Δt had to be zero.

10.5 CONSEQUENCES OF CONSTANCY OF c

Einstein realized that the constancy of c and the principle of relativity could both hold only if space and time were interlinked in ways quite strange to Newtonian physics. He worked out a complete and consistent theory that accurately describes the motions and interactions of matter at high velocities as well as at low.

Moving Clocks Run Slow—Time Dilation

Einstein's theory predicts that a moving clock runs slower than an identical stationary clock. To see that this must be so, consider the "light clock" shown

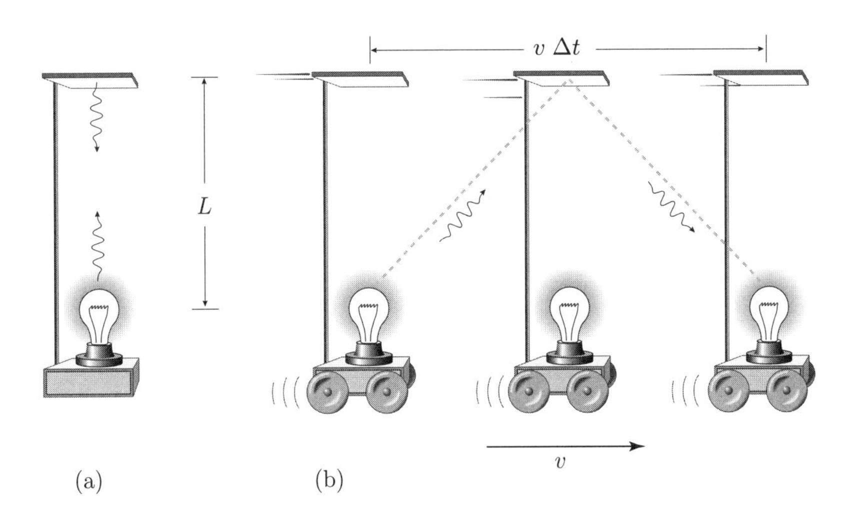

FIGURE 10.11 A light clock that keeps time by counting the number of back-and-forth trips a pulse of light makes between two mirrors separated by a length L. (a) A stationary light clock; (b) a light clock moving with velocity v.

in Fig. 10.11. This imaginary device works by the emission of a pulse of light from near the bottom mirror. The pulse travels to the upper mirror, is reflected, and returns to the bottom one. A "tick" Δt of the clock is the time it takes to make a roundtrip:

$$\Delta t = \frac{2L}{c}.$$

Now imagine an identical clock moving with velocity v parallel to the x-axis, as shown in Fig. 10.11(b). Because of the motion of the clock, a stationary observer sees the light pulse in the moving clock travel over a longer path than the pulse in the stationary clock. Therefore, in the stationary observer's frame a "tick" Δt of the moving clock takes longer than the "tick" of the identical stationary clock. *The moving clock runs slow compared to the stationary clock.*

You can calculate the size of the effect. Imagine two stationary clocks, one at the point where the moving clock emits its pulse and another one a distance $2v\,\Delta t$ away where the light pulse in the moving clock returns to its starting point, i.e., where it completes its "tick." The time Δt on the second stationary clock measures how long the moving clock took to "tick." You can find the size of Δt from the geometry of Fig. 10.11(b). Notice that Δt must satisfy the relationship

$$\Delta t = \frac{2\sqrt{L^2 + v^2\left(\frac{\Delta t}{2}\right)^2}}{c};$$

then solve this equation for Δt.

EXERCISES

21. Show that

$$\Delta t = \frac{2L}{c}\frac{1}{\sqrt{1-\frac{v^2}{c^2}}}.$$

Because $2L/c$ is the "tick" Δt of the stationary clock, you can see that a clock moving with speed v takes a factor of $\frac{1}{\sqrt{1-v^2/c^2}}$ longer to complete its "tick" than does an identical stationary clock; *the moving clock runs slow compared to the stationary clock.*

EXAMPLES

5. Suppose a clock is moving past you at $v = (\sqrt{3}/2)c$. How many minutes would go by on clocks in your reference frame before one minute went by on the moving clock?

First, find the value of $\frac{1}{\sqrt{1-v^2/c^2}}$. For $v = (\sqrt{3}/2)\,c$, you get

$$\frac{1}{\sqrt{1-\frac{3}{4}}} = 2.$$

Therefore, 2 minutes will go by on the stationary clocks while only 1 minute goes by on the moving clock.

The fraction $\frac{1}{\sqrt{1-v^2/c^2}}$ shows up so often in relativity that it is customarily given its own symbol, the Greek letter gamma:

$$\gamma \equiv \frac{1}{\sqrt{1-\frac{v^2}{c^2}}}.$$

Thus for $v = (\sqrt{3}/2)\,c$ we found that $\gamma = 2$ and that a moving clock runs at $\frac{1}{2}$ the rate of an identical stationary clock. In general,

$$\Delta t = \Delta t'\gamma. \tag{8}$$

The phenomenon of moving clocks running slow is called "time dilation." Time dilation has real physical consequences. For example, imagine a beam of radioactive particles such that half of them disintegrate every 1 μs. If that beam is moving past you at $v = (\sqrt{3}/2)\,c$, how long will it take for half of them to disintegrate? Because $\gamma = 2$, in your frame of reference you will find that it takes 2 μs for half of the particles in the moving beam to disintegrate.

EXERCISES

22. Suppose a clock moves past you at $v = 0.8\,c$. How far will it travel in 1 s? How much time will elapse on the traveling clock during that time?

EXAMPLES

6. To see why no one noticed time dilation until the twentieth century, consider the time dilation of a clock in an automobile moving past you at 30 m/s (67 mph). How much would it be after an hour? A year?

First you must find γ. But there is a difficulty that only the binomial expansion can deal with. $v/c = 10^{-7}$, so $\gamma = 1/\sqrt{1-10^{-14}}$, and your hitherto all-powerful hand-held calculator gives you "exactly" $\gamma = 1.000000000$. Because you now know about the binomial expansion, you can write down the answer without using any calculator:

$$\gamma = 1 + \frac{1}{2} \times 10^{-14},$$

so the car's clock will lag behind your measuring clock by 5×10^{-15} times the elapsed time. This would be

$$3600 \times 5 \times 10^{-15} = 18 \times 10^{-12}\ \text{s}$$

after an hour.

EXERCISES

23. What would be the time dilation of a clock moving at 30 km/s relative to Earth?

Time dilation may seem strange enough, but here is something else to think about. Imagine that you switch reference frames, so that you are in the frame of reference of the moving clock. Now the clocks of your former frame of reference are moving with velocity $-v$. What does the principle of relativity tell you about the behavior of those moving clocks when you measure their elapsed time using the clocks of your newly adopted reference frame?

Relativity says—quite correctly—that observers in either frame will measure the clocks of the other frame running slow. How can that be? Take some time and think about it.

Moving Lengths Shrink—Lorentz Contraction

In relativity, length and time are intimately connected. Therefore, once you know that moving clocks run more slowly than stationary clocks, it is less surprising to find that moving lengths differ from stationary lengths. For example, a stick that is 1 m long at rest shortens in the direction of its motion; if it is moving at $(\sqrt{3}/2)\,c$, it will be half as long as when at rest.

This remarkable behavior follows from the constancy of the speed of light for all observers. To see that it must be so consider a rod of length L_0 measured at rest and two observers, one at rest relative to the rod and the

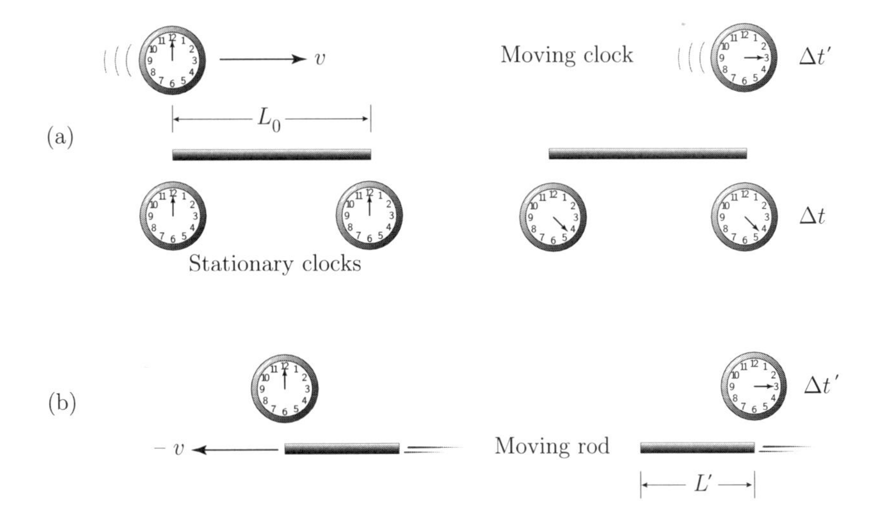

FIGURE 10.12 Measuring the length of a rod moving in the direction of its length. In (a) the length of a stationary rod is found by measuring the time Δt it takes an object (here a clock) to travel from one end of the rod to the other at speed v. In (b) an observer attached to the moving clock sees the rod move past with speed $-v$ and finds its length by measuring the time $\Delta t'$ the rod takes to pass his clock.

other moving past it with a velocity v as shown in (a) and (b) of Fig. 10.12. How does each observer measure the length of the rod?

To measure the length of a rod at rest you can send an object from one end of the rod to the other at a constant speed, call it v, and measure the time Δt that it takes to make the trip. The length of the rod is then

$$L_0 = v\,\Delta t,$$

where Δt is the time interval measured by two stationary clocks positioned one at each end of the rod. This is illustrated in Fig. 10.12(a), where for the sake of the next step in our argument we have the object moving past the rod be a clock. Because this clock is moving, the time that elapses on it will be $\Delta t' = \Delta t/\gamma$.

When the rod is moving at some speed v, you can find its length by timing how long it takes to move past a stationary clock. This is shown in Fig. 10.12(b), where we have moved into the reference frame of the clock that was moving in Fig. 10.12(a) and on which a time $\Delta t'$ elapses as the rod passes by. Therefore, in the reference frame of the clock you will measure the length of the moving rod as

$$L' = v\Delta t'.$$

Replace $\Delta t'$ with $\Delta t/\gamma$ and substitute. This gives

$$L' = \frac{L_0}{\gamma} \equiv L_0\sqrt{1-\frac{v^2}{c^2}}. \tag{9}$$

When $v \neq 0$, γ is *always* > 1; therefore, a moving length is always contracted in the direction of its motion. This effect is called the "Lorentz contraction" after the physicist who first considered its possibility. Like all other predictions of the special theory of relativity, it is a consequence of the constancy of c for all observers and the principle of relativity.

You might ask how such a strange effect could go unnoticed so long. The answer is because the speed of light is so large.

EXAMPLES

7. To see this, consider the Lorentz contraction of a 5 m long automobile traveling at 30 m/s. First find the value of γ. Its difference from 1 is too small to calculate with your calculator. Therefore, approximate γ

$$\gamma = \frac{1}{\sqrt{1-(\frac{v}{c})^2}} \approx 1 + \frac{1}{2}\frac{v^2}{c^2} + \cdots$$

and use the fact that $v/c = 10^{-7}$, so $\gamma = 1 + 5 \times 10^{-15}$. From this it follows that the change in length of the speeding car is 25×10^{-15} m, *i.e.*, 25 fm. This is about the diameter of an atomic nucleus and is 10^7 times less than the wavelength of visible light. Clearly, the effect is not going to be very noticeable at speeds we experience in everyday life.

EXERCISES

24. What would an observer find to be the Lorentz contraction of a 100 m long space ship passing by at 30 km/s?

Notice that in looking at the clock in two different frames of reference in Fig. 10.12, we claimed that the time $\Delta t'$ marked off by the hands of the clock was the same in the different frames. This has to be so. If the hands of the clock read exactly noon, 12:00:00 when the clock and the left-hand end of the rod are at the same point in space, *every observer will agree that this physical event occurs.* And if the hands read 12:00:01.000 when the clock and right-hand end of the rod are at the same point, that physical event must also be the same for all observers regardless of their frames of reference.

The Doppler Effect

It is remarkable that all observers measure the speed of light from any source to be c regardless of whether they and the source are moving towards each other. But surely there must be some effect of the relative motion. And there is. Observers moving with different speeds relative to a light source will measure different frequencies (or wavelengths) of the same light. When a source and an observer are moving toward each other, the observer measures a higher frequency (shorter wavelength) than when they are at rest relative to each other; when they are moving away from each other, the observer measures a lower frequency (longer wavelength) than when they are at rest. This dependence of frequency on the relative motion of source and observer is called the "Doppler effect."

That such an effect must occur follows from time dilation. The demonstration of this is a good way to see whether you understand how to think with the ideas of the special theory of relativity. So let's go through it for the case in which the source and observer are moving directly toward or away from each other.

Consider a source and an observer at rest relative to each other. As shown in Fig. 10.13(a), the source produces a wave of frequency f_0. As a result, it emits a crest every $T_0 = 1/f_0$ seconds, and an observer stationary relative to the source will measure the arrival of those crests every T_0 seconds.

But what if the source and observer are moving toward each other at some speed v? The time interval between the arrival of two successive crests measured by the observer will change for two reasons. First, there will be time dilation. If a time T_0 elapses on the clock of the moving source, a time

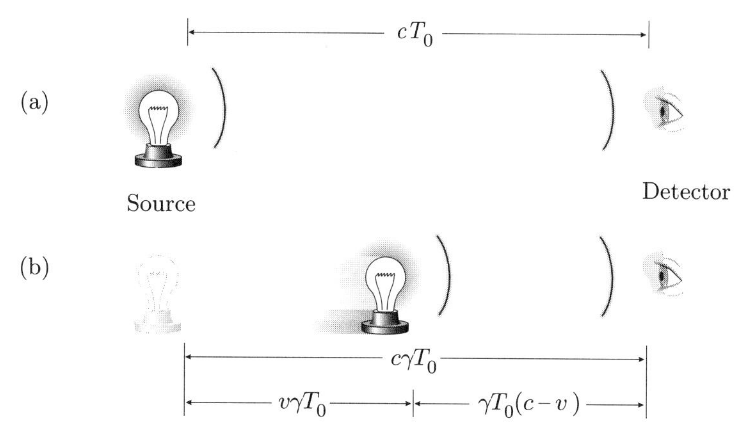

FIGURE 10.13 (a) The distance between crests of a wave moving between a stationary source and observer; (b) the distance between the crests when the source is moving toward the observer with speed v.

γT_0 will elapse on the clock of the observer. The other reason is that the second of two successive crests does not have to travel as far as the first one. During the time between the emission of one crest and the emission of the next one, the source will move closer to the observer, and the second crest will have less distance to travel than the first one. As a result, it will arrive sooner than if the source had not been moving. This is shown schematically in Fig. 10.13(b).

This may be easier to see in terms of distances in the observer's frame. If there were only time dilation, then in a time γT_0 the crest would travel a distance $\gamma T_0 c$, and the next crest would be this distance behind the first one. Because the source moves a distance $\gamma T_0 v$ before the second crest is emitted, the two crests are actually only $\gamma T_0 c - \gamma T_0 v$ apart. The time T measured by the observer between the arrival of one crest and the next is then just this distance divided by the wave's velocity c:

$$T = \frac{\gamma T_0 c - \gamma T_0 v}{c} = \gamma T_0 \left(1 - \frac{v}{c}\right).$$

To see what this means for frequencies, take the reciprocals of both sides of the equation and use the fact that $f = 1/T$ and $f_0 = 1/T_0$. This gives

$$f = f_0 \frac{1}{\gamma \left(1 - \frac{v}{c}\right)},$$

$$f = f_0 \sqrt{\frac{1 + \frac{v}{c}}{1 - \frac{v}{c}}}, \qquad (10)$$

where the last step uses the facts that $\gamma = 1/\sqrt{1 - v^2/c^2}$ and $1 - v^2/c^2 = (1 - v/c)(1 + v/c)$ to get an expression solely in terms of v.

EXERCISES

25. Show that the second version of Eq. 10 follows from the first.

26. Derive the relationship between f and f_0 if the source and observer are moving directly away from each other.

You see here that light from a source moving toward an observer will be shifted up in frequency.

EXERCISES

27. Distant galaxies are observed to emit spectra that have the same patterns as well-recognized atomic spectra but with their wavelengths all shifted toward the red. What does this tell you about the motion of distant galaxies relative to Earth?

How Do Velocities Transform?

We started this chapter talking about a bus, a truck, and an observer in different frames of reference. We did an obvious calculation to find how the velocity v of an object in one frame transforms to its velocity v' in another. We codified the method in Eq. 4 and then announced that the calculation was wrong because it would predict that observers in different reference frames would see a given pulse of light travel with different speeds, which is not what is found experimentally.

If Eq. 4 is wrong, what is correct? Einstein showed that if an object is moving with a speed v in one reference frame, then in a frame moving with a speed V relative to the first, the object will have a speed v' given by the relation

$$v' = \frac{v - V}{1 - \frac{vV}{c^2}}. \tag{11}$$

Here, as in all our examples so far, we are considering only motion in one dimension.

Notice that this equation is like Eq. 4 except for the denominator. And notice that this denominator is not going to be significantly different from 1 unless v and V both begin to approach c.

EXERCISES

28. Suppose you measure the speed of a pulse of light. You find $v = c$. What would an observer in a frame moving with a speed $V = (\sqrt{3}/2)\,c$ find for the velocity v' of this light pulse? You know the answer, but show that Eq. 11 gives it.

29. When we calculated the speed of the bus in the reference frame of the truck, we got +30 mph. Calculate the difference between this answer and the relativistically correct answer.

Something to Think About

Let's conclude this section with a mindbender. Let's see how to put a 10 m long pole into a barn that is only 5 m wide. Imagine a pole 10 m long when at rest moving with a speed of $(\sqrt{3}/2)\,c$ towards the open door of the 5 m

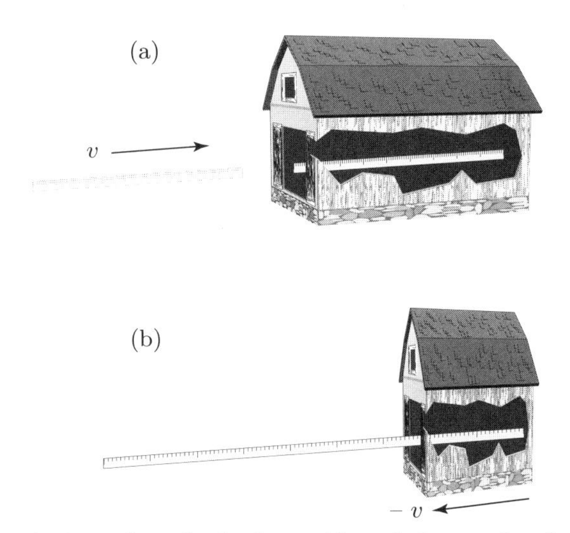

FIGURE 10.14 A 10 m pole and a 5 m barn with a relative speed such that $\gamma = 2$: (a) in the frame of reference of the barn; (b) in the frame of reference of the pole.

wide barn, as shown in Fig. 10.14(a). For this speed (as you should now begin to know by heart) $\gamma = 2$. This means that the pole is contracted to a length of 5 m in the frame of reference of the barn. Will it fit in the barn even ever so briefly before crashing into the back wall?

You can imagine a farmer standing by the barn door. Setting aside certain practicalities of reaction time, inertia of the door, air resistance, and how do you get a pole moving that fast in the first place, you can see that it is in principle possible to shut the door of the barn before the pole goes into the back wall. (If you want a little extra time for closing the door, have the pole move a little faster, so that it is a little shorter than 5 m.)

But what really bends the mind is to imagine riding on the pole so that you are in the pole's frame of reference. Now the pole is 10 m long and the barn is moving toward you at $(\sqrt{3}/2)\,c$, so it is Lorentz contracted by a factor of 2 to 2.5 m. This situation is shown in Fig. 10.14(b). Will the pole fit in the barn? How do you reconcile these two quite different pictures of what is likely to happen? Be assured that the special theory of relativity says that these two different views of the pole entering the barn are correct and consistent.

The crux of the solution is that observers in the two different frames of reference will predict and observe different *sequences* of events: The farmer's clocks will show the barn door closed *before* his clocks show the rod crashed into the wall; the rider's clock will show that the door closed *after* the rod crashed into the wall. However, it would take us too far afield to explore how clocks synchronized in one reference frame are not synchronized in another, with the result that events simultaneous in one frame of reference are not simultaneous in another, so we leave the pole-in-the-barn problem for you to ponder on your own.

PROBLEMS

1. Evaluate each of the following without a calculator and using the kind of series approximation discussed in this chapter.

a. $(1.02)^2$.

b. $\frac{1}{0.96}$.

c. $\sqrt[3]{1.09}$.

d. $\sqrt{0.0408}$.

2. Michelson showed that if there were an ether, the hypothetical medium in which light was thought to travel, then light from a source moving with speed v through this medium would take a time t_1 to travel a distance L and back along the line of motion of the source; and it would take a time t_2 to travel a distance L perpendicular to the line of motion, where

$$t_1 = \frac{2L}{c}\frac{1}{1-\frac{v^2}{c^2}},$$

$$t_2 = \frac{2L}{c}\frac{1}{\sqrt{1-\frac{v^2}{c^2}}}.$$

Show, using the expansion technique, that t_1 is greater than t_2.

3. The He–Ne laser commonly used in physics laboratories emits red light with a wavelength of $\lambda = 633$ nm.

a. What is the frequency of that light?

b. Given that the time for light to travel a distance L parallel to Earth's motion through a hypothetical ether is

$$t_{\parallel} = \frac{L}{c-v} + \frac{L}{c+v} = \frac{2L}{c\left(1-\frac{v^2}{c^2}\right)},$$

while its time to travel perpendicular to Earth's motion is

$$t_{\perp} = \frac{2L}{c\sqrt{1-\frac{v^2}{c^2}}},$$

use the binomial expansion to find an approximate formula for $t_{\parallel} - t_{\perp}$.

c. Find the value of $t_{\|} - t_{\perp}$ for Earth's motion of $v = 30$ km/s if $L = 1$ m. Express your answer as a fraction of the period of the 633 nm red light.

d. What did Michelson and Morley observe when they tried to measure $t_{\|} - t_{\perp}$, and why was their result important?

4. Evaluate without a calculator

a. $1/1.05$.

b. $(1.05)^2$.

c. $(1.05)^{\frac{1}{2}}$.

5. A stick at rest has a length of 2 m. One observer, call her S′, moves with the stick at a speed of $0.6\,c$ past another observer, call him S. The length of the stick is parallel to her direction of motion.

a. When S′ measures the length of the stick, what value does she obtain?

b. When S measures the length of the stick, what value does he obtain?

c. Which of these values is correct?

d. How much time will elapse on the watch of S′ as the stick passes S?

e. How much time will elapse on the watch of S as the stick passes him?

6. In the S frame illustrated in Fig. 10.15 Bob observes three clocks, e′, f′, and g′, moving at $v/c = \sqrt{3}/2$ past three stationary clocks, e, f, and g, 10 light seconds apart. Just as his stationary clocks all read 0, the three moving clocks line up exactly with the stationary clocks, and Bob observes that f′ also reads 0.

a. How far apart are the moving clocks in their rest frame?

b. What does e′ read in seconds when it is coincident with e?

c. What does e read when e′ reads 0?

d. In S′ what does g′ read when f′ reads 0?

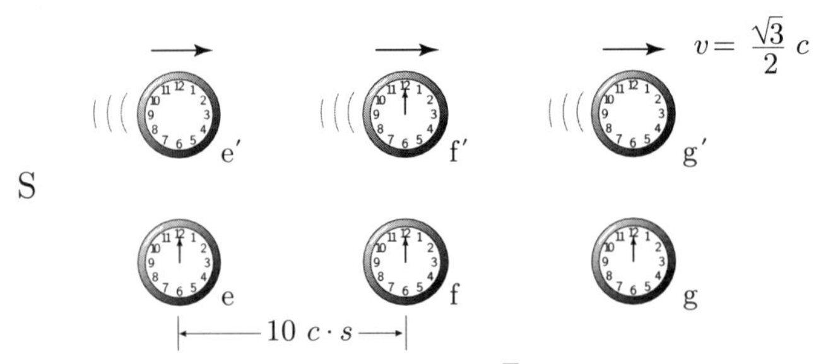

FIGURE 10.15 Three clocks moving at $v/c = \sqrt{3}/2$ past three stationary clocks (Problem 6).

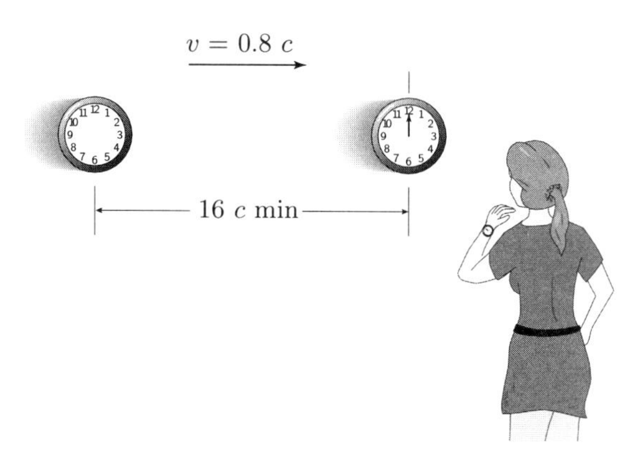

FIGURE 10.16 Two clocks moving past a stationary observer (Problem 8).

e. In S′ what does g read when f′ reads 0?

7. In the S frame of reference object #1 moves from the left with a velocity $v_1 = 0.8c$ while object #2 moves from the right with a velocity $v_2 = -0.8c$. What will be the velocity of object #2 in the rest frame of object #1?

8. Ann observes two clocks moving past her at $0.8c$, as shown in Fig. 10.16. She sees they are 16 light minutes apart and that as the right-hand clock comes by her, it reads 0 and her wristwatch reads 0.

a. What is the proper length between the clocks? ("Proper length" means the length measured in the rest frame.)

b. What does her wristwatch read when the left-hand clock passes her?

c. What does Ann find the left-hand clock to read when her wristwatch reads 0?

d. How much time will have gone by on the right-hand clock when the left-hand one passes Ann? Explain.

CHAPTER 11

Energy and Momentum at High Speeds

11.1 INTRODUCTION

The special theory of relativity shows that the Newtonian concepts of energy and momentum need to be modified if they are to correctly describe bodies moving at high speeds. The modifications lead to the best-known prediction of the theory of relativity: *Energy has mass and vice versa,*

$$E = mc^2,$$

and they also show that the relationship between kinetic energy and momentum that you have frequently used, $K = \frac{p^2}{2m} = \frac{1}{2}mv^2$, is *only an approximation* of the equations that are exact at all speeds. You now need to become familiar with the relativistically correct relationships and how they are used to extract information about atoms and the particles they are made of.

11.2 ENERGY HAS MASS

Einstein used the conservation of momentum and the fact that light exerts pressure to show that energy must have mass.

Light Exerts Pressure

Einstein knew, as did other physicists of his time, that light exerts a force on whatever it strikes. When a light wave is absorbed, it delivers an amount of energy ΔE to an object, and it imparts to the object a change in momentum $\Delta p = \Delta E/c$, where c is the speed of light. This means that light delivering a certain amount of power $\mathcal{P}$ is delivering momentum at a rate of

$$\frac{\Delta p}{\Delta t} = \frac{1}{c}\frac{\Delta E}{\Delta t} = \frac{\mathcal{P}}{c}.$$

But $\Delta p/\Delta t$ is force F, and, spread over a surface area A, it constitutes a pressure $P = F/A$.

It is common to specify intensity of light as the power delivered to a unit area. For example, the Sun delivers about $1.4\,\mathrm{kW\,m^{-2}}$ to the upper atmosphere of Earth. Dividing such a quantity by c gives the force per unit area; this is the pressure exerted by the light.

EXAMPLES

1. Sunlight absorbed by a square meter of collector above Earth's atmosphere results in a pressure of

$$P = \frac{1.4 \times 10^3}{3 \times 10^8} = 4.7 \times 10^{-6}\,\mathrm{Pa}.$$

This is quite small compared to atmospheric pressure of 10^5 Pa, but it is enough to push a spacecraft to the outer reaches of the solar system.

EXERCISES

1. A laser beam 2 mm in diameter carries 0.5 mW of power. How much pressure does this beam exert when it is absorbed?

$E = mc^2$

The most famous result of Einstein's theory, the equivalence of mass and energy, is a straightforward prediction of his special theory of relativity; he also showed that $E = mc^2$ must hold if there is to be a relativistically correct law of conservation of momentum.

Just to remind yourself of an important consequence of the conservation of momentum, imagine a gun and a target mounted rigidly a distance L apart on a cart equipped with frictionless wheels. The situation is shown schematically in Fig. 11.1. When the gun fires a bullet of mass m to the right with speed u, the rest of the apparatus—the gun, target, wheels, etc.—with a total mass M, recoils to the left with velocity v.

The Newtonian form of conservation of momentum says that $Mv = mu$. However, by now you know enough relativity to be uneasy. There may be effects of the relative motion of M and m that make the Newtonian equation only approximate. We can evade this uncertainty by examining the situation

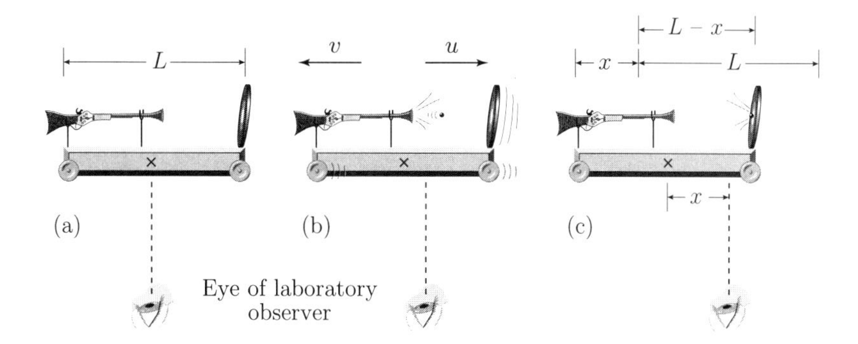

FIGURE 11.1 A gun and target are mounted on a frictionless cart. When the gun fires a bullet of mass m, the apparatus recoils until the bullet stops in the target.

of M and m after they come to rest. This will occur after the time Δt that it takes the bullet to reach the target. During that time the cart travels $x = v\,\Delta t$ to the left, and the bullet travels a distance $L - x = u\,\Delta t$ to the right.

Newtonian physics predicts that $Mv\,\Delta t = mu\,\Delta t$, so that after the transfer of the bullet we must have

$$Mx = m(L - x). \tag{1}$$

Since this result does not depend on the motion of anything, it must also be correct in the special theory of relativity as well as in Newtonian theory. Both theories require that a mass m be transferred a distance $L - x$ in order to conserve momentum.

Now consider the same setup only with the gun emitting a pulse of light of energy E instead of a bullet of mass m. Because light carries momentum E/c, the gun, target, and cart must recoil with momentum $-E/c$. During the time Δt it takes the pulse of light to reach the target, the cart and attachments—still of mass M—recoil a distance $x = v\,\Delta t$, and the light travels a distance $L - x$ to reach the target. Since the light travels with speed c regardless of the recoil velocity, the pulse requires time $\Delta t = (L - x)/c$ to reach the target. In the Newtonian approximation, we have $Mv\,\Delta t = (E/c)\,\Delta t$, and therefore, after the pulse of light has been absorbed by the target and all the parts of the system are again at rest, we have

$$Mx = \frac{E}{c^2}(L - x). \tag{2}$$

Comparing Eqs. 1 and 2 it is apparent that momentum is conserved only if the light pulse carries to the target an amount of mass $m = E/c^2$. Conservation of energy implies that any form of energy can be converted into another; therefore, if light energy E has a mass equivalence of mc^2, so does every other form of energy. The result turns out to be fully general: *Mass and energy are equivalent*. The amount of energy E associated with any mass m is given by Eq. 3:

$$E = mc^2. \tag{3}$$

This famous equation also means that if the energy of an object increases, so does its mass, i.e.,

$$\Delta E = \Delta m\, c^2. \tag{4}$$

Suppose an object sitting on a flat frictionless surface is made to slide. It must have more energy moving than at rest because it has its kinetic energy in addition to any other forms of energy, and more energy means more mass. The equivalence of energy and mass means that an object can have more mass when it is moving than when it is at rest just because it *is* moving. Therefore, for a moving body the mass m appearing in Eq. 3 is greater than the mass m_0 of the body at rest.

It is helpful to distinguish between m and m_0, so we call m_0 "the rest mass" of a particle. It is always the mass that would be measured by an observer in the particle's rest frame. From here on, unless the context tells you differently, m_0 will mean the rest mass and γm_0 or m will represent the relativistic mass. Notice also that Eq. 3 assigns a certain amount of energy to an object even when it is at rest. This is often called the particle's rest energy. For example, an electron has a rest mass of $m_0 = 9.11 \times 10^{-31}$ kg, so its rest energy is

$$m_0c^2 = 9.11 \times 10^{-31} \times (3 \times 10^8)^2 = 8.20 \times 10^{-14}\,\text{J}.$$

Almost never does one express the electron's rest energy in joules; the preferred units are electron volts. Then

$$m_0c^2 = 5.11 \times 10^5\,\text{eV} = 511\,\text{keV}.$$

This is a fact you will use often. Although it is not strictly correct, physicists often do not distinguish between rest energy and rest mass; after all, one is just the other multiplied by a constant. You will often hear physicists say, "The rest mass of the electron is 511 keV."

Einstein's theory predicts that when a body with a rest mass m_0 moves with a speed v, it will have a mass m such that

$$m = \frac{1}{\sqrt{1 - \frac{v^2}{c^2}}} m_0 = \gamma\, m_0. \tag{5}$$

Therefore, Eq. 3 can be written as

$$E = mc^2 = \gamma m_0c^2, \tag{6}$$

and Eq. 4 can be written as

$$\Delta E = \Delta mc^2 = (\gamma - 1)m_0c^2. \qquad (7)$$

EXERCISES

2. Given the validity of Eq. 5, derive Eqs. 6 and 7.

Experimental Evidence for $m = \gamma m_0$

To derive Eq. 5 would take us more deeply into Einstein's theory than we need to go. Instead, let's examine an experiment that shows that mass depends on velocity just as Einstein predicted.[1] The velocity dependence of mass will be difficult to observe until v becomes comparable to c, as you can see by expanding γ:

$$\gamma = \left(1 - \frac{v^2}{c^2}\right)^{-1/2} = 1 + \frac{v^2}{2c^2} + \cdots.$$

The expansion shows that the fractional change in the mass due to its motion, $\Delta m/m = \gamma - 1$, will be less than 10^{-4} until $v/c = 0.014$. This corresponds to $v = 4243$ km/s, nearly 10 million mph and more than 140 times faster than Earth's motion in its orbit.

To find objects moving fast enough to show a measurable effect, Kaufmann used electrons emitted in the radioactive decay of atoms. This choice in 1901 is remarkable when you realize that radioactivity was only discovered in 1896 and the electron in 1897. As did J.J. Thomson, Kaufmann used a combination of electric and magnetic fields to measure e/m. But where Thomson had the E and B fields at right angles to each other, Kaufmann had them parallel. Electrons from a tiny source of radioactive material passed through a set of slits and this arrangement of fields. If the electrons all had the same velocity, they would strike a single point on the detecting photographic plate. However, electrons from radioactive atoms have a wide range of velocities, and passing through the combined E and B fields, they spread out into a nearly parabolic curve. Carefully measuring where they fall on this curve, Kaufmann could determine the particular velocity and charge-to-mass ratio corresponding to each point on the curve. The trajectory of one particular electron is shown in Fig. 11.2 along with a sketch of the parabolic curve formed by electrons with other velocities.

---

[1]The experiment was done by Walter Kaufmann in 1901, (see pp. 502–512 in *The World of the Atom*, edited by H. A. Boorse and L. Motz, Basic Books, New York, 1966) three years before Einstein published his theory.

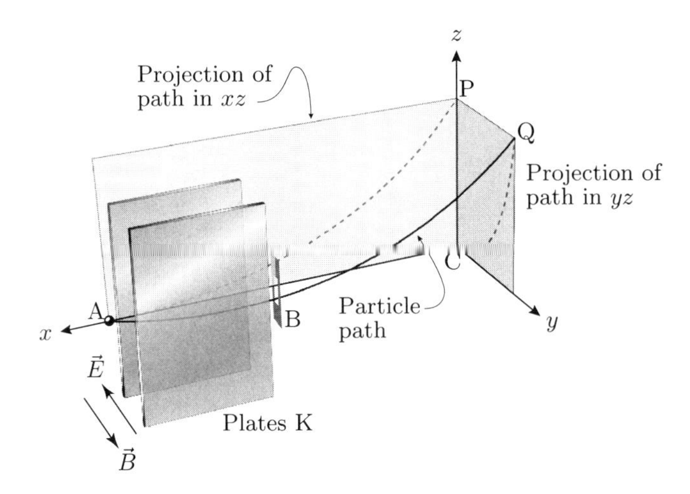

FIGURE 11.2 Kaufmann's arrangement for measuring e/m of electrons from radioactive atoms. Electrons from a point source of radioactive material at A pass between plates K across which is an electric field perpendicular to the plates; the electrons pass through a narrow slit B and strike a screen at Q. Because of a magnetic field parallel to E their trajectory is a circular arc of radius ρ.

Kaufmann's results are shown in Table 11.1. This table shows five measured velocities and the corresponding values of e/m. It is clear that e/m gets smaller as the velocity gets larger. This is just the effect that you would expect if m grows larger with speed. Also included in the table are values of $\gamma = \frac{1}{\sqrt{1-v^2/c^2}}$ corresponding to the measured values of v. If the observed variation in e/m arises because $m = \gamma\, m_0$, then multiplying the measured values of e/m by γ should give the same number. The average of e/m_0 obtained from Kaufmann's data is 1.731×10^{11} C/kg, which is in remarkably good agreement with the currently accepted value of 1.759×10^{11} C/kg.

TABLE 11.1 e/m for Electrons of Different Speeds v

Velocity v (10^8 m/s)	e/m (10^{11} C/kg)	e/m_0 (10^{11} C/kg)	γ	m/m_0
2.815	0.620		2.891	
2.674	0.751		2.205	
2.526	0.927		1.854	
2.366	1.066		1.627	
2.230	1.175		1.495	

EXERCISES

3. Fill in the column of Table 11.1 labeled e/m_0 and calculate the average value of e/m_0 implied by Kaufmann's data. Your results should show that Einstein's prediction that mass will vary as $m/m_0 = \gamma$ is very well born out by Kaufmann's experiment.

4. Another way to show that Einstein's prediction is correct is to compute m/m_0 from Kaufmann's data and compare it to the values of γ calculated for the corresponding values of v. To find m/m_0 you need only divide the accepted value of e/m_0 by the measured value of e/m. Do this and put the values into column 5 of Table 11.1.

The graphical comparison of m/m_0 and γ in Fig. 11.3 shows that the agreement of experiment and theory is very good.

11.3 MOMENTUM AND ENERGY

If the mass of an object depends on its velocity, how does this affect our ideas of momentum and kinetic energy?

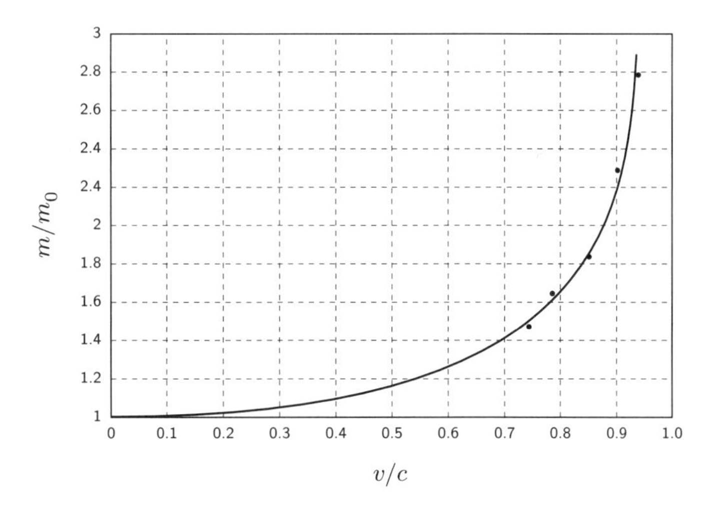

FIGURE 11.3 Comparison of Kaufmann's data (the solid dots) with Einstein's prediction (the solid line).

Relativistic Momentum

Let's begin by asking, What is the momentum of a particle in a frame moving at velocity v? It turns out that only the following specification works for momentum:

$$p = \gamma m_0 v.$$

This is the quantity that must be used to analyze collisions of moving particles, and it is the quantity that is measured when charged particles are deflected by magnetic fields using the relation $p = qBR$ (Eq. 2 discussed in Chapter 7).[2] The combination $\gamma\, m_0$ describes the increase of mass with velocity that is observed in measurements. In Kaufmann's e/m experiment, for example, the momentum determined from the curvature in a magnetic field yielded a larger value of m for a high velocity v and hence a smaller e/m than for a low velocity.

EXAMPLES

2. Suppose that an electron were given a velocity of $v = 0.25c$. What would be its momentum and what change would appear in the value of e/m relative to the low-energy result?

$$\begin{aligned}\gamma &= \frac{1}{\sqrt{1-(0.25c/c)^2}}\\ &= \frac{1}{\sqrt{1-0.0625}} = 1.0328,\\ p &= 1.0328 \times 9.11 \times 10^{-31} \times 0.25 \times 3 \times 10^{8}\\ &= 7.05 \times 10^{-23}\ \text{kg m/s}.\end{aligned}$$

We can get the answer to the second part by noting that if m is multiplied by 1.0328, then e/m must be divided by the same number. So

$$\frac{e}{m} = \frac{1.7587 \times 10^{11}}{1.0328} = 1.7028 \times 10^{11}\ \text{C/kg},$$

or a 3.3% decrease.

[2] Buried in this statement and in the treatment of Kaufmann's data is a nontrivial assumption. We have implicitly assumed that unlike mass, the electric charge is the same in all reference frames regardless of relative velocity. The success of the theory of electrodynamics, which deals with the transformations that determine how electric and magnetic fields appear in different reference frames, shows that this assumption is correct.

3. What would be the radius of curvature of the path of an electron moving with $v = 0.25c$ in a magnetic field of $B = 1$ T?

You can answer this question using Eq. 2 and the value of the momentum p from the previous example.

$$R = \frac{p}{eB} = \frac{7.05 \times 10^{-23}}{1.602 \times 10^{-19} \times 1} = 4.4 \times 10^{-4}\,\text{m} = 0.44\,\text{mm}.$$

Relativistic Kinetic Energy

To handle kinetic energy correctly at high speeds you need to rethink the idea slightly. Kinetic energy is the *additional* energy that an object acquires because it is moving. Therefore, kinetic energy is the difference between the energy of a moving object, mc^2, and its energy at rest, m_0c^2:

$$K = mc^2 - m_0c^2 = (\gamma - 1)m_0c^2. \tag{8}$$

What would be the best approximation of this equation at low velocities, i.e., what should K be when $v/c \ll 1$? You know the answer, but let's see that it is so. Using the binomial expansion you can see that

$$\begin{aligned} K &= m_0c^2\left[\left(1 - \frac{v^2}{c^2}\right)^{-1/2} - 1\right] \\ &\approx m_0c^2\left[1 + \frac{1}{2}\frac{v^2}{c^2} + \cdots - 1\right] = m_0c^2\frac{v^2}{2c^2} \\ &= \frac{1}{2}m_0v^2. \end{aligned}$$

Once again, Newtonian mechanics is the low-velocity approximation to Einsteins's theory; i.e., K is very accurately the old familiar quantity when speeds are small.

The equality of $E = mc^2$ is very general. It applies to all forms of energy, not just to kinetic energy. What effect does potential energy, for example, have on the mass? To be consistent with our principle of energy conservation, we have to conclude that these other forms of energy have mass also. Where would such mass be found? It can only be a part of the rest mass, but of the entire system, not just in a single part of the system. Although it is not always clear where the new energy-associated mass might be, we are forced to conclude that rest mass can include such things as the thermal energy—the $\frac{3}{2}k_BT$ of average random-motion energy you learned about when studying the kinetic theory of gases—and the potential energy. Does this mean that you are a little more massive at the top of a hill (where your potential energy is greater) than at the bottom?

EXAMPLES

4. Find the change in mass, corresponding to the change in gravitational potential energy, of an 80 kg student after he climbs a 60 m high hill. Assume rather unreasonably that there is no loss of material through such things as respiration or perspiration. Is such a mass change detectable?

$$\Delta E = mg\Delta h = 80 \times 9.8 \times 60 = 47040\,\text{J},$$

$$\Delta m = \frac{\Delta E}{c^2} = \frac{47040}{9 \times 10^{16}} = 5.2 \times 10^{-13}\,\text{kg}.$$

It is not likely to be measured. Your gain in weight has nothing to do with climbing hills!

But wait! What about the energy that was used to do the work in climbing the hill? Since it had to come from chemical reactions in the body, all that has been done has been to convert chemical potential energy to gravitational potential energy. Both forms have equivalent mass, so there is no net change at all. Only if some outside agent (a roommate's car, perhaps) lifted you up the hill would your mass change.

EXERCISES

5. Consider an electron and a point charge of $-1\,\mu$C. At first the two particles are at rest very far apart. What will be the change in the mass of this system if they are arranged to be at rest and 10 mm apart?

6. Calculate the change Δm in the mass m when an electron is at rest 1×10^{-12} m away from a *positive* charge of $50e$, compared to being at rest very far away.

7. What is the fractional change of mass found in the previous problem; i.e., what is $\Delta m/m$?

8. In x-ray machines electrons are accelerated through 100 kV or more. What is the speed of 100 keV electrons? What are their momentums? How much error do you make if you use the Newtonian equations instead of the relativistically correct ones?

Two things to remember:
It is common to refer to a particle by its kinetic energy. Thus when the previous problem told you that you have a 100 keV electron, you are expected

to know that this refers to the electron's 100 keV of kinetic energy, not its total energy, which includes its rest energy and is $E = m_0c^2 + K = 611$ keV.

It is also usual to describe the momentum and energy of a moving particle from "our" rest frame. We call this reference frame the "laboratory," or "lab," frame of reference. For example, the curved path of a charged particle in a magnetic field tells you the particle's momentum in the lab frame. Unless it is otherwise stated, all the problems in this book ask for results in the lab frame.

Relation Between Energy and Momentum

It is more convenient to work with momentums and energies directly than it is to first calculate the velocity, find the relativistic factor γ, and then calculate kinetic energy or momentum. The momentum of a particle accelerated to a high energy can easily be measured by bending it in a magnetic field, and its kinetic energy would then be known from the voltage used to accelerate it, $K = qV$.

For such circumstances it is very helpful to have a relationship between momentum p and total energy E. You can find such a relationship from knowledge that $E = \gamma m_0 c^2$ and $p = \gamma m_0 v$. First square each of these equations to get

$$E^2 = \gamma^2 m_0^2 c^4 = \frac{m_0^2 c^4}{1 - v^2/c^2},$$

$$p^2 = \gamma^2 m_0^2 v^2 = \frac{v^2/c^2}{1 - v^2/c^2} m_0^2 c^2.$$

Multiply the second equation by c^2 to give it the same physical dimensions (energy) as the first and then subtract the second from the first. This will give

$$E^2 - p^2c^2 = \frac{1}{1 - v^2/c^2} m_0^2 c^4 - \frac{v^2/c^2}{1 - v^2/c^2} m_0^2 c^4 = \left(\frac{1 - v^2/c^2}{1 - v^2/c^2} \right) m_0^2 c^4.$$

From this you can see that

$$E^2 = p^2c^2 + m_0^2 c^4, \quad \text{or } E = \sqrt{p^2c^2 + m_0^2 c^4}, \tag{9}$$

which, as desired, enables you to find the total energy of a particle of known rest mass m_0 if you know its momentum or to find the magnitude of its momentum if you know its total energy.

EXAMPLES

5. Find the momentum of the electron in Exercise 11.8 without finding γ. Use SI units.

The basic approach is to solve Eq. 9 for p. This gives

$$p = \sqrt{\frac{E^2 - m_0^2 c^4}{c^2}}.$$

Then you need to determine E, and to do this you need to know m_0. From the appendix at the end of the book or from Exercise 17 in Chapter 8 you can find that $m_0 = 9.11 \times 10^{-31}$ kg for an electron. Then

$$\begin{aligned} E &= K + m_0c^2 = qV + m_0c^2 \\ &= 10^5 \times 1.6 \times 10^{-19} + 9.11 \times 10^{-31} \times 9 \times 10^{16} = 9.80 \times 10^{-14}\,\mathrm{J}, \end{aligned}$$

and

$$p = \sqrt{\frac{(9.8 \times 10^{-14})^2 - (8.20 \times 10^{-14})^2}{9 \times 10^{16}}} = 1.79 \times 10^{-22}\,\mathrm{kg\,m\,s^{-1}}.$$

11.4 MASSES IN eV/c^2; MOMENTUMS IN eV/c

Because units of electron volts are widely used in measurements of atomic particles, it is convenient to express their masses and momentums in terms of electron volts.

EXAMPLES

6. To see how to use these units, consider a 100 eV electron. Suppose you want to find its velocity. It is of such low energy that you can use

$$K = \frac{1}{2} m_0 v^2.$$

If you multiply the right-hand side by c^2/c^2, you will get

$$K = \frac{1}{2} m_0 c^2 \, \frac{v^2}{c^2}.$$

In this form the convenience of the units becomes more evident. Knowing that m_0c^2 for an electron is 511 keV, you get

$$100 = \frac{1}{2} \cdot 511 \times 10^3 \cdot \frac{v^2}{c^2},$$

which you can solve for v/c,

$$\frac{v}{c} = \sqrt{\frac{2 \times 100}{511 \times 10^3}} = 1.98 \times 10^{-2},$$

from which it follows that $v = 5.94 \times 10^6$ m/s.

The trick is to bundle together the right combination of m and c to get a quantity identifiable as an energy. Once you get used to it, using eV and suppressing the factor c^2 makes calculations easier and more informative.

EXERCISES

9. What is the velocity of the 200 eV electrons used for measuring e/m in many undergraduate laboratories?

EXAMPLES

7. You can also use this trick to find momentum. Suppose you wanted to know the momentum of a 100 eV electron. Because this is a low-energy electron, you can use the Newtonian approximation

$$K = \frac{p^2}{2m_0}.$$

Multiplying the right side by c^2/c^2 gives

$$K = \frac{p^2c^2}{2m_0c^2},$$

which can be solved for pc:

$$pc = \sqrt{2Km_0c^2}.$$

Now, if K is in eV and m_0c^2 is in eV, you will obtain pc in eV, so

$$pc = \sqrt{200 \times 511 \times 10^3} = 1.011 \times 10^4 \text{ eV},$$

and we would say that $p = 1.01 \times 10^4$ eV/c, or, more compactly, $p = 10.1$ keV/c.

8. Find the momentum of a 1 MeV electron.

This is *not* a low-energy electron, so

$$E = K + m_0c^2 = 1 + 0.511 \text{ MeV} = \sqrt{(pc)^2 + (0.511)^2},$$

$$pc = \sqrt{(1.511)^2 - (0.511)^2} = 1.422 \text{ MeV},$$

and thus $p = 1.422 \text{ MeV}/c.$

Momentum is frequently found by measuring the radius of curvature R of a charged particle's trajectory perpendicular to a magnetic field B. Remember that a particle of charge e and momentum p will bend according to the relationship

$$p = eBR.$$

To use eV units in this equation requires a little ingenuity. Notice that if you multiply both sides by c, the dimensions of the equation become those of energy, i.e., pc has units of energy. At first glance it seems that you must evaluate $eBRc$ using consistent SI units and obtain an answer in joules. This will certainly work, but notice that if you convert the right side of the equation into electron volts, you divide by a conversion factor that has a numerical value just equal to e. This means that if you drop the factor of e and multiply B in tesla by R in meters and c in meters per second, your answer BRc will come out numerically in eV.

EXAMPLES

9. Suppose you wanted to know the radius of curvature with which an electron with momentum of 10 keV/c would bend in a magnetic field of 10^{-3} T. Then

$$R = \frac{pc}{Bc} = \frac{10 \times 10^3}{10^{-3} \times 3 \times 10^8} = 3.37 \times 10^{-2} \text{ m} = 3.37 \text{ cm}.$$

EXERCISES

10. A proton, ($M_0c^2 = 938$ MeV) bends with a radius of curvature of 1 m in a magnetic field of 0.1 T. What are the momentum and the kinetic energy of that proton?

EXAMPLES

10. Suppose we repeat Example 11.9 using units of electron volts. You are given that $K = 100\,\text{keV}$, and you *remember* that for an electron, $m_0 = 511\,\text{keV}/c^2$. The total energy E is the sum of the kinetic energy K and the rest energy, $E = K + m_0c^2$, so $E = 100\,\text{keV} + 511\,\text{keV} = 611\,\text{keV}$.

Equation 9 tells you that $611\,\text{keV} = \sqrt{(pc)^2 + 511^2}$, which you can solve for pc:

$$611^2 - 511^2 = (pc)^2,$$
$$pc = 335\,\text{keV and } p = 335\,\text{keV}/c.$$

EXERCISES

11. Show that the momentum of $335\,\text{keV}/c$ obtained in Example 10 is the same as the value of $1.79 \times 10^{-22}\,\text{kg m s}^{-1}$ obtained in Example 5.

To work easily in eV you need to *know* the masses of the electron and the proton in eV/c^2. It may also be helpful to know the energy equivalent of one atomic mass unit, although for most purposes of this book you can use the proton or hydrogen-atom mass instead. Energy equivalents of these and other useful masses are summarized in Table 11.2. *From now on work all energy and momentum problems in units of eV (or keV or MeV, etc.).* The scale of these units is appropriate to atoms and smaller particles.

TABLE 11.2 Some Masses in Energy Units

Entity	mc^2 (MeV)
electron	0.511
H atom	938.8
proton	938.3
neutron	939.6
1 u	931.5

11.5 WHEN CAN YOU APPROXIMATE?

In the preceding examples we told you when you could use the Newtonian approximation and when you couldn't. What if you have to decide on your own? How can you tell whether to use $K = p^2/(2m_0)$ or $K = E - m_0c^2$? Here we give you some guidance and some rules of thumb. Where the rules come from is left for later courses in physics.

Nonrelativistic Approximations

You will often have to solve one or the other of two related problems:

a. Given the momentum p, find the particle's kinetic energy K.

b. Given the kinetic energy K, find the particle's momentum p.

When v/c is small, you can solve these problems using the approximate Newtonian equations:

$$K = \frac{p^2}{2m_0},$$
$$p = \sqrt{2Km_0}.$$

But how accurate are they? When are they accurate enough?

When the difference between an approximate value and an exact value is a small fraction of the exact answer, the approximation is better than when the difference is a large fraction of the exact answer. Therefore, it is convenient to use this so-called "fractional difference" between the approximate and exact answers as a measure of how good an approximation is. (Fractional difference is often referred to as "fractional error.") Therefore, given an approximate value x_a and an exact value x_e, we want to know

$$\frac{x_a - x_e}{x_e} = \frac{\Delta x}{x_e}.$$

When this quantity is small, the approximation is good.

EXAMPLES

11. What is the fractional difference between 200 and 204? Between 200 and 220?

The answer for the first case is $(204 - 200)/200 = 0.02$, or, as one often says, 2%. For the second case the fractional difference is 0.1, or 10%.

Fractional difference is particularly useful because it is often possible to obtain a compact, simple formula for the fractional difference between an exact equation and an approximate one. Consider the problem of finding p from K. If you use the nonrelativistic equation $p = \sqrt{2Km_0}$, the fractional difference from the answer given by the exact equation will be

$$\frac{\Delta p}{p} = \frac{1}{4}\frac{K}{m_0c^2}$$

as long as $\Delta p/p \lesssim 0.1$. [The symbol $\lesssim$ means "less than or on the order of".] This equation tells you that the accuracy of your approximate answer depends upon the *ratio* of the particle's kinetic energy to its rest energy, $K/(m_0c^2)$.

If this ratio is .1, a value of p calculated from $\sqrt{2Km_0}$ will deviate from the exact value by 2.5%. In other words, if you have a 51.1 keV electron and you calculate its momentum nonrelativistically, you will be making an error of about 2.5%.

EXERCISES

12. What would be the fractional error if the electron has a kinetic energy of 5.1 keV? 102 keV?

For most of what we do in this book a precision of 2.5% is good enough, which gives the following useful rule of thumb:

If $K/(m_0c^2) < 0.1$ you can use nonrelativistic equations to find momentum from kinetic energy.

EXERCISES

13. Suppose you had a 50 MeV proton. Would the Newtonian equations be good enough for calculating the proton's momentum? How much error would you make using the Newtonian equation?

If you know the momentum p and need to find the kinetic energy K, the nonrelativistic equation $K = p^2/(2m_0)$ will give an answer that has a fractional deviation from the relativistically exact answer that is

$$\frac{\Delta K}{K} = -\frac{1}{4}\left(\frac{pc}{m_0c^2}\right)^2,$$

as long as $\Delta K/K \lesssim 0.1$. This means that for an electron with momentum of 162 keV/c the nonrelativistic equation will give an answer that is 2.5% low.

So if 2.5% is good enough precision, you have the following useful rule of thumb:

You may use nonrelativistic equations to find K from p when $pc/(m_0c^2) < 0.3$.

If you would rather remember only one rule of thumb, just remember the most restrictive:

If the ratio of K or pc to the rest energy m_0c^2 *is* < 0.1, *you can use nonrelativistic equations with an error of less than a couple of percent.*

Ultrarelativistic Approximation

There is another extreme that is useful to know about. If the ratios of K or pc to the rest energy m_0c^2 are greater than ≈ 10, you can simply ignore the m_0c^2 terms in Eq. 9. The connections of E, K, and pc become

$$K \approx E \approx pc.$$

This is called the "ultrarelativistic" approximation.

EXAMPLES

12. What is the momentum of a 1 GeV electron? In this case $K/(m_0c^2)$ is about 2000, so the particle is ultrarelativistic. Its momentum is 1 GeV/c.

11.6 SUMMARY

In the study of a moving particle the quantities of most interest are its kinetic energy K, its closely related total energy E, and its momentum $\vec{p}$.

At high speeds, i.e., when $v/c \approx 1$, these quantities behave quite differently than Newtonian physics predicts. This different behavior is related to the fact that inertial mass increases with a particle's speed:

$$m = \frac{m_0}{\sqrt{1 - v^2/c^2}} = \gamma m_0,$$

where v is the particle's speed, m_0 is its mass in its rest frame of reference, and

$$\gamma \equiv \frac{1}{\sqrt{1 - v^2/c^2}}.$$

The relativistically correct formulas connecting total energy E, kinetic energy K, momentum p, and mass m are

$$\vec{p} = m\vec{v},$$
$$E = mc^2,$$
$$E = \sqrt{m_0^2c^4 + p^2c^2},$$
$$K = E - m_0c^2.$$

The following relations are also useful:

$$\frac{v}{c} = \frac{pc}{E} = \sqrt{1 - \frac{1}{\gamma^2}},$$
$$\frac{pc}{m_0c^2} = \sqrt{\gamma^2 - 1},$$
$$\frac{K}{m_0c^2} = \gamma - 1.$$

In the limit of $v/c \ll 1$ the relativistically correct equations are very well approximated by the nonrelativistic Newtonian equations $\vec{p} = m_0\vec{v}$, $K = p^2/(2m_0)$, and $E = K+\text{const}$. For many practical purposes when $v/c < 0.1$ or $K/(m_0c^2) < 0.1$ or $pc/(m_0c^2) < 0.3$, the Newtonian equations are sufficiently accurate and easier to use.

At speeds where $v/c > 0.99$ or when $K/(m_0c^2) > 10$ or so, you can neglect the m_0c^2 terms in the above equations. This is called the ultrarelativistic approximation. Then $K \approx E$ and $p \approx E/c$.

In modern physics the units of energy are electron volts (eV) and their SI mutiples such as keV, MeV, GeV, TeV. They are used as units of kinetic energy and potential energy. By extension they are used as units of mass, eV/c^2, and units of momentum, eV/c. It is fairly common usage to say that the mass of a particle is so many eV, *e.g.*, the mass of a proton is 938 MeV. You are supposed to know that this number includes the factor of c^2.

You will need to know the values of the speed of light $c = 3 \times 10^8$ m/s and the rest masses of the electron (511 keV/c^2) and the proton (938 MeV/c^2).

PROBLEMS

1. A 9.38 MeV proton enters a magnetic field and is bent in a circle of radius $r = 0.2$ m.

a. Is this a relativistic, nonrelativistic, or ultrarelativistic proton? How do you know?

b. What is the extra mass of this proton arising from its motion? Give your answer in the appropriate multiple of eV.

2. Suppose you have an electron traveling with a velocity such that $\gamma = 3$.

a. What is its kinetic energy?

b. What is its momentum?

c. What would be its radius of curvature in a magnetic field of $B = 0.1$ T?

d. What is its velocity?

3. Suppose a stick and S′ are moving past S at a velocity such that $\gamma = 3$. If the stick has a mass of 1 kg when it is at rest,

a. What does S′ measure its mass to be?

b. What does S measure its mass to be?

4. What is the kinetic energy of the stick in the previous question

a. as measured by S′?

b. as measured by S?

5. a. We know that $E = mc^2 = \gamma m_0 c^2$. Use this relation to give a relativistically correct definition of kinetic energy K. Explain why this is a sensible definition.

b. Show by a series expansion that for $v/c \ll 1$ your above definition is well approximated by the usual classical formula for kinetic energy.

c. An electron is accelerated through a potential difference of 10 V. What is the relativistic increase in its mass? Explain.

6. A stationary particle having a rest mass energy of 1400 MeV disintegrates into two particles called "pions", a π^+ and a π^-, that travel in opposite directions, as shown schematically in Fig. 11.4. Each of the pions has a rest-mass energy of 140 MeV.

a. Find the kinetic energy, in appropriate units, of each outgoing particle.

b. What is the value of γ ("gamma") for either pion?

c. When the pions are at rest, they have a lifetime of 0.28 ns. According to an observer in the laboratory, what is the lifetime of the moving pions (in ns)?

7. The rest mass energies of a proton and an electron are 938 MeV and 0.511 MeV, respectively. Calculate the total energy of each of the following particles:

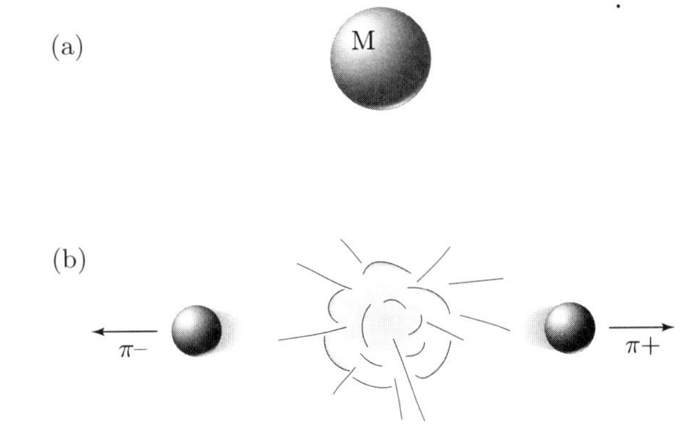

FIGURE 11.4 A particle disintegrates into two pions (Problem 6).

a. a 10 MeV proton.
b. an electron with $\gamma = 1000$.
c. a 500 MeV electron.

Are any of these particles "ultrarelativistic?" If so, which one(s)?

8. There is a particle with the same mass as the electron but positively charged. It is called a "positron." The "Tristan" particle accelerator at Tsukuba, Japan, accelerates both electrons and positrons to energies of 25 GeV and then directs them into head-on collisions in which they annihilate one another; i.e., they turn into a pair of high-energy photons, as illustrated in Fig. 11.5(a) and (b). Give your answers to the following questions in eV.

a. Are these particles nonrelativistic, relativistic, or ultrarelativistic? Justify your answer briefly.
b. Find the momentum of a 25 GeV electron.

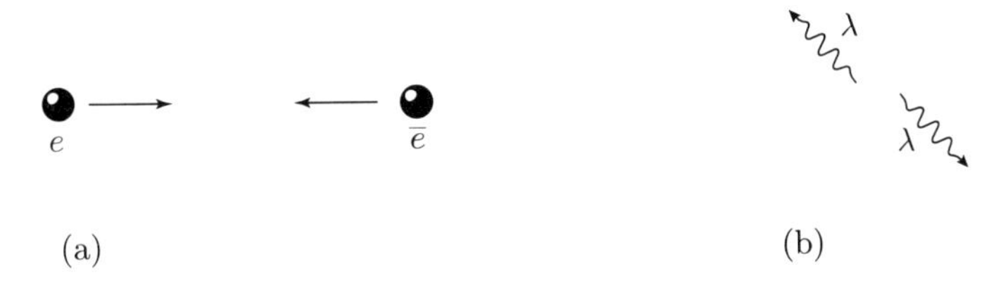

FIGURE 11.5 An electron and a positron collide and become two photons (Problem 8).

c. Suppose a single new particle is formed in the collision of a positron and an electron in this machine. Determine the momentum, kinetic energy, and rest mass of such a particle.

d. If instead of forming a new particle, all of the electron and positron energy is used to produce two photons of equal wavelength traveling in opposite directions (See Fig. 11.5(b)):

i. Estimate the wavelength of each photon.

ii. Why are two photons, rather than one, generated? (Hint: consider the conservation laws.)

9. What magnetic field will bend a 1.02 MeV electron in a circle with a radius of 1 m?

10. What magnetic field will bend a 1.02 MeV proton in a circle with a radius of 1 m?

11. What magnetic field is required to bend a 1 GeV electron in a circle 10 m in radius?

12. What magnetic field is required to bend a 1 GeV proton in a circle 10 m in radius?

13. What is the velocity of a 1.02 MeV electron? Express your answer as a fraction of the speed of light c.

14. An electron in a magnetic field of 1 kG (0.1 T) is bent in an arc with a radius of 1 m. What is the electron's kinetic energy?

15. For each of the preceding six questions, justify using the nonrelativistic approximation, the ultrarelativistic approximation, or the exact relativistic equations.

16. The rest mass energies of the electron and proton are respectively 511 keV and 938 MeV. For the cases i–iv below:

a. Determine whether the particles are nonrelativistic or ultrarelativistic. Justify your answer.

b. Calculate the kinetic energies (in eV) for each case, using the appropriate approximate formulas in each case *and also* the exact relativistic formulas.

i. An electron with a momentum of 10 keV/c.

ii. An electron with a momentum of 10 GeV/c.

iii. A proton with $\gamma = 50$.

iv. A proton with $v/c = 0.05$.

17. A proton and an antiproton each traveling with a momentum of 10 GeV/c collide head-on in a high-energy physics experiment. The proton and antiproton have equal rest masses (938 MeV/c^2), and charges of opposite sign but equal magnitude.

a. What is the total energy before the collision?

b. If they collide to form a new particle, determine its momentum, charge, and rest-mass energy.

c. The new particle is unstable and decays into two pions, π^+ and π^-, both of which have rest masses of 140 MeV/c^2, and opposite but equal-magnitude charges. Find the resulting kinetic energy of each pion.

18. What is the speed, relative to the speed of light, of an electron with kinetic energy equal in value to its rest mass?

19. A typical nuclear reactor produces about 2.5×10^9 J of thermal energy in 1 s.

a. What mass does this energy correspond to?

b. If 200 MeV of energy is released per fission, how many fissions occur per second in this nuclear reactor?

c. What fraction of the mass of a ^{235}U nucleus is converted into thermal and radiant energy when it undergoes fission?

20. Find the energy released in the deuterium–deuterium fusion reaction

$$^2\text{H} + ^2\text{H} \rightarrow ^3\text{He} + ^1\text{n}.$$

The rest masses of ^{2}H, ^{3}He, and the neutron ^{1}n are 2.014102 u, 3.016029 u, and 1.008665 u respectively.

21. Consider the decay $^{55}\text{Cr} \rightarrow\ ^{55}\text{Mn} + e^-$, where e is an electron. The nuclei ^{55}Cr and ^{55}Mn have masses of 54.9279 u and 54.9244 u, respectively. Calculate the mass difference of the reaction. What is the maximum kinetic energy of the emitted electron?

22. In the electron–positron collider at Cornell University, electrons and positrons acquire oppositely directed momenta of 4.0 GeV/c before they collide head-on.

a. Suppose that after the two particles collide they unite to form a new particle *at rest*. What is the rest mass of the new particle?

b. Suppose the new particle decays into a proton and an antiproton that move in opposite directions. What is the kinetic energy of either particle?

c. What is the velocity relative to the speed of light and the momentum of either particle?

C H A P T E R

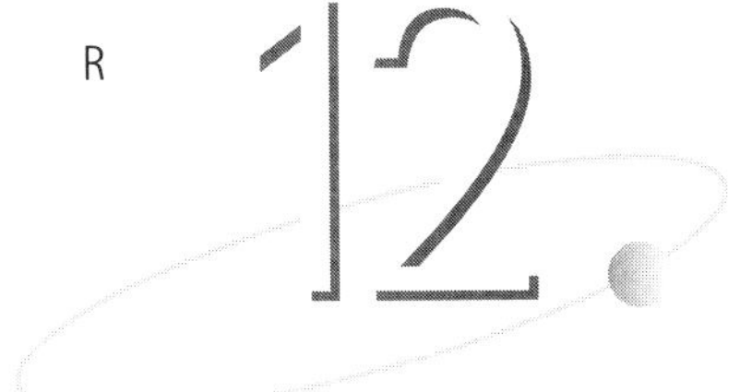

The Granularity of Light

12.1 INTRODUCTION

The discovery of the electron swiftly led to better understanding of the nature of matter. This in turn led to a revolution in the understanding of the nature of light. The most surprising outcome was the discovery that under many circumstances light and other forms of electromagnetic radiation behave like particles instead of waves. There are two outstanding examples of light behaving like particles. One example is called "the photoelectric effect" and the other "the Compton effect."

12.2 THE PHOTOELECTRIC EFFECT

Discovery of the Photoelectric Effect

In 1887 Heinrich Hertz did experiments that for the first time produced and detected electromagnetic radiation in a controlled way. On one side of a room he generated radio waves by means of a high-frequency current sparking across a gap. The resulting waves crossed the room and induced sparking in a properly adjusted detecting apparatus. He observed that this induced spark was much larger when the metal tips of the spark gap were illuminated with light from the sparks of the generator. When the light from the generating sparks was passed through glass before reaching the detecting spark gap, the spark became smaller, but when the light came through a quartz plate, the induced spark remained large. Hertz knew that glass absorbs ultraviolet light and that quartz does not, so he concluded that it was ultraviolet light causing the spark in the detector to become larger.[1]

In 1888, shortly after Hertz's observations, Hallwachs found that a negatively charged zinc plate would discharge when illuminated with ultraviolet light, while a positively charged plate would not.[2] This and other evidence showed that this "photoelectric effect" consisted in the emission of negative electricity from the metal when illuminated with light of suitably short wavelength. In 1889 Elster and Geitel showed that the photoelectric effect could be produced using *visible* light on metals that were amalgams of mercury with the alkali metals cesium, sodium, and potassium, as well as with zinc.

As we have seen, it was another decade before techniques were developed that gave reliable measurements on charged particles such as the electron. Understanding of the photoelectric effect did not progress much until it was possible to measure directly the actual charges released when light struck these metals.

Properties of the Effect

In 1899, two years after discovering the electron, J.J. Thomson, in England, and Philipp Lenard, in Germany, measured the charge-to-mass ratio q/m of the negative charges emitted in the photoelectric effect. They found q/m to be similar to the ratio already measured by Thomson for cathode rays, i.e., electrons, and concluded that the photoelectric charges must also be electrons.

Figure 12.1 is a diagram of the apparatus Lenard used to study the electrons emitted when light strikes a metal. Electrons produced using light are often called "photoelectrons." He used the apparatus both to measure the q/m ratio of the photoelectrons and to learn some things about their energy. The apparatus is in some ways like the one used to measure e/m. Here the cathode rays are produced when ultraviolet light from S shines through the window Q and strikes the cathode C, releasing charges. The flow of charges initiated by the light is often called the "photocurrent." The charges emitted at C have some initial energy K_i, and they acquire more kinetic energy by passing through a potential difference V maintained between C and A. The charges that pass through the hole in A proceed on with kinetic energy

$$\frac{1}{2}mv^2 = e\,V + K_i.$$

In Fig. 12.1 the dashed circle shows where a uniform magnetic field B can be applied directed out of the plane of the paper. When the field is off,

---

[1] Heinrich Hertz, *Electric Waves*, MacMillan & Co., London and New York, 1893.
[2] W. Hallwachs, *Ann. d. Phys.* **33**, 301 (1888).

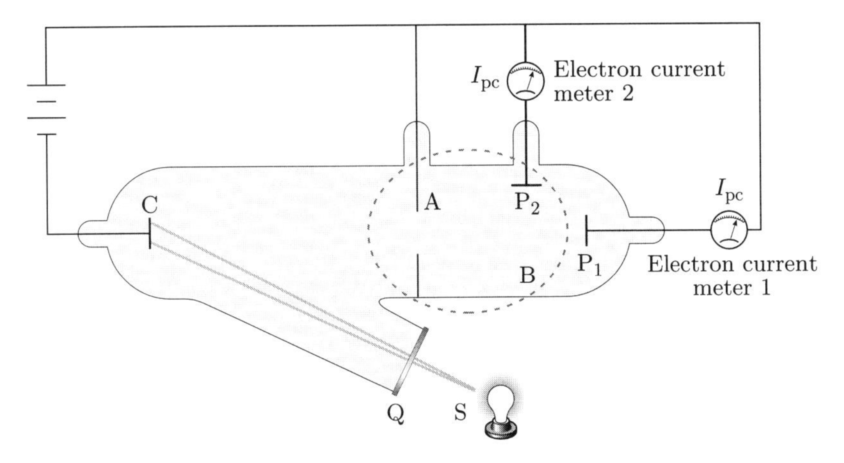

FIGURE 12.1 Lenard's apparatus for measuring q/m and K_{max} of photoelectrons. The dashed oval curve surrounds the region where a magnetic field is applied perpendicular to the plane of the page.

charges coming through the hole A strike the anode P_1 and the photocurrent is read there; when the field is on and properly adjusted, the charged particles bend in a circle of radius R onto P_2. One can tell when B is correct because then the maximum photocurrent is read from P_2. R is measured from the geometry of the apparatus. Because $mv = eBR$, it is easy to combine the energy equation with this to show that

$$\frac{e}{m} = \frac{2V}{B^2R^2}. \tag{1}$$

Equation 1 is the same as Eq. 6 derived in Chapter 8. Lenard measured q/m to be 1.2×10^{11} C/kg and concluded that the emitted charges were electrons.

EXERCISES

1. How does this value compare with the value of e/m obtained by Thomson as described in Chapter 8? With the currently accepted value? Why do you suppose people measuring values of e/m that differed by $\approx 50\%$ or more all agreed that they were seeing the same particle?

Lenard observed that some photocurrent flowed for positive voltages up to 2.1 V applied to C. He realized that this meant that some of the electrons were emitted with enough energy to escape the attraction of plate C even when it was positively charged. When the voltage on C became too positive, that is, greater than 2.1 V, no more charge flowed. This result meant that the

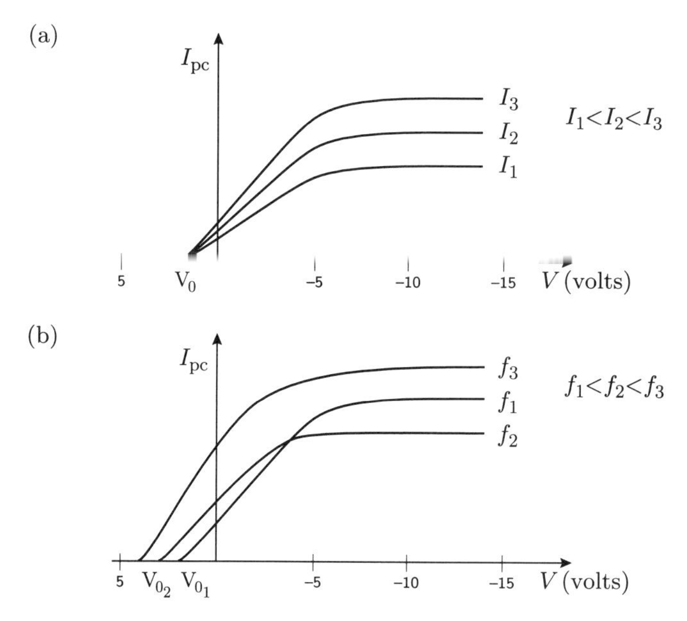

FIGURE 12.2 Photocurrent I_{pc} vs. cathode voltage V (a) for different intensities I_1, I_2, I_3 and (b) for different frequencies f_1, f_2, and f_3 of incident light. Increases in intensity increase the photocurrent but do not change the voltage V_0 corresponding to K_{max}; only increases in frequency increase K_{max}.

photoelectrons had a *maximum* kinetic energy of 2.1 eV. Lenard's most surprising observation was that an increase in the intensity of the incident light led to a proportional increase in the number of photoelectrons but did not change their maximum energy! Only when Lenard varied the frequency of the incident light did the maximum energy change. Later, more refined experiments confirmed these results and showed that the higher the frequency of incident light, the higher would be the maximum energy of the emitted electrons. In Fig. 12.2 schematic plots of photocurrent I_{pc} vs. the cathode voltage V show the effects of changing the intensity and also the frequency of the illuminating light.

EXERCISES

2. When you shine ultraviolet light on a zinc plate connected to a charged electroscope, it discharges when negatively charged but not when positively charged. Use the discovery that the charges emitted in the photoelectric effect are electrons to explain these observations.

There was another surprising property of the photoelectrons. They began to flow with no delay as soon as the cathode was illuminated. Why was this unexpected? It was unexpected for the same reason that it was a surprise to find that the maximum energy of the electrons did not depend on the intensity of light. Recall that the model of light as a wave implies that light emitted from a source spreads out through space, becoming ever weaker just like a water or sound wave.

The fact that the intensity of light from a point source of light decreases as $1/r^2$ as you move a distance r from the source is easily understood from the geometry of a sphere. As an example, suppose we use the Sun, which emits a total of $W_0 = 3.9 \times 10^{26}$ J/s in the visible spectrum. By the time this radiation reaches Earth, 1.5×10^{11} m from the Sun, its energy is spread over a sphere of area

$$4\pi r^2 = 4\pi(1.5 \times 10^{11})^2 = 2.8 \times 10^{23}\,\mathrm{m}^2,$$

so that measurements of the solar energy incident above the Earth show that the amount of energy incident per second per unit area (which is how we define intensity) is

$$\frac{3.9 \times 10^{26}}{2.8 \times 10^{23}} = 1390\,\mathrm{J\,s^{-1}\,m^{-2}}.$$

As the sunlight travels further and further, the area of the surrounding sphere grows as r^2, so the energy is spread over a larger area, and its intensity drops as $1/r^2$.

Thus, at the planet Jupiter, which is 5.2 times further from the Sun than Earth, the intensity of sunlight striking the outer atmosphere of Jupiter is

$$\frac{1390}{(5.2)^2} = 51\,\mathrm{J\,s^{-1}\,m^{-2}},$$

and so on, the light from our Sun ever decreasing to the ends of space.

To see what is surprising about the fact that photoelectrons are emitted as soon as light strikes the metal, let's make some estimates of how quickly an atom could accumulate energy from a wave and get enough to eject the photoelectron. We could use sunlight to produce the photoelectric effect, selecting out a narrow band of frequencies, say between 700.0 THz and 700.3 THz (a deep blue light). This range contains about 7×10^{-5} of the total energy in the sunlight, and only half of that reaches the ground. Therefore, the intensity of the light for our experiment might be in the range of

$$3.5 \times 10^{-5} \times 1390 = 0.049\,\mathrm{J\,s^{-1}\,m^{-2}}.$$

A beam of light with this intensity is very easy to see.

From Lenard's (and later) research we know that it takes different amounts of work to release an electron from different metals. Suppose we have a metal in which it takes about 2.9 eV to release an electron. How long do you expect it to take light with an intensity of 0.049 $\mathrm{J\,m^{-2}\,s^{-1}}$ to supply the necessary 2.9 eV to an atom? If the energy of light is spread smoothly over space, then the rate that energy is supplied to an atom is $W_a = I\pi r^2$, where πr^2 is the cross-sectional area of the atom of radius r and I is the intensity of the light. For an atom with a typical radius of 0.2 nm,

$$W_a = 0.049\pi(0.2 \times 10^{-9})^2 = 6.2 \times 10^{-21}\,\mathrm{J\,s^{-1}},$$

or

$$W_a = 0.038\,\mathrm{eV\,s^{-1}}.$$

From this result it follows that it would take about 76 seconds before the incident light could supply the 2.9 eV the atom needs to release an electron.

EXERCISES

3. Complete the calculation to show that it would take about 76 seconds, or well over a minute, to release one electron under the assumptions discussed above.

Einstein's Explanation: $E = hf$

The essentially instantaneous emission of an electron was an extraordinary result, totally at odds with classical physics. Sir William Bragg described the strangeness of the effect very well when he said:

> It is as if one dropped a plank into the sea from a height of 100 feet, and found that the spreading ripple was able, after travelling 1000 miles and becoming infinitesimal in comparison with its original amount, to act on a wooden ship in such a way that a plank of that ship flew out of its place to a height of 100 feet.

Think about this image. When you understand it, you will understand why the photoelectric effect was so remarkable. No matter how low the intensity of the light, energy was either delivered instantly and completely to the atom or not at all.

The idea of the continuous, smooth spreading of electromagnetic energy must simply be wrong. Einstein was the first to accept that possibility. He generalized an idea of Max Planck and asserted that light (both the visible part and all other forms of electromagnetic energy) carries energy in indivisible amounts, or "quanta." Each quantum of light of frequency f has an energy of hf, where h is a constant known as "Planck's constant" with the value 6.63×10^{-34} J s.

EXERCISES

4. Find the value of h in eV s, units that will often be useful in this book.

Your answer $h = 4.15 \times 10^{-15}$ eV s means that light of frequency 700 THz consists of many packets, or quanta, each carrying an energy of $700 \times 10^{12} \times 4.15 \times 10^{-15} = 3.7$ eV.

If that is the case, said Einstein, then we can understand that each electron is emitted immediately following the absorption of a single quantum of light. It is then evident that reducing the intensity of the light just reduces the number of quanta, but not the time it takes a particular quantum to be absorbed and produce the emission of an electron.

Also, it is clear from conservation of energy that no electron can come out with more energy than the quantum of light brought to it in the first place. In fact, experimental evidence showed that even the electrons emitted with maximum energy did not have the full energy supplied by the "photon," as a quantum of light is often called. Einstein understood that it would take some energy to break the electron loose from the surface; we call this energy the "work function" of the metal and often represent it by the lowercase Greek letter phi, ϕ. Einstein summarized his ideas by a simple statement of the conservation of energy,

$$K_{max} = \frac{1}{2}mv_{max}^2 = hf - \phi, \tag{2}$$

which says that the maximum kinetic energy a photoelectron can have is the amount carried in by the photon less the energy used to break the electron loose.

Experimental Verification of Einstein's Equation

Precise experimental verification of Einstein's equation was difficult for practical reasons. The energies of emitted photoelectrons depend sensitively on the state of the metallic surface. Metal surfaces always have oxide layers, and usually surfaces have thin films of oil or other contaminants. It is difficult to prepare surfaces so that from experiment to experiment the surface is in the same state. Millikan[3] partially solved this problem by using a machine tool in the vacuum to shave off a thin layer of metal to make

[3]R.A. Millikan, A direct photoelectric determination of Planck's "h," *Phys. Rev.* **7**, 355–388 (1916); Einstein's photoelectric equation and contact electromotive force, *Phys. Rev.*7(Second series), 18–32 (1916).

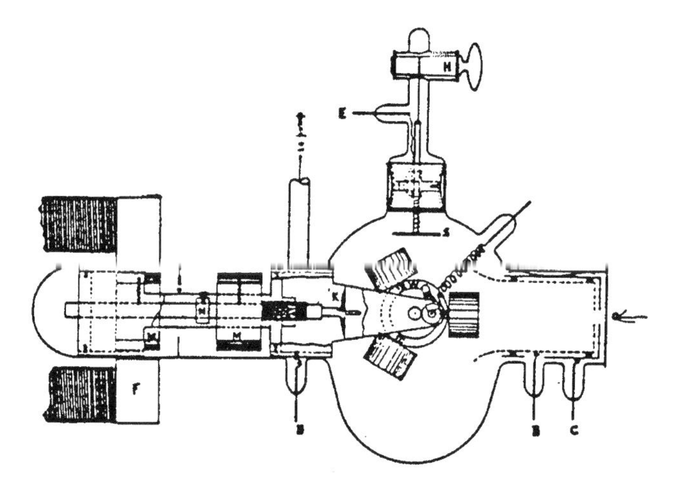

FIGURE 12.3 Diagram of Millikan's photoelectric effect apparatus. *Taken with permission from R.A. Millikan, Phys. Rev. Vol. 7, 355–388 (1916), ©1916 The American Physics Society*

a surface in a well-defined state of cleanliness. Even then, it was hard to get reproducible results because a vacuum is never perfect, and a surface will quickly oxidize and pick up contaminants after it is cleaned. Millikan's results were convincing because he found an independent way to determine ϕ, and then he showed that there was a way to prepare the surface so that ϕ remained unchanged over several weeks. The surface was quite dirty, but it was reproducibly dirty. That was enough to make his measurements meaningful.[4]

Millikan measured the maximum kinetic energy of photoelectrons emitted from sodium metal illuminated with various wavelengths of light. His apparatus, shown schematically in Fig. 12.3, was entirely contained in vacuum. The wheel W would rotate one of the three metal cylinders—one of sodium metal, one of lithium, and one of potassium—under the knife K, which then shaved off a layer of metal to clean the surface. The cylinder was then rotated around to face a window O through which came a beam of monochromatic light.

The energy of the electrons was determined by measuring the photocurrent versus the electric potential difference between the sodium metal and the wire mesh B and collector C. This electric potential difference is the cathode voltage, V, and when it is positive, it impedes the flow of the electrons

[4]It is interesting to note that this is another situation where inadequate vacuum resulting from the limits of the technology of the day caused much confusion, which was cleared up only by ingenious design of the experiment and by carrying it out very carefully.

from the sodium cathode to the mesh and the anode. The cathode voltage was varied until the photocurrent vanished. The value of V_0 at which electrons just cease to flow corresponds to the maximum electron energy, *i.e.*, K_{max}, because

$$\frac{1}{2}mv_{\text{max}}^2 = eV_0$$

(after correcting for the voltage produced by the battery action of the different metals of which the apparatus was made).

Millikan made careful measurements using several different wavelengths of light on lithium, sodium and potassium. His data for sodium are given in Table 12.1. These data illustrate an important difficulty of experiments that try to determine the point at which a quantity becomes zero. Accurate location of this point is usually difficult because there are always a few electrons around from stray light or other sources, and they obscure the zero point. The answer is found by measuring V at points where the photocurrent is not zero and then determining V_0 by extrapolation, as shown in Fig. 12.4. There is always some guesswork in such extrapolation because the data points that will best inform you about where the zero point is are also the least reliable data points because they are small and more like the obscuring background. Millikan's extrapolated values are given in the last row of Table 12.1.

TABLE 12.1 Experimental Values of Photocurrent I_{pc} vs. Cathode Voltage V for Sodium Illuminated with Six Different Wavelengths of Light[a]

546.1 nm		433.9 nm		404.7 nm		365.0 nm		312.6 nm		253.5 nm	
V_0 (V)	I_{pc} (mm defl)	V_0 (V)	I_{pc} (mm defl)	V_0 (V)	I_{pc} (mm defl)	V_0 (V)	I_{pc} (mm defl)	V_0 (V)	I_{pc} (mm defl)	V_0 (V)	I_{pc} (mm defl)
0.253	28	0.829	44	0.934	82	1.353	67.5	1.929	52	2.452	68
0.305	14	0.889	20	0.986	55	1.405	36	1.981	29	2.568	38
0.358	7	0.934	10	1.039	36	1.458	19	2.034	12	2.672	26
0.410	3	0.986	4	1.091	24	1.510	11	2.086	5	2.777	16.5
				1.143	10	1.562	4			2.882	8
				1.196	3						
0.46 V		1.03 V		1.21 V		1.59 V		2.13 V		3.03 V	

[a] Adapted from Millikan, A direct photoelectric determination of Planck's "h," *Phys. Rev.* **7** 355–388 (1916). Three entries from his table have been corrected to agree with his graphs, and every entry has been corrected for 2.51 V of contact potential. The bottom row of the table contains the values of V_0 at which I_{pc} goes to zero as determined by Millikan's extrapolation. The current was read as millimeters of deflection (mm defl) on the scale of an electrometer.

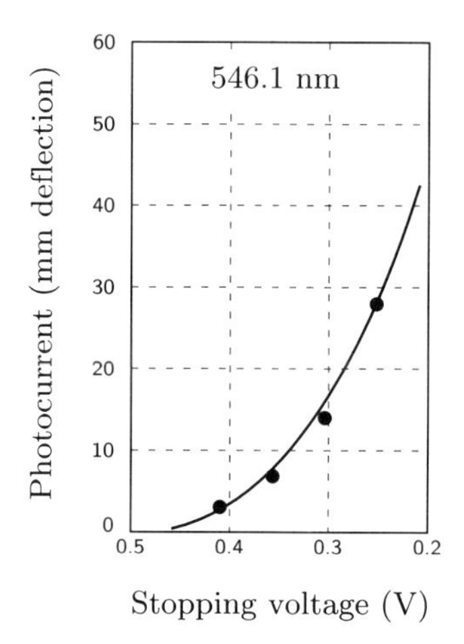

FIGURE 12.4 Use of extrapolation to determine the cathode voltage V_0 at which the photocurrent $I_{pc} \to 0$ for 546.1 nm light on sodium.

EXERCISES

5. Try some extrapolation yourself. Find the zero points for 433.9 and 404.7 nm light. How well do your values agree with Millikan's?

EXAMPLES

1. How can you find the work function from Millikan's data? For 546.1 nm light the cathode voltage at which the photocurrent becomes zero (found by extrapolation) is 0.46 V; therefore, the maximum energy K_{max} of electrons from the photoelectric effect induced with this light is 0.46 eV. Because the energy of a 546.1 nm photon is 1240/546.1 = 2.27 eV, the work function must be 2.27 − 0.46 = 1.81 eV.

EXERCISES

6. Calculate the work function using Millikan's data for λ = 433.9, 404.7, 365.0, 312.6, and 253.5 nm. How do your answers compare among themselves? With the value(s) given in Table 12.2?

7. Consider Eq. 2. In a plot of K_{max} of the photoelectrons vs. the frequency of the incident light, what functional dependence and shape of curve do you expect? What should be the slope?

TABLE 12.2 Work Functions of Some Metals

Metal	Work Function (eV)
Silver	4.73
Aluminum	4.20
Calcium	2.7
Cesium	1.9
Potassium	1.76 to 2.25
Sodium	1.90 to 2.46
Rubidium	1.8 to 2.2
Copper	4.1 to 4.5
Iron	4.72
Nickel	5.01
Platinum	6.30

8. Use Millikan's data to determine values of K_{max}. Plot them versus frequency. Do the slope and intercept agree with what you would expect from Eq. 2? Your results should show excellent agreement between Millikan's measurements and Einstein's predictions.

So Einstein was right. Electromagnetic radiation has its energy in packets. It is quantized. Light of frequency f has a smallest indivisible amount of energy hf. Light possesses a kind of atomicity; it consists of photons.

EXAMPLES

2. What is the maximum velocity of photoelectrons emitted when 250 nm light strikes a clean aluminum plate? First find hf and ϕ and then use Eq.2 to find the maximum kinetic energy of an emitted electron. For Al, $\phi = 4.20$ eV (see Table 12.2). For 250 nm light,

$$hf = \frac{hc}{\lambda} = \frac{1240}{250} = 4.96\,\text{eV}.$$

In this calculation we have used $hc = 1240\,\text{eV nm}$ in order to carry through the calculations in electron volts. It is convenient to know that $hc =$

1240 eV nm, for then you can divide by $\lambda = 250$ nm without converting from nanometers to meters.

Therefore,

$$\frac{1}{2}mv_{\max}^2 = 4.96 - 4.20 = 0.76\,\text{eV}.$$

And now you can find $v_{\max}$ by rewriting the above to use mass in units of eV/c^2,

$$\frac{1}{2}mv_{\max}^2 = \frac{1}{2}mc^2\left(\frac{v_{\max}}{c}\right)^2 = 0.76\,\text{eV},$$

which you can solve for $v_{\max}/c$:

$$\frac{v_{\max}}{c} = \sqrt{\frac{2 \times .76}{mc^2}} = \sqrt{\frac{1.52}{511000}},$$

so

$$v_{\max} = 1.73 \times 10^{-3}c = 5.2 \times 10^5\,\text{m}\,\text{s}^{-1}.$$

12.3 PHOTOMULTIPLIER TUBES: AN APPLICATION OF THE PHOTOELECTRIC EFFECT

Within five years of its discovery the photoelectric effect was used to measure ultraviolet radiation from the Sun. Since then, many instruments for detecting and measuring light have been based on the photoelectric effect. One of these, the photomultiplier tube, is widely used in industry and in research. This device can detect individual photons; it can directly demonstrate the granularity of light.

How the Photomultiplier Tube Works

The photomultiplier uses two different effects: the photoelectric effect and electron multiplication. When a photon strikes a surface coated with a material with low work function, an electron is sometimes emitted. This is just the photoelectric effect, and it represents a conversion of light energy into an electric current, i.e., into the motion of an electric charge.

A single electronic charge is difficult to detect, and it is desirable to amplify it. This is done by placing nearby a second surface at a voltage positive with respect to the first. The electron emitted from the first surface is then accelerated toward the second surface. It gains enough energy so that when it strikes the second surface, it causes the emission of several electrons. This process is called "secondary emission," and it multiplies the electrons.

The multiplication can be repeated by placing a third surface nearby with a voltage positive relative to the second. Then if the surfaces are properly shaped and arranged, the electric fields associated with the voltage difference between them will direct all the electrons emitted from the second surface onto the third one. Each impacting electron causes the emission of several electrons, and the multiplication repeats, as shown in Fig. 12.5.

Parts of a Photomultiplier Tube

The surfaces that emit and collect electrons are called the "electrodes" of the photomultiplier. The first electrode, where the photoelectric effect occurs, is called the "photocathode." The last electrode, where the multiplied electrons emerge as a current, is called the "anode." (You can see that these names are adapted from Faraday's names for the parts of an electrolytic cell.) The electrodes between the photocathode and the anode are called "dynodes." It is at the dynodes that electron multiplication occurs.

An electron from the photocathode passing down a string of dynodes each 100 volts higher than the preceding one can give rise to an overall amplification from 10^3 to 10^8. Exactly what amplification occurs depends upon the voltage between the dynodes, the material of which the dynodes are made, the total number of dynodes in the string, and the particular geometric arrangement of the dynodes. Figure 12.5 shows schematically one possible arrangement of photomultiplier electrodes.

EXERCISES

9. What is the overall amplification of a photomultiplier tube that has 9 dynodes and an electron multiplication factor of 6 at each dynode?

Among the many practical considerations in the design of photomultiplier tubes is finding photocathode materials that can emit an electron for long wavelengths of light, i.e., materials with especially low work functions. Even with enough energy, not every photon causes the emission of an electron. Therefore, we look for materials in which production of a photoelectron is both energetically possible and maximally probable. Some alkali metals produce only one photoelectron for every 1000 incident photons, while some mixtures of alkalis, e.g., the multialkali Na K Sb Cs, can produce as many as 300 electrons for every 1000 photons. This ratio is very important and is called the "quantum efficiency." The two examples here illustrate quantum

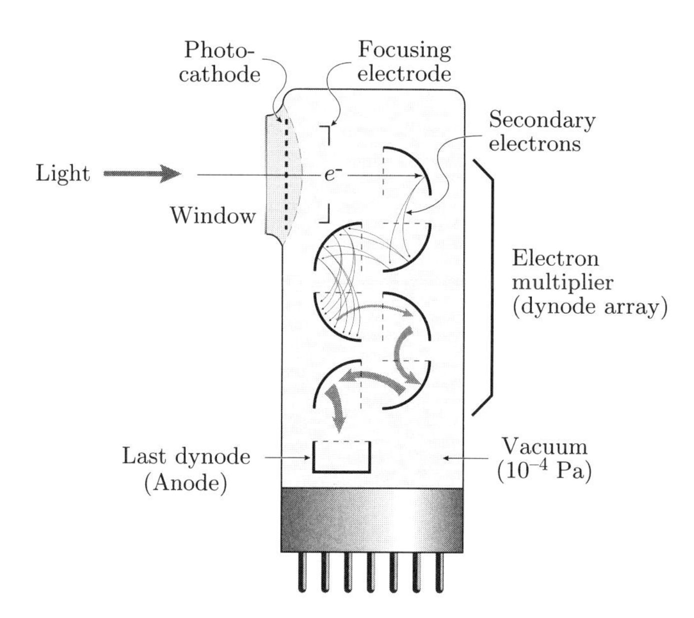

FIGURE 12.5 Schematic diagram of electrodes and electron multiplication in a photomultiplier tube.

efficiencies of 0.1% and 30%. A quantum efficiency of 30% is about as large as you can get with visible light.

Scintillation Counting of Radioactivity: A Useful Application

When an energetic charged particle or a high-energy photon strikes a piece of matter, it may impart all or some of its energy to an electron in the matter. The electron's energy will often go to make a small flash of light, and if the crystal is transparent, as is, for example, sodium iodide doped with thallium, NaI(Tl), this flash can be detected. A small flash of light is called a "scintillation." By detecting and counting the scintillations, one detects and counts incident particles. Even better, the size of the light flash is proportional to the amount of energy the particle or incident photon leaves in the crystal. By comparing the amounts of light in different scintillations one compares the energies of incident particles. The scintillator converts the energy of an invisible charged particle or high-energy photon into a pulse of low-energy visible photons.

The photomultiplier tube is very effective for counting such scintillations. The tube (see Fig. 12.5) is arranged to view the crystal into which the incident particles are directed. Photons from a scintillation strike the photocathode of the photomultiplier, which emits a tiny pulse of electrons. These are then amplified, and there appears at the anode an electrical pulse signaling the presence of an incident particle; the pulse size is proportional to the energy deposited in the crystal by the detected particle.

Incident high-energy photons, say 1 keV or more, can release electrons inside the crystal by the photoelectric effect; i.e., the incident photon may be absorbed by an atom, which then emits an electon. The electron, if it does not escape from the crystal, interacts with other atoms of the crystal and produces the scintillation that can be observed with a photomultiplier. There are several ways that an incident particle can energize an electron in the crystal, but however the electron gets its energy, there is then a probability that some of it will go to produce a scintillation. Notice that a low-energy photoelectric effect is used to detect and measure a higher-energy one.

12.4 SUMMARY

Electromagnetic radiation comes in discrete packets of energy called photons. For radiation of frequency f, the energy of a single photon is hf, where h is an experimentally determined constant called "Planck's constant." Its value is 4.15×10^{-15} eV s, but it is especially useful to *remember* that $hc = 1240$ eV nm.

Einstein hypothesized this granularity of light in order to explain the photoelectric effect. He showed that if light came in discrete quanta, then radiation of frequency f upon being absorbed into a surface could release electrons with a maximum kinetic energy

$$\frac{1}{2}mv_{\text{max}}^2 = hf - \phi,$$

where ϕ is the work function of the surface. Millikan's experimental measurements verified Einstein's predictions.

PROBLEMS

1. You have been put in the basement of a building from which you can escape only by riding up to the ground floor on a coin operated elevator that requires $2. Someone passes you a $5 bill through the bars of the basement window. Fortunately, the elevator's money slot makes change. There is also in the elevator a video game that requires $0.50 per game.

a. What is the maximum amount of money that you can have left when you reach the ground floor?

b. What is the minimum amount of money that you can have left when you reach the ground floor?

c. How is this situation analogous to the photoelectric effect?

2. Light is directed onto a metal surface for which the work function is 2 eV. If the light's frequency f is such that $hf = 5$ eV, what is the maximum energy with which an electron can be emitted from the surface?

3. Light with wavelength of $\lambda = 450$ nm shines on a cesium sample, and a photoelectric current flows. With a retarding voltage of 0.85 volts the current goes to zero.

a. What is the maximum kinetic energy of electrons emitted by the cesium?

b. Find the work function for cesium.

4. A marvelous new metal, phonium, is found to have a work function of 1 eV.

a. If a photon of 3 eV energy strikes phonium, what is the maximum energy with which an electron can be emitted?

b. Light of wavelength 620 nm strikes phonium. What is the maximum energy of electrons emitted by the photoelectric effect under these circumstances?

c. What are two features of the photoelectric effect that led Einstein to postulate that light energy comes as multiples of some smallest, indivisible packet?

d. What is the maximum-energy photon that can be produced when electrons that have been accelerated through 40 kV are stopped by a sudden collision with a tungsten anode?

5. The schematic diagram in Fig. 12.6 shows an apparatus for studying the photoelectric (PE) effect. When light of wavelength 620 nm shines on the cathode, electrons are emitted. The current I of electrons that reach the anode across a potential difference V is measured on the ammeter A. (In what follows neglect any contact potential.)

a. What did Einstein conclude about the nature of light in order to explain the various features of the PE effect?

b. As the voltage V is varied from negative to positive values, the current I changes as shown in the graph above.

i. Why is the current of emitted electrons not zero when $V < 0$?

ii. What is the maximum kinetic energy of an electron emitted when 620 nm light strikes the cathode?

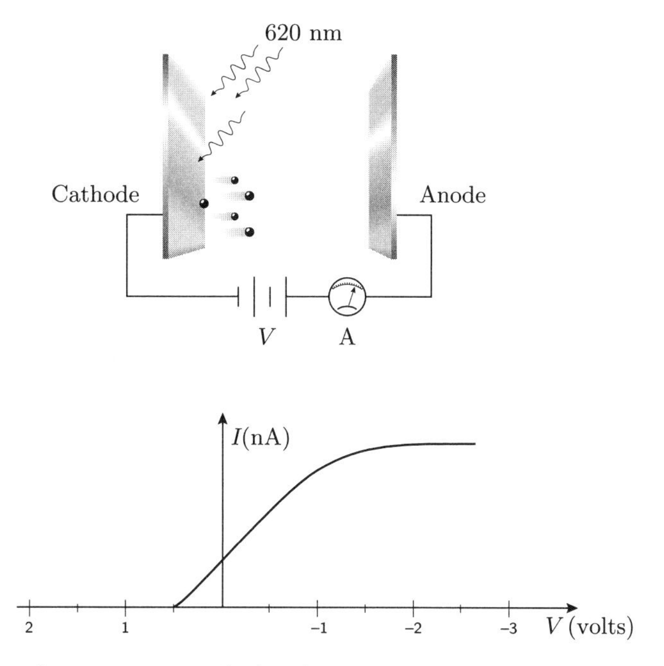

FIGURE 12.6 Photocurrent vs. cathode voltage (Problem 5).

c. On the same graph, sketch how the current I varies as V is changed when the cathode is illuminated with 310 nm light.

d. What is the work function (in eV) of the cathode?

6. Visible light from a mercury lamp is composed mainly of five wavelengths: 615 nm, 579 nm, 546 nm, 435 nm, and 405 nm. The lines are colored blue, violet, red, green, and yellow (in no particular order). When passed through a diffraction grating, the colors separate as shown in Fig. 12.7 (only one side of the pattern is shown).

a. Identify the color and wavelength of λ_1 in the figure.

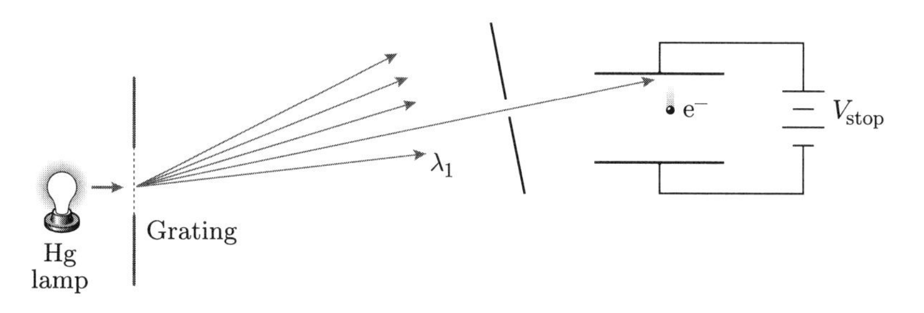

FIGURE 12.7 A photoelectron is just prevented from reaching the anode (Problem 6).

b. By appropriate placement of the apparatus, one of the five wavelengths is selected and illuminates the cathode in a photoelectric experiment. If the violet line is used, the stopping potential $V_{\text{stop}} = 0.762$ Volts. What is V_{stop} for the blue line?

c. What is the stopping potential when red light from the mercury lamp is used? Explain your result carefully.

7. An electric discharge causes atomic hydrogen to emit photons with energies of 1.89, 2.55, and 2.86 eV. In an experiment these are passed through a 1 mm wide slit and then through a diffraction grating that has 5000 rulings per centimeter. The photons then go on and strike a screen 1 m away.

a. What are the wavelengths of these photons?

b. Describe what appears on the screen and where. Be quantitative. Also give colors and tell how you know them.

c. Explain why these results suggest that light is a wave?

8. Millikan's photoelectric data for a lithium cathode are:

λ (nm)	433.9	404.7	365.0	312.5	253.5
V_0 (V)	0.55	0.73	1.09	1.67	2.57

Make a plot of the stopping potential versus frequency and find:

a. The value of Planck's constant.

b. The work function of lithium.

c. The cutoff frequency below which no electrons are emitted from the lithium cathode.

9. The stopping potential for photoelectrons emitted from a photocathode surface illuminated by light of wavelength 491 nm is 0.63 V. When the wavelength is changed to a new value, the stopping potential is found to be 1.43 V.

a. What is the new wavelength?

b. What is the work function of the surface?

c. Use Table 12.2 to identify the material of which this photocathode might be made.

C H A P T E R

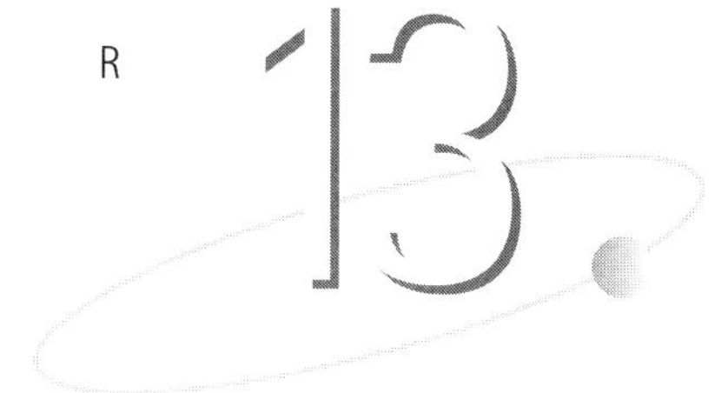

X-Rays

13.1 INTRODUCTION

As the nineteenth century came to a close, three discoveries revolutionized physics, set the stage for remarkable technological developments, and ushered in a new century and a new physics. In 1895 Roentgen discovered x-rays; in 1896 Becquerel discovered radioactivity; in 1897 J.J. Thomson discovered the electron. Each discovery became a new tool with which physicists explored the atom and made the discoveries that have led to the extraordinary technology that underpins our society today.

The pattern of exploration is easy to see. A new particle or radiation is discovered. Its properties are determined. A technology for its production is developed. As soon as it is under control, it is used to probe matter further. For example, the discovery of the electron led to the realization that atoms have structure of which electrons are part. Electrons became probes with which to explore the atom.

This same pattern applies to x-rays. In this chapter we discuss the technology for producing and measuring x-rays. A later chapter will discuss how x-rays are used to probe and reveal basic features of atomic structure.

13.2 PROPERTIES OF X-RAYS

Roentgen discovered x-rays while studying cathode rays. He noticed that certain materials outside and a little distance away from the cathode-ray tube would fluoresce, i.e., give off light, when cathode rays were striking the wall of the tube. He realized that some new form of radiation was traveling from the tube to make the material fluoresce. He did not then know that it was a particularly short-wavelength kind of electromagnetic radiation, so he called the mystery radiation "x-rays."

He quickly found that x-rays would cause fluorescence in many materials, that they would darken photographic plates, and that they would cause air to become electrically conducting. Fluorescence, photography, and ionization are still the principal ways to detect and measure x-rays.

13.3 PRODUCTION OF X-RAYS

You can produce x-rays by sufficiently rapid acceleration of any kind of charged particle. Electrons, because of their small mass, are the easiest and most practical to accelerate. The simplest way to accelerate electrons abruptly is to bring them to a sudden stop by directing a beam of them onto a metal target. The "Coolidge tube" shown in Fig. 13.1 works this way. Electrons are boiled off a hot cathode C, formed into a beam, and accelerated to a potential of some tens of thousands of volts, 10 kV to 100 kV, and then directed onto a water-cooled metal target T. The target is often made of copper or tungsten or molybdenum. The collision produces x-rays. The target is set at an angle so that x-rays can come off at a right angle to the tube without being absorbed by the target material.

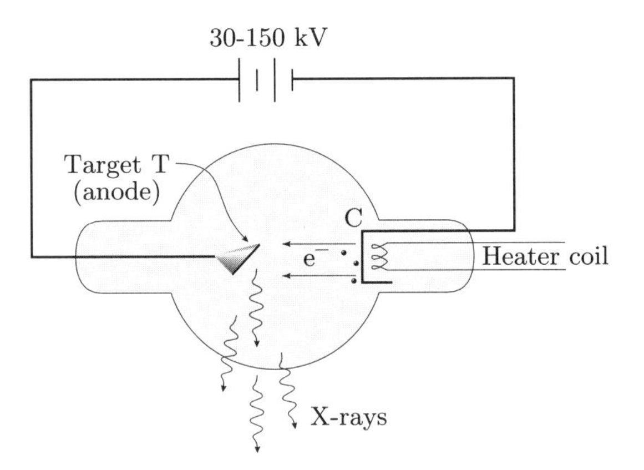

FIGURE 13.1 Diagram of a Coolidge tube for the production of x-rays. Electrons are boiled off from the cathode C, which is heated by the current produced by the power supply B. Electrons from C are accelerated toward target T by voltage from a high-voltage source. X-rays are produced when the electrons strike the target.

EXAMPLES

1. You can see why water cooling might be needed. An electron current of 10 mA at 50 kV delivers $10 \times 10^{-3} \times 50 \times 10^3 = 500$ W of power to the target. Water cooling is a practical way to carry off this heat.

EXERCISES

1. How much power would be delivered to the target by a 75 kV, 10 mA beam of electrons?

13.4 X-RAYS ARE WAVES

The wave nature of x-rays remained hidden for more than 15 years after their discovery. As you know from Chapter 9, the test for wave nature is to look for interference. The results were ambiguous. When x-rays passed around a sharp edge, faint fuzziness could be observed that might be diffraction. The problem was that if x-rays were waves, their wavelengths were so short that they did not exhibit much diffraction in the sizes of slits one could make in a machine shop and use in a laboratory. Indeed, the unsuccessful attempts at observing diffraction implied that the wavelength must be shorter than 0.1 nm, more than a thousand times less than the wavelength of visible light.

In 1912 von Laue suggested that nature provides slits or gratings with dimensions small enough to diffract waves as short as the x-rays might be.[1] From indirect evidence of shapes and sizes it appeared that crystalline matter was composed of atoms laid out in simple patterns with regular spacings of the order of a few tenths of nanometers. Von Laue realized that such arrays should act as three-dimensional diffraction gratings for x-rays, and he predicted that when a beam of x-rays of many different wavelengths passed through a crystal, interference would cause a single incident beam to emerge as a sheaf of beams the way a single beam on a diffraction grating emerges as a fan of beams. He predicted that x-rays passed through a crystal would make a pattern of dots on a photograph placed to intercept the emerging sheaf of beams.

---

[1]W. Friedrich, P. Knipping, and M. von Laue, "Interference phenomena with Röntgen rays," *Sitzungsberichte d. Bayer. Akademie der Wissenschaften*, 303–322 (1912); an abridged translation is given in *The World of the Atom*, edited by Henry Boorse and Lloyd Motz, Basic Books Publishers, New York, 1966, pp. 832–838.

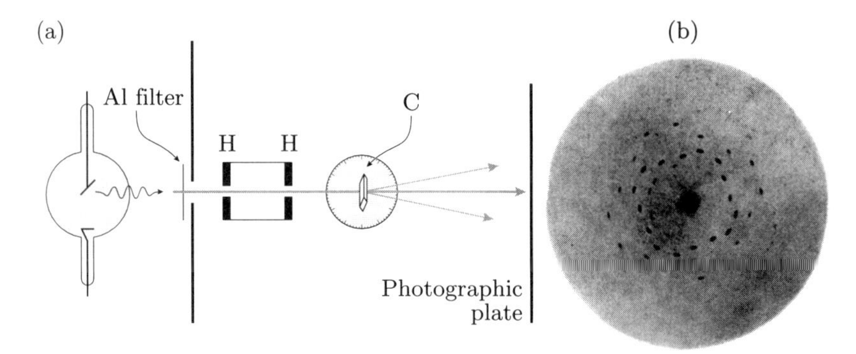

FIGURE 13.2 (a) Friedrich and Knipping's apparatus for producing and detecting x-ray diffraction: X-rays are collimated by apertures in the lead plates H and then diffracted by crystal C. (b) An example of the diffraction data they obtained: The black dots are places where diffraction concentrated the x-rays on the photographic plate. *By permission of Oxford University Press from M. Siegbahn,* The Spectroscopy of X-Rays, *p. 14, Oxford University Press, 1925*

Acting on von Laue's idea, Friedrich and Knipping did the simple experiment diagrammed in Fig. 13.2(a). They used a crystal of copper sulfate (because it was easy to obtain) and a beam of x-rays containing a broad range of wavelengths. The very first results confirmed von Laue's prediction. Fig. 13.2(b) shows the pattern Friedrich and Knipping obtained. Each dot in the pattern corresponds to a diffracted beam caused by constructive interference from some regular array of atoms in the three-dimensional crystal.

Nowadays, the pattern of dots produced when a beam of x-rays containing a broad range of wavelengths diffracts from the crystal is called a "Laue pattern," and the dots are called "Laue spots." For this work von Laue received the Nobel Prize in physics in 1914. The experiment established that x-rays are waves and confirmed that crystals are three-dimensional, ordered arrays of atoms.

13.5 THE BRAGG LAW OF CRYSTAL DIFFRACTION

The work of von Laue, Friedrich, and Knipping established the wave nature of x-rays, but it did not provide a tool for further exploring their properties. Such a tool was developed by W.H. Bragg and W.L. Bragg, father and son.[2]

[2]W.H. Bragg and W.L. Bragg, "Reflection of x-rays by crystals," *Proc. Roy. Soc.* (London), Series A, **88** (1913), 428–438; reprinted in part in *The World of the Atom*, pp. 845–852. See also *X-Rays and Crystal Structure*, W.H. Bragg and W.L. Bragg, G. Bell and Sons, Ltd., London, 1915.

They recognized that occurrence of Laue spots could be thought of as the result of reflections from parallel planes of atoms in the crystal. They pointed out (see Fig. 13.3) that when x-radiation of a single wavelength was incident on a crystal at a grazing angle θ with respect to planes of atoms in the crystal, a beam would result from constructive interference and would emerge at a grazing angle of θ relative to those planes of atoms whenever the wavelength λ, the grazing angle of incidence θ, and the spacing between the planes d satisfied the relation

$$2d \sin\theta = n\lambda. \tag{1}$$

Here n is called the "order" and can be any integer but is usually 1 and seldom greater than 2 or 3. Because the angle of emergence of the diffracted beam equals the angle of incidence as in reflection, the equation is called the "Bragg law of x-ray reflection" or, more often, just the "Bragg law." It is as though each plane of atoms in the crystal acts as a semitransparent mirror for x-rays.

Bragg's law can be derived by the following argument. In Fig. 13.3(a) is shown a line perpendicular to the incident wavefront striking a stack of parallel planes of atoms. If part of that wavefront reflects from the top plane and part from the next plane, the two will reinforce each other constructively if their difference in path length is an integer multiple of λ. For the two rays shown in Fig. 13.3(a) the path difference is $2d \sin\theta$. Bragg's law follows immediately.

It is important to realize that many different sets of planes of atoms can be imagined in the same crystal. This point is illustrated by the different sets of lines connecting the dots in the diagram of Fig. 13.3(b). Each set of planes can act as the source of diffracted x-rays. Of course, the spacings of these different sets of planes are different, and the angles of incidence of incoming x-rays vary with respect to the different sets of planes. The intensity of the diffracted beams drops off as the density of atoms in a plane decreases. The many spots in the Laue photograph (Fig. 13.2) arise because the incident beam contains a continuum of wavelengths, and the various planes of the crystal select out those wavelengths that satisfy Eq. 1 and reflect them as beams that are recorded on the film.

EXAMPLES

2. Suppose x-rays strike a crystal plane at a grazing angle of 20°, and suppose the spacing between the planes of the crystal is 0.2 nm. What wavelength will emerge at 20° to the crystal? The first-order value of λ is just

$$\lambda = 2d \sin\theta = 2 \times 0.2 \sin 20^\circ = 0.137\,\text{nm}.$$

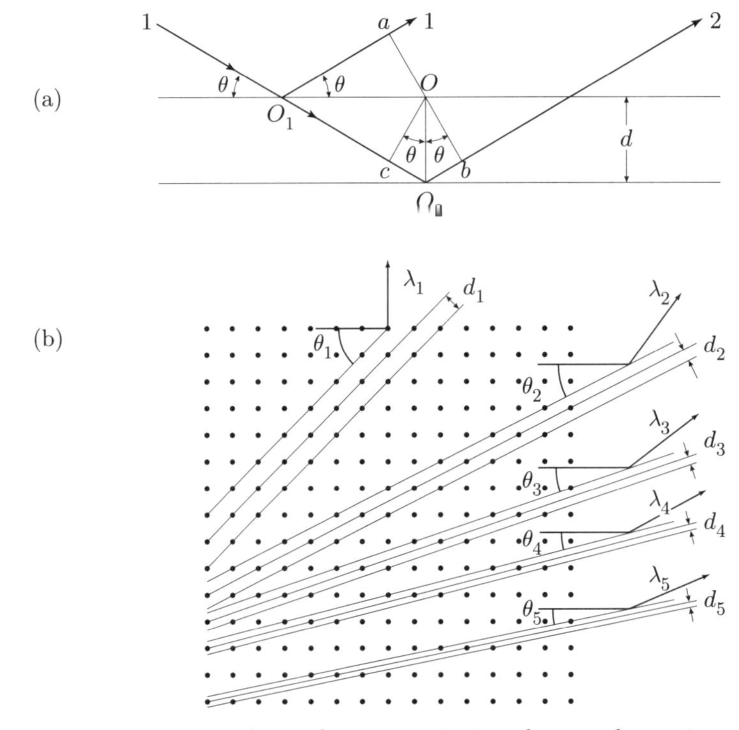

FIGURE 13.3 Bragg conditions for constructive interference of x-rays in a crystal. (a) Geometry for derivation of the Bragg law; (b) illustration of the presence of different sets of Bragg planes; in this illustration five different planes can satisfy the Bragg law for five different wavelengths and so produce five Laue spots.

There might also be a second-order or a third-order wavelength present.

EXERCISES

2. Suppose the crystal in the above example is rotated to 30° with respect to the incident beam and the detector is rotated to 30° with respect to the crystal. What is the wavelength of the x-rays that are detected?

Powder Diffraction Patterns

If you use x-rays of a single wavelength, you can obtain diffraction patterns from powdered samples instead of from single crystals. A powdered sample of crystalline material consists of many small crystals randomly oriented relative to each other; foils of hammered metal may also consist of many small randomly oriented crystals. When such a sample is irradiated with x-rays, the

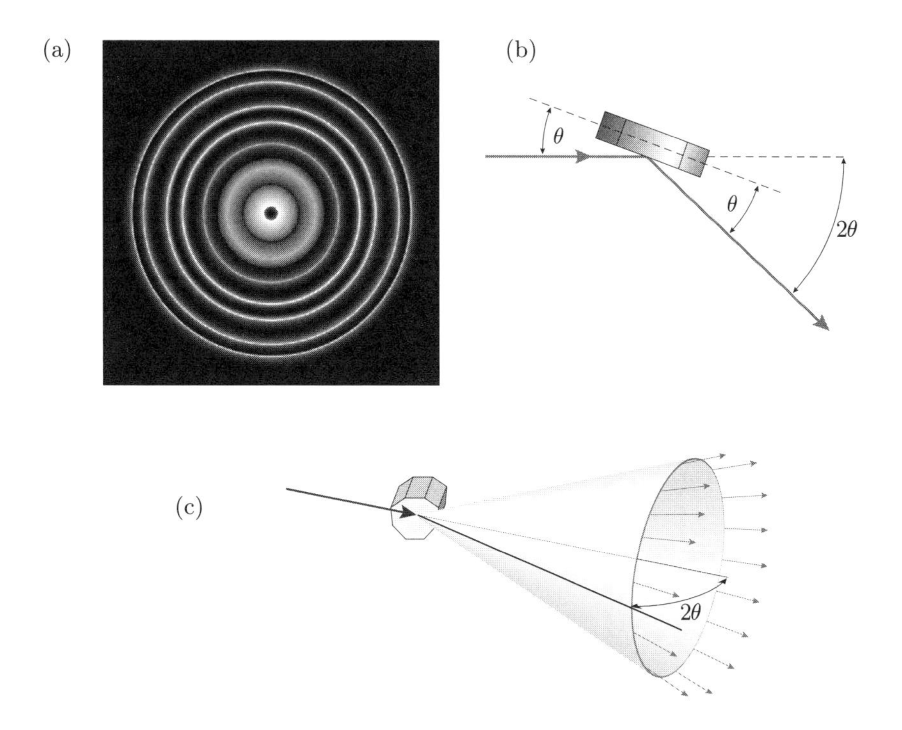

FIGURE 13.4 Why small randomly oriented crystallites produce ring-shaped x-ray diffraction patterns: (a) example of Debye–Scherrer diffraction; (b) schematic illustration of a crystallite satisfying the Bragg law for the incident x-ray wavelength; (c) randomly oriented crystallites will produce a cone of radiation at an angle 2θ to the axis of the incident beam; this forms the circular pattern on the photograph.

outgoing diffracted x-rays form a pattern of concentric rings like that shown in Fig. 13.4(a) and not the array of spots observed in the Laue experiments. Such diffraction from many small randomly oriented crystallites is called "Debye–Scherrer diffraction."

Each ring of the Debye–Scherrer diffraction pattern (Fig. 13.4) corresponds to a different spacing between planes in the crystal being irradiated. To see how a ring might arise, consider a single set of parallel atomic planes within a tiny crystal. If the crystal is tipped so that the incoming beam is incident at the angle θ satisfying the Bragg law, a beam will be diffracted and emerge at an angle 2θ relative to the incident beam and at an angle θ relative to the plane of the crystal, as illustrated schematically in Fig. 13.4(b). For the particular spacing d there will be a few other orientations of the crystal relative to the beam that satisfy the Bragg law. These orientations correspond to the higher orders, i.e., $n = 2, 3$, etc.

If the crystal can be tipped only about an axis perpendicular to the plane of the page, as in Fig. 13.4(b), the outgoing Bragg reflected beam will lie only in that plane. However, if the crystal were rotated about an axis parallel to the incident beam, the beam would go off at some angle out of the plane of the page. Therefore, from a sample consisting of a large number of tiny crystallites oriented every which way, there will be outgoing beams in a conical shape around the incident beam, as shown in Fig. 13.4(c). When these conical shapes strike a photographic plate some distance away, they produce circles like those shown in Fig. 13.4(a).

EXERCISES

3. Explain what pattern might appear on the photograph if the x-rays had a continuum of wavelengths? Why would this not be useful?

13.6 A DEVICE FOR MEASURING X-RAYS: THE CRYSTAL SPECTROMETER

The Braggs realized that diffraction from the planes of a single crystal spread x-rays out in space according to their wavelengths, and that this property could be the basis for constructing a spectrometer for x-rays. To make their x-ray spectrometer they arranged a crystal at an angle θ relative to the beam coming from an x-ray tube and placed a detector at an angle θ relative to the crystal, as shown in Fig. 13.5. According to Eq. 1, only those x-rays with wavelength $\lambda = 2d\sin\theta$ or, less likely, some simple fraction, e.g., $\lambda/2$, $\lambda/3$, will undergo Bragg reflection through the angle θ. Thus the device selects out from all the different wavelengths of incident x-rays the particular wavelengths that will produce a beam at angle θ. As θ is changed (by rotating the crystal), the wavelength of the Bragg reflected beam changes as determined by the Bragg equation. Then the detector is rotated to the new value of θ to intercept the outgoing beam of the new wavelength. By successive rotations of crystal and detector, through θ and 2θ respectively, you can measure the intensity of x-rays as a function of λ.

Determining the Spacing of Atoms in Crystals

Notice that to extract useful numbers you need to know either the crystal spacing d or the wavelength λ. If you know the crystal spacing, you can use the crystal spectrometer to measure x-ray wavelengths. If you know the wavelength, then you can use those x-rays to measure the spacings of various crystals and so learn how crystalline solids are put together. Physicists do both: They use crystals to study x-rays; they use x-rays to study crystals. But how do they get started? It is necessary to measure some crystal spacing or

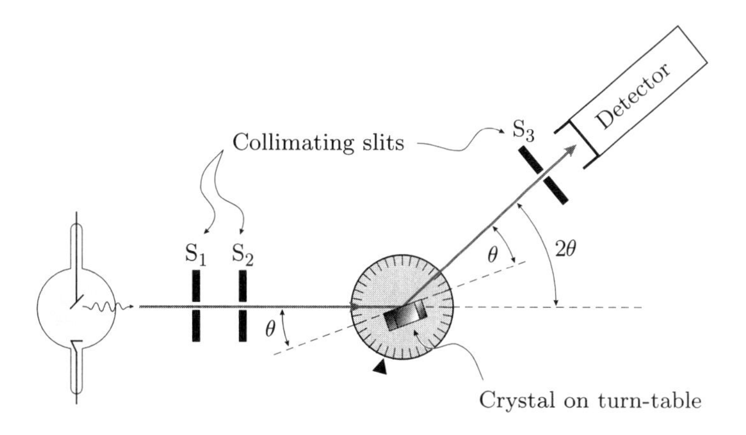

FIGURE 13.5 Schematic diagram of a Bragg spectrometer.

some x-ray wavelength. After that you can use one to measure others. Where do you get the first measurement?

From its Laue pattern the Braggs were able to infer that the Na and Cl atoms of a crystal of table salt sit on the corners of a cube equally spaced a distance d from one another. For such a simple structure the volume occupied by a single atom is d^3, and a numerical value of d^3 can be calculated from the molecular weight M, the density ρ of salt, and Avogadro's number N_A.

A mole of NaCl has a mass of $M = 35.453 + 22.99 = 58.443$ g, and the density of salt is $\rho = 2.163\ \mathrm{g\,cm^{-3}}$. Therefore, the volume of a mole of NaCl is

$$\frac{58.443}{2.163} = 27.01\ \mathrm{cm}^3.$$

Because 1 mole of NaCl contains 2 moles, i.e., $2N_A$, of atoms, each atom occupies a volume

$$d^3 = \frac{27.01}{2 \times 6.022 \times 10^{23}}\ \mathrm{cm}^3,$$

from which it follows that

$$d = 2.820 \times 10^{-8}\ \mathrm{cm} = 0.2820\ \mathrm{nm}.$$

It is common, especially in older books, to measure atomic-sized lengths in angstroms (Å) where $1\ \text{Å} = 10^{-8}\ \mathrm{cm} = 10^{-10}\ \mathrm{m} = 0.1\ \mathrm{nm}$. In these units the spacing between the planes of a rock salt crystal is $d = 2.82\ \text{Å}$. In modern textbooks, however, it is customary to use nanometers for atomic-sized dimensions, and we will do that.

Knowing the spacing of a rock-salt crystal, you can make quantitative measurements of the yield of x-rays as a function of wavelength, and you can find spacings of other crystals.

EXAMPLES

3. For example, one might take x-rays from a tungsten target, form them into a beam with slits, as in Fig. 13.5, and allow them to strike a crystal of rock salt. The x-rays emitted at an angle of, let's say, 25° will then have a wavelength of $\lambda = 2d \sin\theta = 2 \times 0.2820 \times \sin 25° = 0.2384\,\text{nm}$.

In this way the Bragg spectrometer is used to measure unknown x-ray wavelengths.

EXAMPLES

4. The x-rays coming from the rock-salt crystal in Example 13.3 are allowed to strike a calcite crystal ($CaCO_3$). The calcite crystal is rotated until there is a strong reflection from it. This occurs when the angles of grazing incidence and reflection are each 23.1°. From this measurement we deduce that the spacing of the planes of atoms in the calcite crystal is

$$d = \frac{0.2384}{2 \sin 23.1°} = 0.3036\,\text{nm}.$$

So now a known x-ray wavelength is used to find an unknown spacing of atoms, i.e., the x-rays are used to probe the structure of crystalline matter.

EXERCISES

4. At what angle would such a beam of x-rays reflect strongly from a quartz crystal? From a mica crystal?

It is important to keep in mind that there will be other sets of planes within the crystals, and these will also give rise to x-ray beams when the Bragg law is satisfied. The theory of these is well worked out, but we will not concern ourselves with it here. The simplest cases are enough to show how x-rays can be used to find new details of atomic structure.

The spacing of the calcite crystal is useful information because calcite was a crystal frequently used in early x-ray spectrometers. Its spacing and those of some other historically important crystals are given in Table 13.1.

TABLE 13.1 Spacing Between Planes of Commonly Used Crystals at 18°C

Crystal		d (nm)
Calcite	$CaCO_3$	0.3036
Rock salt	NaCl	0.2820
Quartz	SiO_2	0.4255
Mica	SiO_2	0.9963

13.7 CONTINUUM X-RAYS

Using known crystal spacings you can make a spectrometer from any convenient crystal and measure how much of each different wavelength is present in the x-rays. Ulrey used a Bragg spectrometer with a calcite crystal to measure the intensity of x-rays from a tungsten target as a function of the angle of orientation θ of the crystal. This angle is related to the wavelength by the Bragg law, Eq. 1.

Remember that in a Bragg spectrometer (Fig. 13.5) the detector is rotated to view an exit angle θ corresponding to the angle of grazing incidence θ. In other words, the detector must always be at an angle of 2θ relative to the axis of the incident beam of x-rays in order to register the intensity of the x-rays diffracted through the Bragg angle θ. Setting the crystal and detector at many different angles, Ulrey obtained the data shown in Table 13.2: the intensity distribution of the x-rays produced when energetic electrons strike a tungsten (W) anode in an x-ray tube.

Ulrey's data[3] are plotted in Fig. 13.6. They show two particularly interesting features: a continuous, smooth variation of intensity from long wavelengths down to some short wavelength and a sudden termination, or "cutoff," of the curve at that short wavelength.

The continuum is to be expected from the simple picture of how x-rays are produced. Not all electrons striking the target undergo the same acceleration; some are accelerated suddenly, some less so. The distribution of wavelengths should follow the distribution of accelerations. The largest intensity corresponds to the acceleration that occurs with greatest probability.

[3] C.T. Ulrey, "Energy in the continuous x-ray spectra of certain elements," *Phys. Rev.* **11**, 401–410 (1918).

TABLE 13.2 Intensity of X-rays vs. Angle of Grazing Exit θ from Calcite[a]

X-Ray Intensity (relative units)	θ (degrees)	X-Ray Intensity (relative units)	θ (degrees)
1.9	2.45	6.4	6.24
5.0	2.83	5.6	6.62
6.8	3.30	4.8	6.91
8.6	3.59	4.1	7.19
9.5	3.87	3.2	7.57
9.9	4.25	2.6	7.86
9.9	4.53	2.3	8.14
9.6	4.91	2.0	8.52
9.1	5.29	1.7	8.81
8.0	5.58	1.5	9.19
7.0	5.86		

[a] Taken from Ulrey, *Phys. Rev.* **11**, 401 (1918).

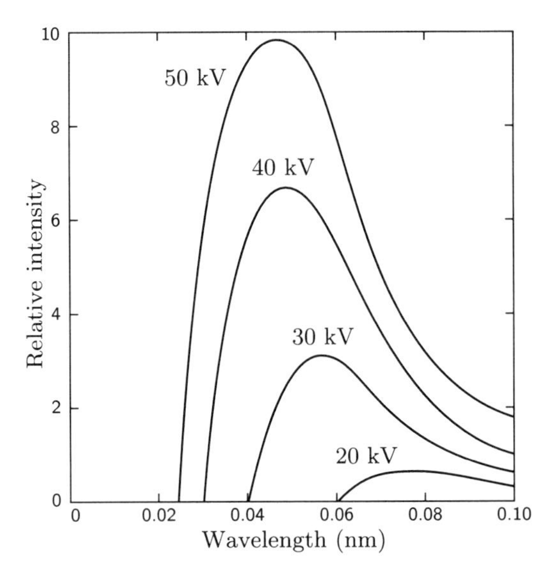

FIGURE 13.6 Relative intensity of continuum x-rays from electrons striking a tungsten target. Each curve is labeled with the voltage through which its corresponding electrons were accelerated. *Taken with permission from C.T. Ulrey, Phys. Rev. 11, 401–410 (1918) ©1918 The American Physical Society*

TABLE 13.3 X-Ray Tube Voltage and Cutoff Wavelengths

Tube Voltage (kV)	λ_{cutoff} (nm)
20	0.062
30	0.0413
40	0.031
50	0.0248

But the short wavelength cutoff warns us that this simple picture is incomplete. Although we might expect extremely sudden decelerations to be less probable, there is no reason in classical physics why their probability should suddenly drop to zero. Classical physics leads us to expect that the continuum will peter out at short wavelengths, not stop suddenly.

13.8 X-RAY PHOTONS

The short-wavelength cutoff is explained by the granular nature of electromagnetic radiation. Just as there are photons of visible light, so also are there x-ray photons, and x-ray energy comes in packets $E = hf = hc/\lambda$. Because x-ray photons cannot have energy larger than the maximum kinetic energy of the incident electrons, electrons of charge e accelerated through a voltage V cannot produce photons with energies greater than eV. Therefore, there will be a shortest possible λ, and it will be

$$\lambda_{\text{cutoff}} = \frac{hc}{eV}.$$

In Fig. 13.6 the curves are labeled with the accelerating voltage applied to the x-ray tube. As expected, the value of the cutoff wavelength becomes smaller as the accelerating voltage is increased. Let's calculate what the cutoff wavelength should be for a given accelerating voltage.

EXAMPLES

5. Electrons accelerated through 30 kV cannot deliver more than 30 keV of energy to the target. Therefore, the shortest possible x-ray wavelength is $hc/(eV) = 1240/30{,}000 = 0.0413$ nm. This agrees well with Ulrey's data.

EXERCISES

5. Predict the cutoff wavelengths for the curves labeled 40 kV and 20 kV in Fig. 13.6. How do your answers compare with Ulrey's data?

6. Using the data in Table 13.2 and the fact that Ulrey's spectrometer used a calcite crystal:

(a) Calculate the wavelengths corresponding to angles of reflection, and then plot the intensity of the x-radiation vs. λ.

(b) What is the energy of the photon that corresponds to the wavelength at which the intensity is a maximum?

(c) By extrapolation of your graph determine the short-wavelength cutoff.

(d) What is the voltage difference through which these electrons passed before they struck the tungsten anode of the x-ray tube?

7. Suppose you used a rock-salt crystal instead of calcite in your crystal spectrometer. At what angle of the detector relative to the crystal would the detected intensity be a maximum?

8. Ulrey observed that continuous x-ray spectra generated with electrons produced with the x-ray tube voltages shown in Table 13.3 exhibited sharp cutoffs at the wavelengths shown in the table. Predict what sort of a curve you will get if you plot f (or hf) vs. V. Explain your prediction.

9. In a color-television tube electrons pass through a potential difference of 30 kV before they strike the phosphorescent screen at the end of the tube. What is the maximum energy that an x-ray photon can have coming from this tube? What is the wavelength of that photon?

10. In the previous problem what (approximately) is the most probable energy of photon that will be produced by the TV tube?

11. Using the insights you generated to answer Exercise 8, derive a formula relating the voltage V of an x-ray tube to the maximum frequency f_{max} of the emitted radiation. Why is this formula familiar to you?

13.9 THE COMPTON EFFECT

Introduction

You have just been reminded that x-rays do not always behave like waves. Indeed, studies with the crystal spectrometer and other techniques for measuring x-rays showed that in some circumstances they behave like hard, featureless particles. Photons can collide with electrons and bounce off them like tiny BBs.

Bragg scattering is an important special case of the interaction of x-rays with matter, but it is not the only kind of x-ray scattering. A beam of x-rays directed at a crystal will result in the weak emission of x-rays at all angles relative to the incident beam. In the early 1920s, the American physicist Arthur Holly Compton studied this kind of x-ray scattering. He scattered x-rays of a single well-defined wavelength from a small piece of carbon and observed that at angles other than 0° some of the scattered radiation came out with its wavelength unchanged and some came out with its wavelength increased. He explained the change in wavelength as the result of a billiard-ball-like collision of an x-ray photon with a nearly free electron of the carbon atoms. Such a change in wavelength (or energy) of a photon after it scatters from a charged particle is called the "Compton effect."

The discovery and explanation of the Compton effect decisively persuaded physicists of the granularity of light. This was important because the idea that electromagnetic radiation, of which light is one example and x-rays are another, has a corpuscular nature was by no means entirely accepted even after Einstein explained the photoelectric effect in terms of the absorption of photons.

Compton Scattering

Figure 13.7 shows a diagram of the apparatus Compton used to exhibit the effect. X-rays of a well defined wavelength of 0.0708 nm came from the molybdenum anode of a specially designed x-ray tube and struck a small block of graphite at the point labeled R. The x-rays scattered from the carbon in all directions; collimating slits S_1 and S_2 selected those scattering through some particular angle θ. For the case shown at the left in Fig. 13.7 the scattering angle is $\theta = 90°$.

To measure the wavelengths of the x-rays scattered through an angle θ from the graphite, Compton let them strike a calcite crystal, which he rotated through a succession of angles. At each angle he rotated the detector to the angle that would look at outcoming radiation that satisfied the conditions of the Bragg law (Eq. 1). In this way Compton knew the wavelength of the radiation he was detecting, and he could measure its intensity from the

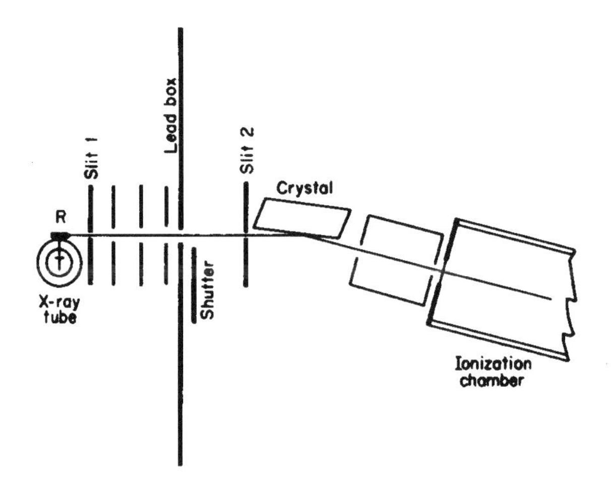

FIGURE 13.7 Diagram of Compton's apparatus. *Taken with permission from A.H. Compton, Phys. Rev. 22, 409–413 (1923) ©1923 The American Physical Society*

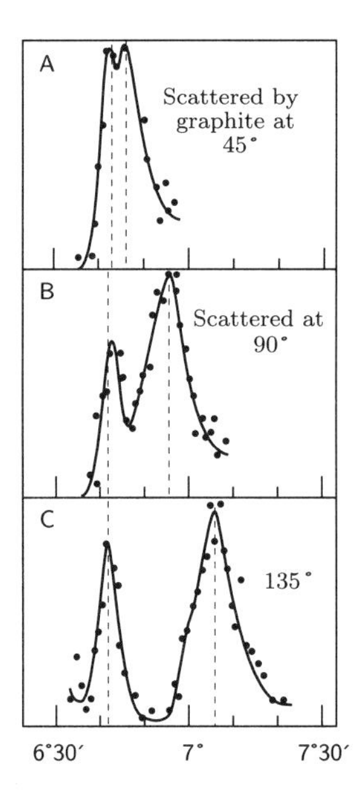

FIGURE 13.8 Compton's data showing the appearance of the longer-wavelength x-rays as the scattering angle is increased. *Taken with permission from A.H. Compton,* Phys. Rev. *22, 409–413 (1923) ©1923 The American Physical Society*

amount of ionization produced in the chamber of his detector. The result was a map of the intensity of the scattered radiation as a function of wavelength. Figure 13.8 shows Compton's data for scattering angles of 45°, 90°, and 135°.

Notice that Compton plots the angle of the calcite crystal along the x-axis. To find the wavelengths, you need to use Bragg's law.

EXAMPLES

6. For calcite $d = 0.3036$ nm. From the scale of Fig. 13.8 you can see that the primary peak occurs at $6^\circ\ 42' = 6.7^\circ$. The wavelength of the x-rays making this peak must then be

$$\lambda = 2 \times 0.3036 \sin 6.7^\circ = 0.0708\ \text{nm}.$$

The figure shows that a second peak emerges and moves toward larger Bragg angles as the scattering angle is increased. Larger Bragg angles correspond to longer wavelengths and thus to lower photon energies.

EXERCISES

12. Use the data in Fig. 13.8 to find the wavelength of the lower energy photons when the scattering angle is 90°.

Table 13.4 shows the data Compton obtained when he scattered 0.0708 nm radiation and looked at x-rays coming out at 135°.

EXERCISES

13. Make a graph of the data in Table 13.4 and draw a smooth curve though them. Label the peak corresponding to the incident photons; label the peak arising from Compton scattering. By how much is the wavelength shifted?

The presence of the shifted peak is the Compton effect. Compton explained the effect as due to elastic scattering, i.e., scattering like that of hard spheres bouncing off each other. One of the "spheres" was an electron in the carbon; the other was a packet of electromagnetic energy—the photon. Elastic scattering of a photon from an unbound charged particle is called "Compton scattering."

The fact that Compton's explanation worked so well led to general acceptance of the concept of the particle-like photon with energy hf, where h is Planck's constant and f is the frequency of the radiation.

TABLE 13.4 Relative Intensity of X-Rays Scattered from Carbon

$\theta = 135^\circ$ Intensity	Wavelength (nm)	$\theta = 135^\circ$ Intensity	Wavelength (nm)
4.3	0.0688	13.8	0.0734
0.5	0.0689	20.0	0.0739
[illegible]	0.0697	[illegible]	0.0745
8.0	0.0703	25.6	0.0750
13.0	0.0707	18.9	0.0754
7.8	0.0713	10.8	0.0760
3.0	0.0719	4.8	0.0765
2.8	0.0725	2.0	0.0773
11.0	0.0729		

Derivation of the Energy Change of a Compton Scattered Photon

Figure 13.9(a) is a diagram of the scattering process as Compton imagined it. A photon of energy hf enters from the left and strikes a stationary electron of rest mass m. After the collision, the electron recoils away with momentum p_e at an angle ϕ with respect to the line of motion of the incident photon, and a photon of energy hf' travels away at an angle of θ; the electron stops inside the target and is not measured. Because the electron has gained energy from the collision, the outgoing photon's energy of hf' will be less than that of the incoming photon hf.

Although the data in Table 13.4 are the wavelengths λ and λ', it is more convenient to discuss Compton scattering in terms of the corresponding frequencies f and f'. Our goal is to derive a relationship between f' and f and the scattering angle θ. We do this using the relativistically correct expressions for conservation of energy and momentum.

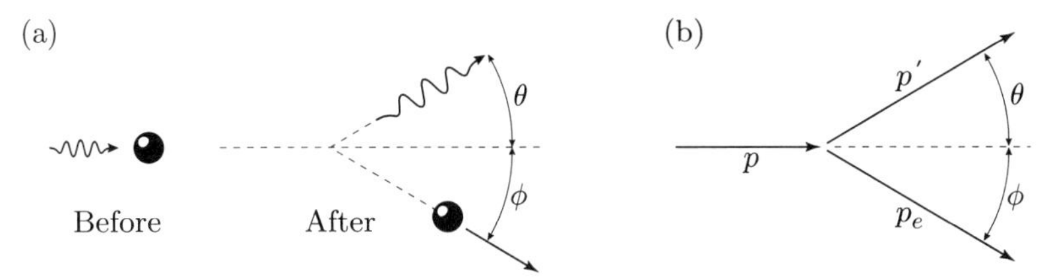

FIGURE 13.9 (a) Schematic representation of a Compton scattering event. (b) Momentum diagram of Compton scattering.

Start by noticing that the relation of photon energy to photon momentum is

$$E = hf = \sqrt{0 + p^2c^2}$$

because $m_0 = 0$ for a photon. Therefore, *for a photon*,

$$E = hf = pc.$$

EXERCISES

14. Show from the above equation that for a photon, $p = h/\lambda$.

Now use this simple relationship of photon energy and photon momentum to write down the total energy of the photon *and* electron before the collision and after. Before the collision, the energy is just the photon's energy plus the rest energy of the electron: $pc + m_e c^2$. After the collision, the total energy is the photon's somewhat lower energy $p'c$ plus the electron's larger energy, as shown below. Because energy is conserved, the total energy before the collision equals the total energy after:

$$pc + m_e c^2 = p'c + \sqrt{m_e^2 c^4 + p_e^2 c^2}, \tag{2}$$

where p_e is the momentum of the recoiling electron and m_e is its rest mass.

To find the relation of the scattering angle of the photon and its momentum after scattering, use conservation of momentum as well as conservation of energy. Conservation of momentum gives two more equations that you can use to eliminate p_e from Eq. 2 so that you can find the photon's momentum and thus its energy.

In Fig. 13.9(a) a photon travels toward positive x and scatters from an electron as shown. The electron recoils at an angle ϕ while the photon goes off at an angle θ relative to the x-axis. Figure 13.9(b) shows that there are two momentum equations, one for vertical, or y, momentum and the other for horizontal, or x, momentum. In the y direction the total momentum is initially zero, so after the collision the amount of momentum up must equal the amount down, and

$$p' \sin\theta = p_e \sin\phi. \tag{3}$$

The amount of momentum in the x direction before the collision equals the total amount in the x direction after, so

$$p = p' \cos\theta + p_e \cos\phi. \tag{4}$$

Our aim is to find p' of the outgoing photon in terms of the incident p and the scattering angle θ. This means that we need to eliminate both ϕ and p_e from Eqs. 2, 3, and 4.

Whenever you have an angle α appearing in trigonometric functions that you want to eliminate from a set of equations, the first approach is to see whether you can combine the trig functions in such a way that you can use the identity $\sin^2\alpha + \cos^2\alpha = 1$.

Rewrite Eq. 1 as

$$p - p'\cos\theta = p_e \cos\phi \tag{5}$$

and square it to get

$$p^2 + p'^2\cos^2\theta - 2pp'\cos\theta = p_e^2\cos^2\phi.$$

Square Eq. 3 to get

$$p'^2\sin^2\theta = p_e^2\sin^2\phi.$$

Add the two equations and use the trig identity to eliminate ϕ:

$$p_e^2 = p^2 + p'^2 - 2pp'\cos\theta. \tag{6}$$

EXERCISES

15. Show that Eq. 6 is correct by deriving it as above but include any missing steps.

Replacing p_e^2 in Eq. 2 with the expression given by Eq. 6, you reach the goal of eliminating the electron momentum. From this result you get an explicit expression for p' in terms of the incident p and the outgoing scattering angle θ:

$$\begin{aligned}(pc - p'c + m_ec^2)^2 &= m_e^2c^4 + p_e^2c^2,\\ (pc - p'c + m_ec^2)^2 - m_e^2c^4 &= p_e^2c^2 = p^2c^2 + p'^2c^2 - 2pp'c^2\cos\theta,\\ pp'(1-\cos\theta) &= m_ec(p - p'),\\ p' &= \frac{m_epc}{p(1-\cos\theta) + m_ec},\end{aligned} \tag{7}$$

which is what we wanted to find.

EXERCISES

16. Do this entire derivation from start to finish.

17. From Eq. 7 derive the following expression for f' in terms of f and θ,

$$hf' = \frac{hf}{\frac{hf}{m_e c^2}(1 - \cos\theta) + 1}, \tag{8}$$

which is very convenient for calculating how the energy of a photon changes when it scatters.

It is often helpful to express these photon energies relative to the rest energy of the electron. Notice that if you define the ratio of the photon energy to the electron rest energy as

$$x = \frac{hf}{m_e c^2},$$

you can write Eq. 8 in the compact form

$$x' = \frac{x}{x(1 - \cos\theta) + 1}. \tag{9}$$

EXAMPLES

7. Remembering that the rest mass energy of an electron is 511 keV, suppose a 256 keV photon undergoes Compton scattering through 180°. What will be the energy of the outgoing photon? What will be the energy of the recoiling electron?

For this case $x = \frac{1}{2}$. Since $\cos\theta = -1$, it follows from Eq. 9 that $x' = \frac{1}{4}$. This means that $hf' = 511/4 = 128$ keV. Because the electron must have the remaining part of the energy, its kinetic energy in this particular case will be 128 keV.

Another virtue of this version of the Compton scattering equation (Eq. 9) is apparent if you want to know about Compton scattering from particles other than an electron. Suppose you want to know what happens to a photon that scatters off a proton, $m_p c^2 = 938$ MeV.

EXAMPLES

8. To see what happens when a 256 keV photon scatters 180° from a proton, calculate $x = 0.256/938 = 2.729 \times 10^{-4}$, from which it follows that $x' = 2.728 \times 10^{-4}$. There is hardly any effect.

EXERCISES

18. If a 100 keV photon scatters from an electron through an angle of 90°, what will be the recoil energy of the electron?

19. What is the maximum energy that an electron can acquire from a 600 keV photon by means of Compton scattering?

The Compton effect is customarily described as a change in wavelength rather than as a change in frequency. To get it in that form divide both sides of the equation just preceding Eq. 7 by pp' and also by m_e. This gives

$$\frac{1-\cos\theta}{m_e} = c\left(\frac{1}{p'} - \frac{1}{p}\right) = c\left(\frac{\lambda'}{h} - \frac{\lambda}{h}\right),$$

from which it follows that

$$\lambda' - \lambda = \frac{h}{m_e c}(1 - \cos\theta). \tag{10}$$

When people speak of the "Compton scattering equation" or the equation for "Compton wavelength shift," they mean Eq. 10, which gives the change in wavelength of the scattered photon as a function of its angle of scattering.

Equation 10 shows the curious fact that the change in wavelength due to Compton scattering does not depend upon the frequency (i.e., energy) of the incident photon.

EXERCISES

20. Is the frequency shift of the scattered photon independent of the frequency of the incident photon? Explain.

21. Table 13.4 shows actual Compton scattering data. What is the energy of the 0.07078 nm incident photon? Calculate the shift in wavelength you would expect when that photon scatters through 135°. How well does your answer compare with the data shown in Table 13.4?

Compton Scattering and the Detection of Photons

Radioactive materials often emit energetic photons called, for historical reasons, "gamma rays." Radioactive cesium emits a gamma ray with an energy of 662 keV. Obviously, this is not a visible photon.

EXERCISES

22. How do you know that this photon is not visible?

As previously noted, such photons, although not visible themselves, can cause other materials to emit visible light. We described in Chapter 12 how a photon can deposit energy in a crystal of sodium iodide [NaI(Tl)] and cause the crystal to give off a flash of light that can be detected with a photomultiplier tube. By measuring the relative intensity of the scintillations you can measure the relative amounts of energy left by the photons in the crystal.

A common measurement of photons is to count the number of flashes that appear in the crystal and sort the counts according to how bright the flashes are.

The graph in Fig. 13.10 was obtained after hundreds of thousands of photons passed through a NaI(Tl) crystal. Of these, somewhere between 1% and 10% produced flashes of light. The flashes were counted and sorted according to their brightness. The graph shows the number of flashes on the y-axis and the brightness along the x-axis. Because brightness is proportional to energy deposited in the crystal, it is convenient to label the brightness scale with the corresponding photon energy, and that is done here.

If flashes came only from complete absorption of incoming single-energy photons, there would be only one size of flash. In Fig. 13.10 you can see that many of the flashes correspond to photons of 662 keV. This group of photons is labeled "photopeak" on the graph. The energies of the emitted photons are very closely the same, but because the flashes of light that they produce vary somewhat in brightness, the peak in the graph is wide. We say that the "resolution" of the instrument is not perfect. The name "photopeak" always refers to the peak corresponding to the full energy of the incident photon.

Figure 13.10 shows that other things are going on besides absorption of 662 keV photons. The experimenter detected flashes of light over quite a wide range of brightness. These correspond to the absorption of energy all the way from 0 keV up to an apparent 800 keV. Two features show up particularly: the bump labeled "backscatter" and the sudden drop from a constant level of counts labeled "Compton edge."

It turns out that there are several ways for a photon to deposit energy in a crystal, and these depend on the energy of the photon. For example, between 250 keV and 1.02 MeV a photon is most likely to do either of two things. It may produce a photoelectron in the crystal and leave all of its energy in the crystal. Or it may Compton scatter off a nearly free electron of some atom and leave the crystal; the energy left in the crystal is then whatever was given to the electron by the Compton scattering. This will produce a smaller flash

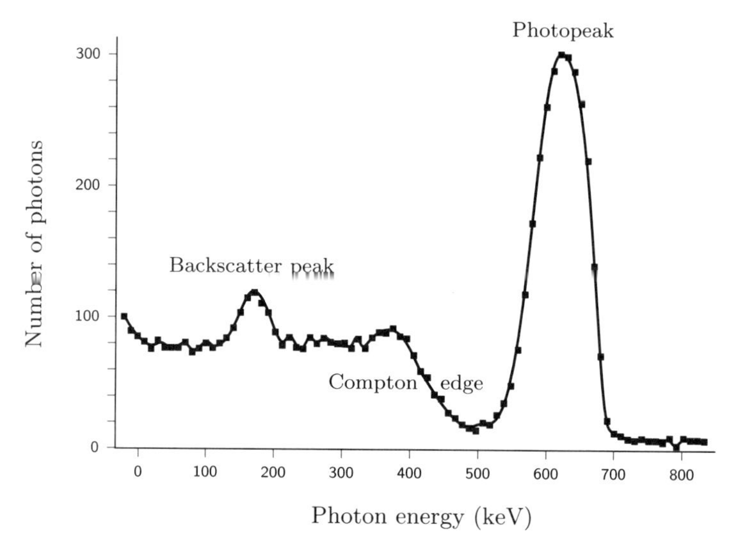

FIGURE 13.10 Distribution of energy deposited in a NaI(Tl) crystal by 662 keV photons emitted following the radioactive decay of ^{137}Cs.

of light than total absorption because the energy left in the crystal is smaller. It is in this way that most of the lower-energy light pulses occur.

You can calculate the energies at which the backscatter peak and the Compton edge occur by using your knowledge of Compton scattering. The Compton edge occurs when a photon enters the crystal and undergoes 180° Compton scattering from an electron of one of the crystal's atoms and then leaves the crystal. As a result of scattering off the electron, the photon leaves the crystal with reduced energy. The electron remains behind with increased energy, which goes to produce a flash in the crystal.

EXERCISES

23. Under what circumstances will a photon that is Compton scattered out of the crystal leave the most possible energy with the electron from which it scattered?

24. Calculate the maximum energy that an escaping 662 keV photon will impart to an electron. How does your answer correspond to the energy of the graph in the region labeled "Compton edge"?

25. To what do you attribute the counts observed at energies below the Compton edge?

The backscatter occurs because the NaI crystal is surrounded by heavy material, and there is a fair probability that some photons will enter the crystal, pass through it unabsorbed, and then scatter off electrons of surrounding metal (usually aluminum, but it does not make much difference). These photons scatter back into the crystal (hence the name, backscatter) where they are absorbed. The crystal is usually arranged so that only those photons that bounce directly back are likely to reenter the crystal. Of course, photons that have scattered through 180° are reduced in energy, and the flashes of light that they can produce will not be as bright as those produced by the full-energy photons.

EXERCISES

26. What is the energy of a photon that has Compton scattered 180° off Al? Off Si? Off Fe?

27. Compare your answer to the previous problem with the energy corresponding to the "backscatter peak" in the graph. Why should they be the same?

13.10 SUMMARY

Useful Things to Know

X-rays are usually produced by sudden acceleration of charged particles. In practical circumstances x-rays are made when energetic electrons are brought to a sudden halt by collision with a metal electrode.

The existence of Laue diffraction spots and Debye–Scherrer diffraction rings shows that x-rays are waves with wavelengths a thousand times smaller than those of visible light.

Diffraction of x-rays by crystals obeys the Bragg law

$$2d\sin\theta = n\lambda,$$

where d is the spacing between successive planes of atoms in the crystal, θ is the angle of grazing incidence, λ is the wavelength of the x-rays, and n is an integer $n = 1, 2, \ldots$. In any given crystal there will many sets of planes with different spacings. The Bragg law holds for each set.

The Bragg law is the basis for the design and operation of the crystal spectrometer. With this device it is possible to measure x-ray wavelengths and map out the intensity of x-ray emissions as a function of wavelength.

X-rays also exhibit particle-like behavior. To understand the short-wavelength cutoff of the continuous x-ray spectrum, it is necessary to assume that x-rays are photons each with energy $E = hf$. Furthermore, the Compton effect, i.e., the decrease in energy of x-rays scattered from free or weakly bound electrons as the scattering angle increases, can be understood as resulting from the scattering of one small structureless object by another.

Some Important Things to Keep in Mind

If the wavelengths of x-rays are thousands of times smaller than those of visible light, they can in principle be used to probe the structure of objects thousands of times smaller than the smallest things we can see with our eyes. But notice that x-ray photons are going to be thousands of times more energetic than visible ones; the energy scale shifts from eV to keV.

We have already used x-rays to find out two interesting things about matter. Bragg diffraction shows that crystals are made of regular arrays of atoms. We did not emphasize that fact much, but it is quite important. Also, notice that the Compton effect provides quite direct evidence that the electron is a constituent part of the atom.

There is every reason to expect that x-rays can yield interesting information about structures as small as 0.1 nm, that is, as small as atoms. Furthermore, nothing in principle prevents us from making x-rays of energies of 100 keV or higher. These would have wavelengths that could probe even smaller structures.

Be able to convert from energy to wavelength and vice versa without hesitation.

EXAMPLES

9. For example, what is the wavelength of a 100 keV photon? Because $E = hf = hc/\lambda$, a 100 keV photon has a wavelength $\lambda = hc/E$. Never forget that $hc = 1240$ eV nm. Then if you know E in units of eV, you can immediately get that $\lambda = \frac{1240}{100\times10^3} = 0.0124$ nm.

EXERCISES

28. What is the wavelength of a 12.4 keV photon?

PROBLEMS

1. The Bragg Law for crystal diffraction of x-rays is $2d \sin\theta = \lambda$. Use Fig. 13.11 to show:

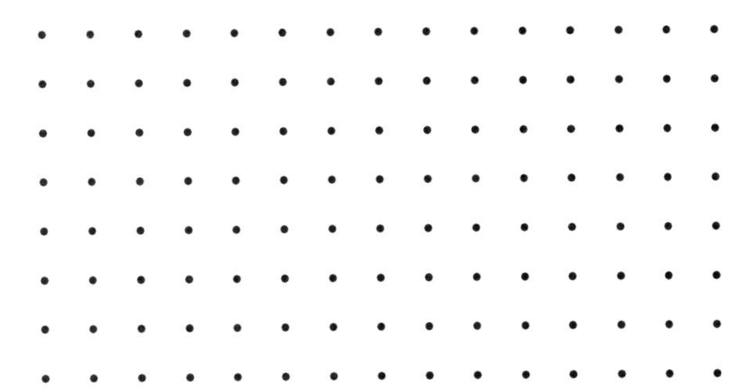

FIGURE 13.11 Representation of a crystal lattice (Problem 1).

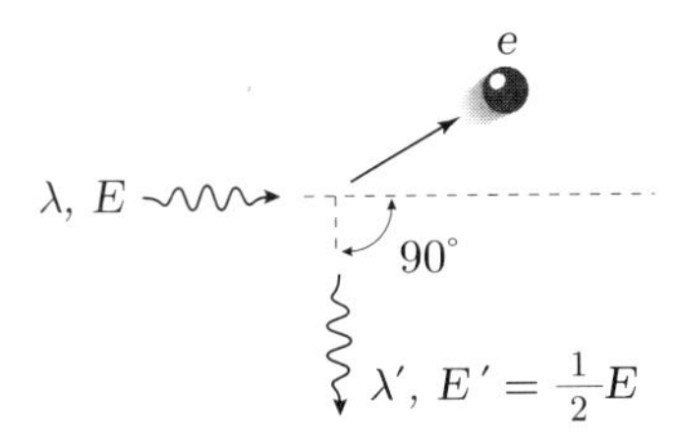

FIGURE 13.12 Compton scattering event (Problem 3).

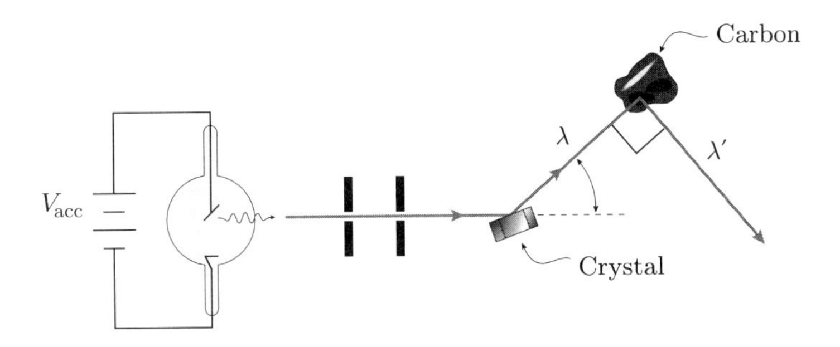

FIGURE 13.13 Compton scattering arrangement for Problem 4.

a. What the symbols d and θ refer to.
b. What is meant by Bragg diffraction.

2. a. What is the Compton effect?
   b. How is the Compton effect important to our understanding of the nature of electromagnetic radiation?

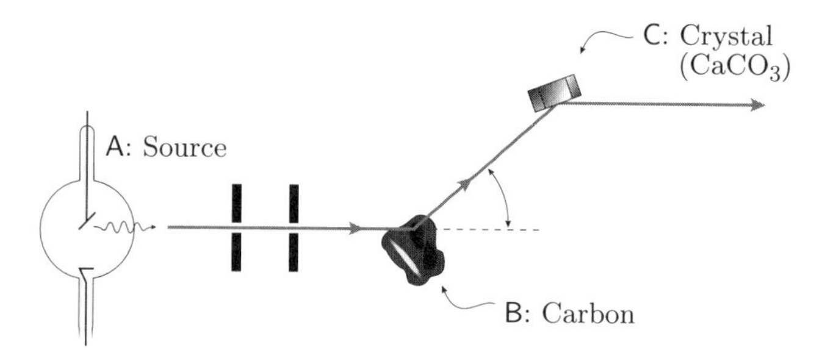

FIGURE 13.14 A Compton scattering arrangement in which the incident x-rays have a single well-defined wavelength (Problem 5).

3. A photon Compton-scatters from an electron through an angle of 90°, as shown in Fig. 13.12. In doing so, it loses half its initial energy:

a. What is the photon's initial wavelength?

b. What are the initial and final energies of the photon?

c. Find the final wavelength and energy (λ', E') for a photon with the values of λ and E found in part (b) after being scattered through 180°.

d. What is the final energy of the *electron*?

4. Monochromatic x-rays of wavelength $\lambda = 0.243$ nm are produced in the following way: Electrons bombard a metal target in a standard x-ray tube, and the resulting radiation is reflected from a crystalline material, as shown in Fig. 13.13.

a. What is the purpose of the crystal?

b. What is the minimum accelerating voltage V_{acc} that could have produced x-rays of this λ?

c. The reflected x-rays go on to collide with a carbon block C, where their wavelength is shifted to λ' by the Compton effect. For the direction shown in the drawing, calculate the change in wavelength.

d. Does the x-ray gain or lose energy in this process? Where does the energy go? What is the most striking assumption made by Compton in his analysis of this phenomenon?

5. Write a short description of the experiment depicted in Fig. 13.14. Include the following:

a. Identify and explain the *purpose* of items A, B, and C in Fig. 13.14.

b. Discuss the significance of the experiment. Why were the experimental results, and the analysis of those results, important and surprising?

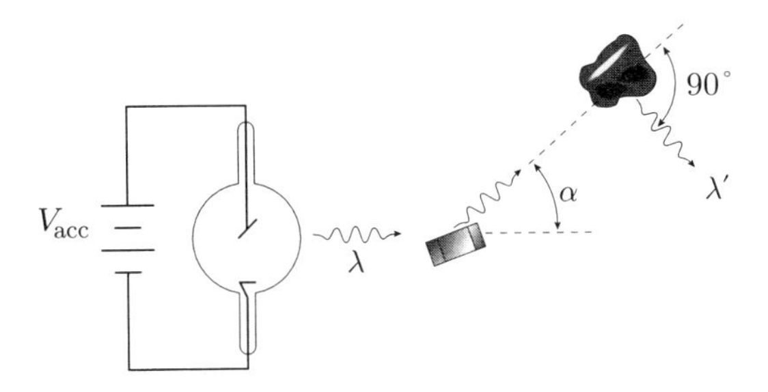

FIGURE 13.15 Compton scattering arrangement for Problem 6.

6. X-rays from an x-ray tube are Bragg reflected as shown in Fig. 13.15 from the planes of a calcite crystal spaced 0.3036 nm apart. Photons of wavelength $\lambda = 0.3654$ nm emerge from the crystal at an angle α with respect to their initial direction (see Fig. 13.15) and are Compton scattered through an angle of 90° as they pass through a second material.

a. Find the angle α that the diffracted photon forms with its initial direction.

b. What is the minimum accelerating voltage V_{acc} that would produce a photon of the given wavelength?

c. How much energy was lost by the photon, and where did the energy go?

7. A 50 kV accelerating voltage is applied to an x-ray tube as shown in Fig. 13.16. When electrons from the cathode crash into the target, K_α radiation of wavelength $\lambda_{K_\alpha} = 0.0723$ nm is emitted.

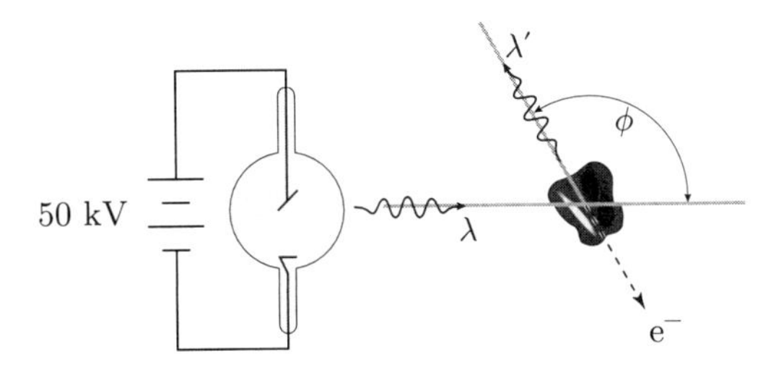

FIGURE 13.16 X-ray generation and Compton scattering geometries for Problem 7.

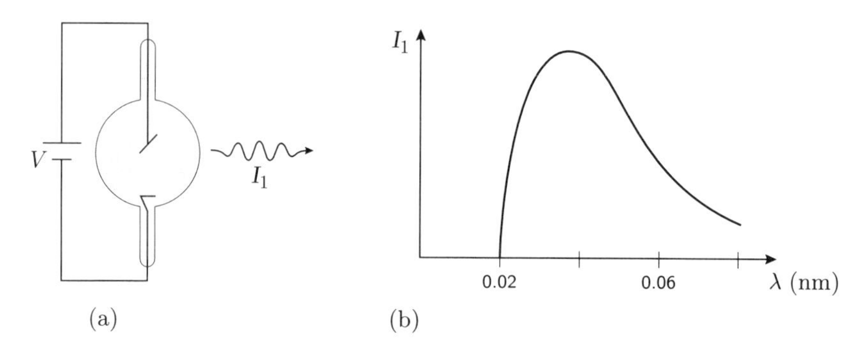

FIGURE 13.17 Intensity distribution of x-rays vs. λ for Problem 9(a).

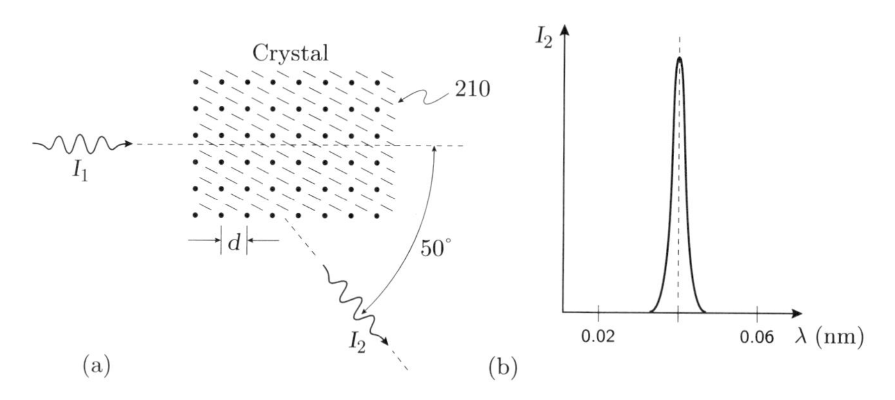

FIGURE 13.18 Schematic of Bragg crystal and diffracted x-ray intensity for Problem 9(b).

a. What is the atomic number Z of the target material? Hint: Recall that the energy levels of an electron orbiting a nucleus of charge $+Ze$ are given by

$$E_n = -13.6\frac{Z^2}{n^2}\text{ eV}.$$

The K_α photons strike a carbon block and undergo Compton scattering, as shown in Fig. 13.16. If the photon is scattered through an angle $\phi = 180^\circ$, the scattered photon has a wavelength $\lambda = 0.0771$ nm.

b. Find the kinetic energy and the momentum imparted to the electron in the scattering event. Use appropriate units. In what direction is p?

8. Figure 13.8 shows three of Compton's scans obtained using calcite as the crystal in the analyzing Bragg spectrometer.

a. Use the information on the x-axis to calculate the wavelength of the incident radiation and the wavelength of the scattered radiation for each plot.

b. From the incident wavelength and the scattered angle use Compton's equation to calculate the wavelength of the scattered radiation for each plot.

c. Compare the results of (a) and (b).

d. For each case calculate the energy gained by the scattered *electron*.

9. a. An electron tube (Fig. 13.17(a)) generates x-rays of intensity $I_1(\lambda)$, as shown in Fig. 13.17(b). What is the voltage applied to the electron tube?

b. The x-rays from the tube are incident on a crystal. It is observed that the x-rays that undergo Bragg diffraction emerge at the angle shown in Fig. 13.18(a). The intensity of the emerging beam $I_2(\lambda)$ is shown in Fig. 13.18(b). What is the spacing d between the planes of the crystal from which the x-rays have diffracted?

c. The x-rays of part (b) are incident on a block of graphite, as shown in Fig. 13.19(a). Photons Compton scattered through 130° are observed to have a distribution of intensity I_3 vs. wavelength, as shown in Fig. 13.19(b). *Calculate* the wavelength λ' of the scattered x-rays.

d. How much energy is gained by electrons that Compton scatter photons 130°as in part (c)?

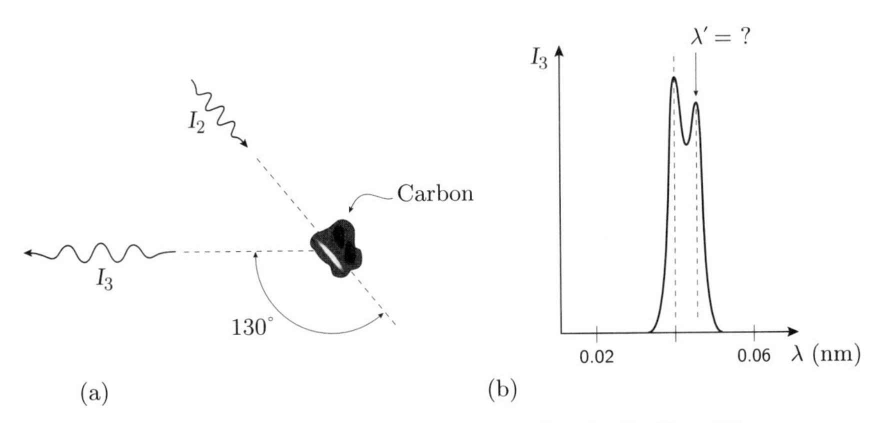

FIGURE 13.19 Compton scattering and Compton effect for Problem 9(c).

C H A P T E R 14

Particles as Waves

14.1 INTRODUCTION

We have seen that waves can act like particles. The energy of light waves is packaged in quanta, which we call photons. Photons can scatter from electrons like tiny hard objects. If this behavior is not surprising or mysterious enough, we now find that particles can act like waves. Electrons, protons, neutrons, and atoms of all kinds can exhibit interference; they show all the same kinds of patterns that waves do.

14.2 THE DE BROGLIE WAVELENGTH

The first clue that particles might have wavelike behavior came when Louis de Broglie noted that (as you saw in Chapter 13) the momentum of a photon is

$$p = \frac{hf}{c} = \frac{h}{\lambda} \tag{1}$$

and suggested that perhaps particles might obey the same equation. In other words, he proposed that particles might behave as waves with a wavelength of

$$\lambda = \frac{h}{p}. \tag{2}$$

This wavelength of particles is called the "de Broglie wavelength."

Notice that photons and particles that have the same energy K have different wavelengths. The de Broglie wavelength of the particle differs from the photon wavelength because the particle has a rest mass and the photon does not.

EXAMPLES

1. To illustrate this point let's calculate and compare the momentums of a 10 keV photon and a 10 keV electron. From Eq. 1 you see that the photon's momentum is 10 keV/c. To calculate the momentum of a 10 keV electron you may use the nonrelativistic expression connecting momentum p and kinetic energy K:

$$\frac{p^2}{2m_e} = K.$$

Now use our favorite trick of multiplying and dividing the left side by c^2 and solving for pc. This will give you

$$pc = \sqrt{2Km_ec^2} = \sqrt{2 \times 10 \times 511} = 101.1\,\text{keV}.$$

The 10 keV electron's momentum is more than 10 times greater than that of a 10 keV photon, and therefore the electron's wavelength is one-tenth the photon's.

EXAMPLES

2. Using this calculated momentum, you can find the wavelength of a 10 keV electron:

$$\lambda = \frac{h}{p} = \frac{1240}{101.1 \times 10^3} = 0.0123\,\text{nm}.$$

EXERCISES

1. Find the wavelength of a 50 keV photon and compare it to the de Broglie wavelength of a 50 keV electron.

2. What energy of photon will have the same wavelength as a 50 keV electron?

14.3 EVIDENCE THAT PARTICLES ACT LIKE WAVES

De Broglie's idea that particles might have a wavelength has been strongly confirmed for all kinds of particles: electrons, neutrons, protons, atoms, subnuclear particles, etc. What is the evidence for such a wavelength? How can particles be made to exhibit wavelike behavior? Let's consider G.P. Thomson's experiments, Davisson's and Germer's experiment, Stern's experiment, and a modern version of the double-slit interference done with electrons.

G.P. Thomson's experiment

G.P. Thomson, the son of the J.J. Thomson who discovered the electron, took a very straightforward approach. If particles are waves, then a beam of electrons passing through a polycrystalline foil should make a diffraction pattern of concentric rings like the pattern made by x-rays when they diffract through a polycrystalline sample, as described in Chapter 13.

To see whether electrons would produce such a pattern, Thomson used the apparatus shown schematically in Fig. 14.1. Electrons with kinetic energy of about 40 keV were produced in the tube A and then formed into a beam by passage through a tube B of bore 0.23 mm and length 6 cm. They struck a thin, polycrystalline gold foil at C. The resulting pattern of the electrons could be viewed on a phosphorescent screen E or recorded on a photographic plate D that could be lowered in front of the screen. The distance from the foil to the screen was 32.5 cm.

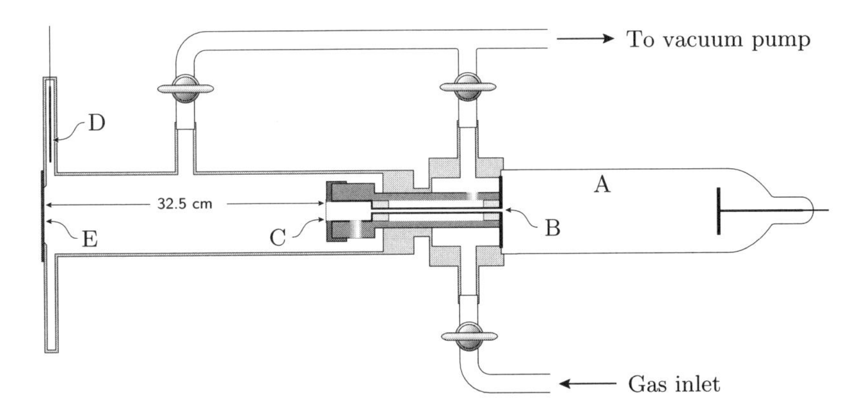

FIGURE 14.1 Schematic diagram of G.P. Thomson's apparatus for the study of electron diffraction.

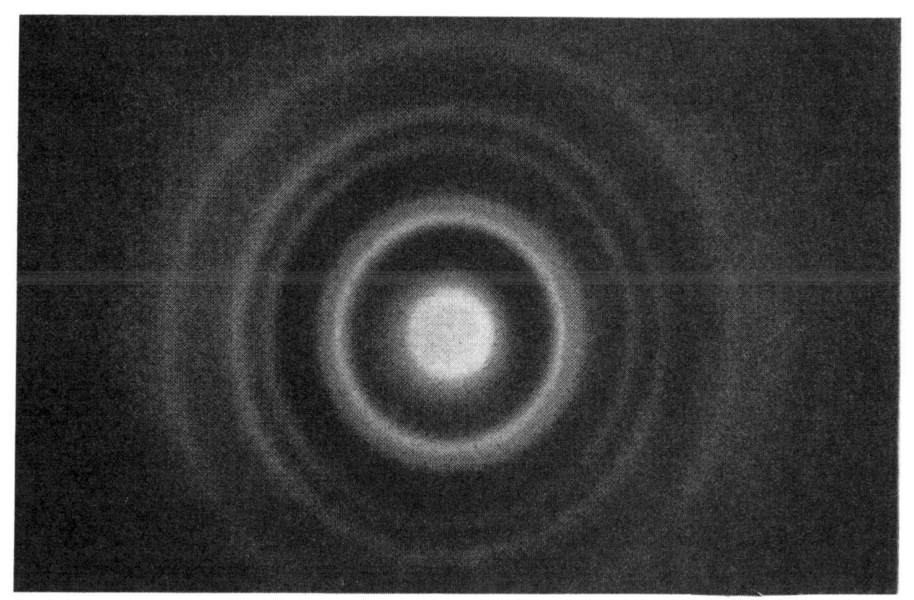

FIGURE 14.2 The diffraction pattern from electrons on a polycrystalline gold foil observed by G.P. Thomson.

The electrons made the pattern shown in Fig. 14.2; it is reproduced from Thomson's publications.[1] The rings are clearly analogous to those seen in the Debye–Scherrer patterns of x-ray diffraction from polycrystalline samples.

Do these results agree with de Broglie's predictions? To answer this question, we will compare the pattern with the predictions of the Bragg law for crystal diffraction:

$$2d \sin\theta = n\lambda. \tag{3}$$

First let's see why there are so many rings. Remember from the discussion of x-ray diffraction in Chapter 13 that there should be rings for each order, $n = 1, 2, 3, \ldots$, of the Debye–Scherrer pattern, and for each different set of crystal planes. Therefore, a pattern of concentric rings makes sense at least qualitatively.

To interpret the results quantitatively we need to know the various possible spacings d for a gold crystal. Refer back to Fig. 13.4. This figure shows you that the value of d depends on which set of planes you are considering. For labeling the sets of planes there is a notation that we introduce here without much explanation.

---

[1]G.P. Thomson, "The diffraction of cathode rays by thin films of platinum," *Nature* **120**, 802 (1927); "Experiments on the diffraction of kathode rays," *Proc. Roy. Soc. London* **A117**, 600–609 (1928); *The Wave Mechanics of Free Electrons*, McGraw-Hill, New York, 1930.

Crystal planes are specified by a set of three small integers called Miller indices. They are usually written in parentheses in the form $(h k \ell)$. For example, the familiar, obvious parallel planes shown in Fig. 14.3(a) are denoted by (100). The planes that pass through a crystal at 45° as shown in Fig. 14.3(b) are labeled (110).

You can use the Miller indices of a crystal plane and the lengths of the sides of the smallest unit of the crystal to find the spacing between adjacent planes. It is particularly simple to do this for the planes of cubic crystals. For cubic crystals the spacing d between the planes is related to the edge b of the elementary crystal cube by the expression

$$d = \frac{b}{\sqrt{h^2 + k^2 + \ell^2}},$$

where $(h\,k\,\ell)$ are the Miller indices.

Table 14.1 lists values of b, the length of the edge of the basic cube, for a number of different cubic crystals. This quantity b is usually called the "lattice constant."

EXAMPLES

3. What is the spacing between adjacent (111) planes of a gold crystal? Table 14.1 gives the lattice constant for gold $b = 0.407$ nm. This means that the spacing between adjacent (111) planes is $0.407/\sqrt{3} = 0.235$ nm.

EXERCISES

3. What is the spacing between the (110) planes of gold?

TABLE 14.1 Lattice Constants for Some Common Cubic Crystals

Element	Lattice Constant b(nm)
Al	0.404
Au	0.407
Ni	0.352
Cu	0.361

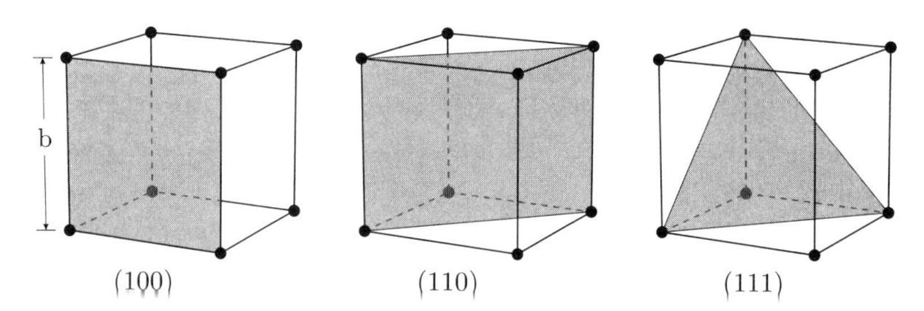

FIGURE 14.3 Miller indices of some important planes in a cubic crystal—the atoms are located at the corners of a cube

4. What is the spacing between the (331) planes of gold?

There is one more point to make about spacings between planes of a crystal. One often talks about planes like (200). Notice that the spacing between such planes is $d = b/2$. But how can that be? If the spacing between planes is $d = b$, there should not be any atoms halfway between these two planes. And there aren't. Bragg reflection from (200) planes is the same thing as second-order Bragg reflection from the (100) planes. You can see that this is so in two ways.

First, look at Eq. 3. For (200) planes, $d = b/2$, and the equation becomes

$$\frac{2b}{2}\sin\theta = \lambda,$$

or

$$2b\sin\theta = 2\lambda,$$

which is the same as Eq. 3 for (100) with $n = 2$; i.e., it is the second-order result.

Second, you can imagine the pattern of wavelengths between adjacent planes. For first order, one wavelength just fits between the planes. For second order, two wavelengths just fit. The notation (200) pretends that there is a second plane of atoms halfway between the real ones. The effect is to put two wavelengths between the planes of real atoms, i.e., it generates the second-order Bragg reflection. It is customary to deal with higher orders in the Bragg equation by using only $n = 1$ and then including the higher orders by appropriate choice of Miller indices and the corresponding value of d.

Now we can get back to Thomson's experimental proof of the validity of de Broglie's idea. One of his tests was to observe how the radius of one ring, the one produced by the (200) planes, varied in diameter as he varied the voltage through which the electrons were accelerated. He obtained the data shown in Table 14.2.

TABLE 14.2 Diameter of the (200) Diffraction Ring vs. Electron Voltage

Acceleration Voltage V (kV)	Diameter D (cm)	$DV^{1/2}$ (cm keV)$^{1/2}$
24.6	2.50	12.4
31.8	2.15	
39.4	2.00	12.6
45.6	1.86	
54.3	1.63	12.0
61.2	1.61	12.6

Thomson showed that if the diameter D of any given diffraction ring varies as the inverse of the square root of the accelerating voltage V, then the data are consistent with de Broglie's prediction. Why should this be so? The argument has two parts. First, we show that de Broglie's equation $\lambda = h/p$ predicts that the wavelength will be inversely proportional to $V^{1/2}$. Then we show that the ring diameter D will be proportional to λ.

The first part has three steps. First, the kinetic energy K of the electrons is proportional to the acceleration voltage V because $K = eV$. But K is also proportional to the square of the momentum, because

$$\frac{p^2}{2m} = K = eV.$$

Therefore, it must follow that the momentum of the electrons is proportional to the square root of the accelerating voltage, i.e.,

$$p \propto V^{1/2}.$$

Consequently, if de Broglie is correct and λ is proportional to $1/p$, it follows that $\lambda \propto 1/V^{1/2}$.

The second part of the argument uses the Bragg law, $2d \sin\theta = \lambda$. If θ is small, then $\sin\theta$ is approximately $r/(2L)$, where r is the radius of the diffraction ring and L is the distance from the gold foil to the photographic plate. But the radius r is proportional to the diameter D, so for fixed values of d and L,

$$D \propto \lambda,$$

from which it follows that

$$D \propto \frac{1}{V^{1/2}}$$

if de Broglie is correct.

Do the data agree with this prediction? There are two standard ways to find out.

EXAMPLES

4. One is to compute directly with the numbers in Table 14.2 and see whether they obey the expected relationship. If $D \propto 1/V^{1/2}$, then the product $DV^{1/2}$ should be a constant. Is it? You can see from the entries in the third column of Table 14.2 that agreement is pretty good. It convinced most physicists. Thomson was awarded a Nobel Prize for this work.

EXERCISES

5. Calculate the values for $DV^{1/2}$ that are missing from Table 14.2.

6. Functional relationships are often verified by plotting measured quantities in some form that will give a straight-line graph if the expected relationships are correct. How would you plot the data of Table 14.2 to determine whether de Broglie was correct? Do it.

Thomson also analyzed the diffraction rings that came from the gold foil as shown in Fig. 14.2. He assigned the five successive rings from the center outwards to the following indices: (111), (200), (220), (113)+(222), and (331)+(420). In the last two cases rings from two different sets of indices fell too close to one another to be distinguished.

EXERCISES

7. Measure directly from the circles in Fig. 14.2 and show that the progression of circle diameters is what you would expect for the diffraction of de Broglie waves.

8. Thomson says that the photographic plate was 32.5 cm from the gold foil. Use that number and the fact that the diameter of the (200) diffraction ring was 2.5 cm for 24.6 keV electrons to find the lattice constant of the gold foil. How does your result compare with the value given in Table 14.1?

Warning: Remember that the angle of the diffraction ring relative to the incident beam is twice the diffraction angle θ.

The Experiment of Davisson and Germer

The first experiment to show the wave nature of electrons was done by C.J. Davisson and L.H. Germer, working in the laboratories of the Western Electric Company, the manufacturing arm of the American Telephone and Telegraph Company. Their original interest was to study the emission of secondary electrons from the electrodes of vacuum tubes (these were produced when the electrons the tubes were intended to control struck the electrodes). When Davisson directed a beam of electrons at various samples, he observed that about 1% of the incident electrons were reflected back from them. He realized that the reflected electrons might be used as a direct probe of the structure of the atom just as E. Rutherford (about whom more later) used alpha particles to probe atomic structure.[2]

Davisson and Germer observed that when a beam of 54 eV electrons struck the (111) surface of a nickel crystal perpendicularly, the electrons diffracted just as light would diffract from a reflection grating. The rows of atoms on the crystal surface acted as the lines of the grating, and the distance between these rows of atoms corresponded to the grating spacing D. A diffraction maximum in the intensity of reflected electrons occurred just where the grating equation $D \sin\theta = \lambda$ predicted for waves of wavelength $\lambda = h/p$.

For their experiment Davisson and colleagues used the apparatus shown schematically in Fig. 14.4. It was designed to find how the intensity of the outgoing electrons varies as a function of the angle between the incident beam and the scattered electrons. They could measure the intensity of electrons at any scattering angle they chose by moving the detector to that angle. They could also rotate the target and change the angle at which incident electrons hit the target's surface.

An electron gun made a beam of electrons with energy that could be varied from a few eV to around 200 eV. This beam was directed onto a target of single-crystal nickel, and electrons reflected from the target were detected and measured by catching them in a metal box.

The cut and orientation of the crystal determine which rows of atoms will be exposed to the incident electrons. The spacing between the rows of atoms affects the experimental results because it is the grating spacing. For the experiments of Davisson and Germer the face-centered cubic crystal was

[2]For a short readable history of the Davisson–Germer experiment, read Richard K. Gehrenbeck, "Electron Diffraction Fifty Years Ago," *Physics Today* **31**, 1, 34–41 (1978).

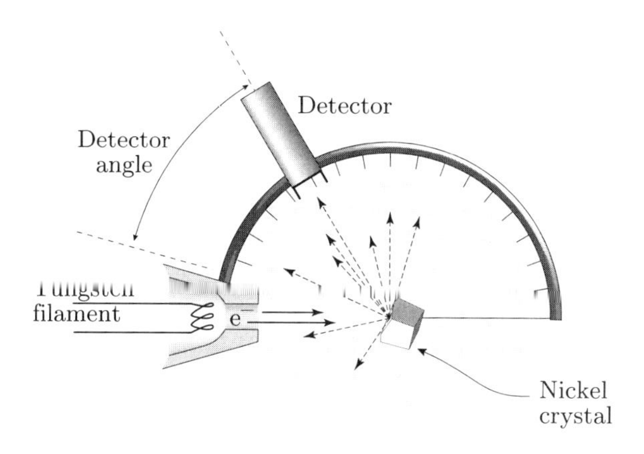

FIGURE 14.4 Schematic diagram of the apparatus of Davisson and Germer.

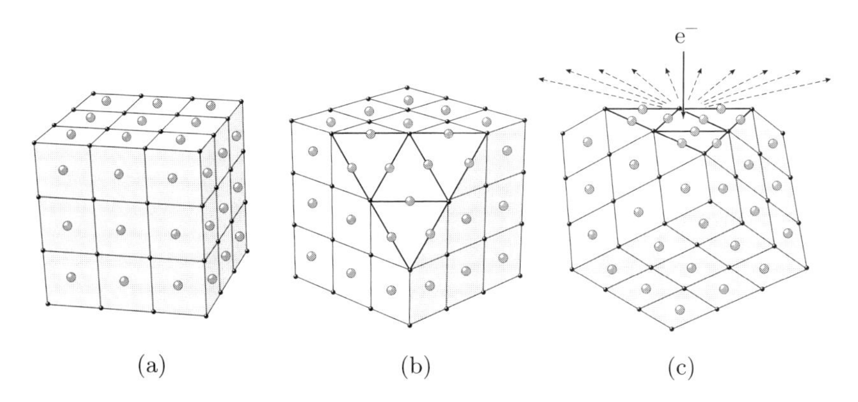

FIGURE 14.5 The arrangement of atoms on the (111) face of a Ni crystal. *Taken from J. Frank. Instit. Vol. 206, C.J. Davisson, "Are Electrons Waves?" (1928) with permission from Elsevier Science*

cut at right angles to the cube's diagonal, i.e., at 45° to each face, as shown in Fig. 14.5, exposing the (111) face. Their electron beam came in normal to that face. For this orientation the rows of atoms on the (111) face are $D = 0.215$ nm apart, and this is the grating spacing.

Davisson and Germer looked at the variation of intensity of the scattered electrons as a function of scattering angle for different energies of the incident beam. Their results are shown in Fig. 14.6, where the electron intensity is plotted on polar graphs. Each point on these graphs shows the intensity of the scattered electrons and the angle at which it was measured. The intensity is given by the length of the radial distance from the origin to the curve, and the angle at which this intensity occurs is the angle that a line from the origin to the curve makes with the vertical axis. The progression of the data is striking. At all incident energies the intensity of electrons scattered directly backwards with a reflection angle of 0° is high and then drops off rapidly at larger reflection angles. But notice what happens as the electron energy is increased. For 40 eV incident electrons the intensity of the reflected electrons drops off smoothly as the angle is increased from 0° to 90°. For 44 eV incident electrons a bump appears in the number of electrons reflected at 50°. The bump grows to a maximum as the energy is increased to 54 eV and then diminishes as the incident energy is increased further.

This bump is the expected interference maximum. According to the grating equation it should occur at 50° for a wavelength of

$$\lambda = 0.215 \sin 50^\circ = 0.215 \times 0.766 = 0.165\,\text{nm}.$$

How does this wavelength compare with de Broglie's prediction? An electron with kinetic energy $K = 54\,\text{eV}$ has a momentum given by

$$pc = \sqrt{1.02 \times 10^6 \times 54} = 7.43 \times 10^3\,\text{eV}.$$

The de Broglie wavelength of such electrons is then

$$\lambda = \frac{hc}{pc} = \frac{1240}{7430} = 0.167\,\text{nm}.$$

This degree of agreement is convincing. Many other measurements provide equally good agreement. Davisson was awarded a Nobel Prize for this work.

You might wonder why we treated electrons as reflected from a diffraction grating rather than as waves interacting with a three-dimensional crystal lattice in the way that x-rays interacted. The answer is that because of their electrical charge and low energy, the electrons penetrate no more than a few layers of atoms into the crystal surface, and therefore grating diffraction is a better description of what happens to such low-energy electrons than Bragg reflection.

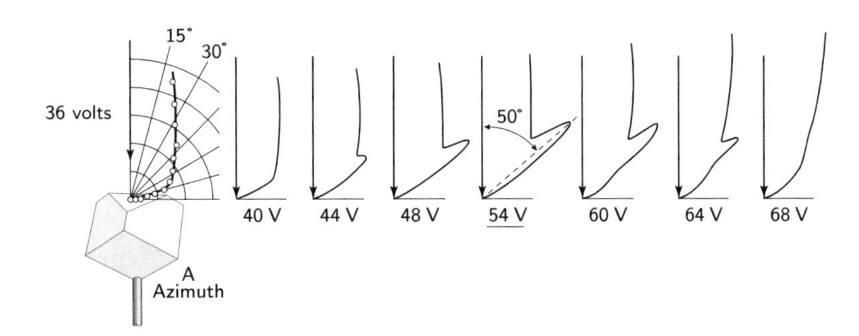

FIGURE 14.6 This succession of polar plots of the intensity of electrons reflecting off the surface of a nickel crystal shows that a diffraction maximum occurs at 50° when the electrons are accelerated through 54 V. *Taken from J. Frank. Instit. Vol. 206, C.J. Davisson, "Are Electrons Waves?" (1928) with permission from Elsevier Science*

"Double-Slit" Interference with Electrons

A striking example of wave behavior of electrons is the generation of a double-slit interference pattern by Möllenstedt and Düker.[3] Their apparatus is shown schematically in Fig. 14.7(a). For two slits they used a thin metal-coated quartz fiber and applied to it a voltage of about 10 V. This device slightly deflected the 50 nm wide beam of 19.4 keV electrons coming from the assembly of cathode, anode, and demagnifying electrodes in such a way as to make the electrons behave as though they had come from the two sources. (In optics this effect can be obtained with a device called a biprism, so that is the label given to the electrical equivalent in Fig. 14.7(a).) The electrons then formed the interference pattern on a photographic plate as shown in Fig. 14.7(b). Notice how strikingly it resembles the optical double-slit interference pattern described in Chapter 9. Figure 14.7(b) shows the negative of a photograph, so the bright lines are where the electron waves interfere constructively, and the black spaces represent destructive interference. *Electrons interfere just like photons.*[4]

EXERCISES

9. What is the wavelength of 19.4 keV electrons?

10. Suppose the effective separation of the sources of interfering 19.4-keV electrons was 6 μm and the effective distance from these sources to the photographic plate was 11 cm. What would be the distance between adjacent fringes on the photograph.

11. Suppose the fringes of the previous problem were magnified by a factor of 160 before they struck the plate. And then suppose the photograph was enlarged by a factor of 20 when it was printed. What then would be the spacing between the fringes? Compare your answer with the spacing of the fringes shown in Fig. 14.7(b).

---

[3]G. Möllenstedt and H. Düker, "Fresnel interference with a biprism for electrons," *Naturwiss.* **42**, 41 (1955); H. Düker, "Interference pattern of light intensity for electron waves using a biprism," Z. *Naturforsch.* **10a**, 256 (1955).

[4]There is a certain amount of art that goes into a picture like Fig. 14.7(b). If only the slits were producing the spacing of the electron fringes, they would be about 160 nm apart and very difficult to see. To make the pattern visible, a cylindrical electrical lens was used to magnify electrically the fringe-to-fringe dimension by a factor of 160. To suppress the effects of irregularities in the shape of the fiber, magnification of the dimension along the fringe was limited to between a factor of 5 and 10. Then the overall pattern was magnified optically a factor of 20 to make the picture with visible fringes.

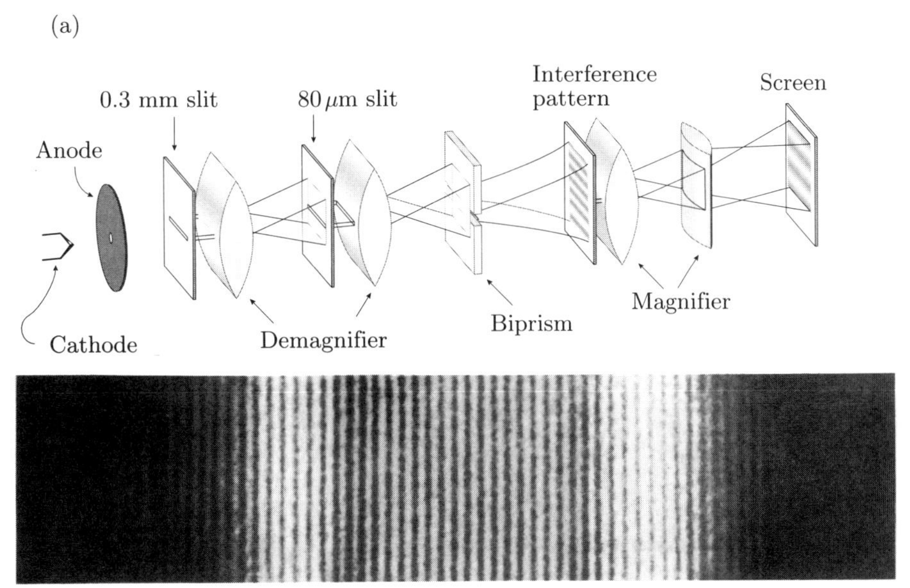

FIGURE 14.7 Double-slit interference of electrons: (a) apparatus; (b) interference pattern. *Taken from H. Düker,* Z. Naturforsch. *10a, 256 (1955) with permission of the publisher*

Waves of Atoms

The wavelike behavior of particles is quite general. In 1931 Estermann, Frisch, and Stern[5] showed that a beam of helium atoms would diffract from the surface of a crystal of LiF and that the wavelength of the atoms was what the de Broglie relationship predicted.

Since 1932, when the neutron was discovered, this neutral particle with a mass slightly larger than that of a proton has been widely used for diffraction studies of crystals and other matter. Figure 14.8 shows a comparison of diffraction data taken using the same wavelengths of x-rays and neutrons.[6] The target was powdered copper. The counter angle 2θ is twice the diffraction angle. The peaks occur at the angles where Debye–Scherrer maxima occur. In this case the diffraction patterns are not recorded on a flat screen, which would show the Debye–Scherrer circles, but with a detector that moves in an arc around the sample so that it crosses the circles and measures them as intensity maxima at their corresponding angles. The similarity of the two spectra is striking.

The energy of the neutrons was measured to be 0.07 eV. To understand the resulting diffraction patterns you need to know the neutrons' wavelength.

---

[5] I. Estermann, and O. Stern, "Diffraction of molecular rays," *Zeit. f. Physik* **61**, 95–125 (1930).

[6] C.G. Shull and E.O. Wollan, "X-ray, electron and neutron diffraction," *Science* **108**, 69–75 (1948).

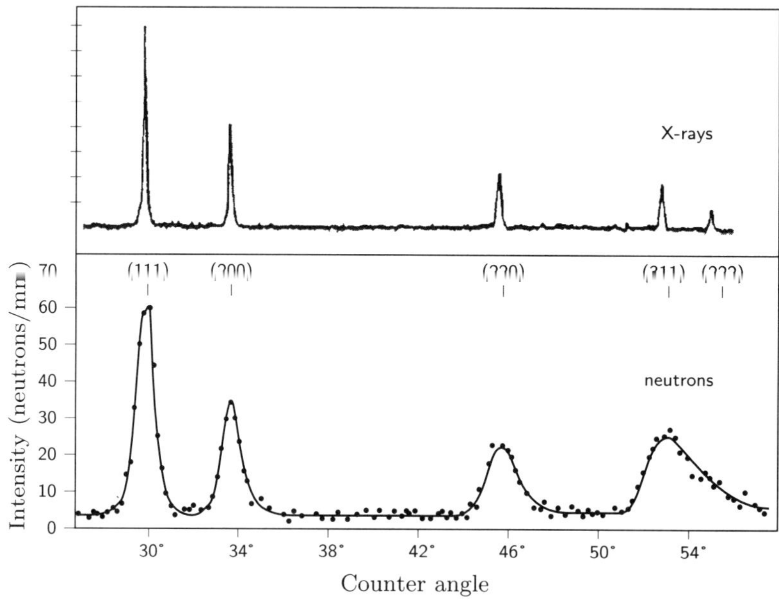

FIGURE 14.8 Bragg diffraction patterns for $\lambda = 0.1$ nm neutrons and x-rays in powdered copper. *Taken with permission from C.G. Shull and E.O. Wollan, "X-Ray, Electron, and Neutron Diffraction," Science 108, 2795, 69–75 (1948) ©1948 American Association for the Advancement of Science*

EXAMPLES

5. To find the neutron wavelength from the de Broglie formula $\lambda = h/p$, first find the momentum p from the kinetic energy $K = 0.07$ eV. For such a low energy you can find the product pc from the nonrelativistic relationship

$$pc = \sqrt{2mc^2K} = \sqrt{2 \times 939 \times 10^6 \times 0.07} = 1.146 \times 10^4 \text{ eV}$$

using the fact that for a neutron, $mc^2 = 939 \times 10^6$ eV. This value of pc implies that the wavelength is

$$\lambda = \frac{hc}{pc} = \frac{1240}{1.146 \times 10^4} = 0.11 \text{ nm}.$$

This is a wavelength just about the size of an atom, and therefore just right for probing the structure of something made up of atoms, such as a crystal or a molecule.

EXAMPLES

6. Now, if these 0.07 eV neutrons are undergoing Bragg diffraction in the copper crystallites, at what angle should there be a maximum due to reflections from the (111) planes?

Use the Bragg law to answer this question, and solve $2d\sin\theta = \lambda$ for $\sin\theta$. First find d. From Table 14.1 note that the lattice constant for Cu is 0.361 nm, so the spacing d between the (111) planes is

$$\frac{0.361}{\sqrt{3}} = 0.208\,\text{nm},$$

which yields

$$\sin\theta = \frac{\lambda}{2d} = 0.260,$$

so that $\theta = 15.1°$, and there should be a peak at the counter angle $2\theta = 30.2°$. There is!

EXERCISES

12. Calculate the counter angle at which you would expect to see a diffraction maximum from the reflection of 0.07 eV neutrons from the (200) plane of copper. Compare your result to the value given in Fig. 14.8.

13. Think of a nice way to show graphically that all the peaks in the neutron diffraction spectrum of Fig. 14.8 are consistent with Bragg diffraction. Do it.

14.4 SUMMARY AND CONCLUSIONS

All the objects we have thought of as particles—electrons, helium atoms, neutrons, protons—exhibit the diffraction and interference that characterizes waves. They act like waves with a wavelength λ given by the de Broglie relationship

$$\lambda = \frac{h}{p},$$

where p is the momentum of the particle and h is Planck's constant.

Some Useful Things to Know

Diffraction and interference are basic tools for working with particles. Consequently, you need to be able to find a particle's wavelength given its energy or momentum in order to predict how a beam of such particles will interact with an array of other particles, e.g., in a crystal lattice according to the Bragg law. Fairly often, this means that you must find the momentum p when you are given the kinetic energy K.

Before you can find the momentum of a particle from its kinetic energy, you must also be able to tell whether you can get by with the nonrelativistic relationship between momentum p and kinetic energy K of a particle of mass m,

$$pc = \sqrt{2mc^2 K},$$

or whether you need to use the relativistically correct relationship

$$pc = \sqrt{2Kmc^2 + K^2}.$$

The rule of thumb we developed in Chapter 11 is that the nonrelativistic formula will be accurate to 1% if $K/(mc^2) \leq 0.1$. If $K/(mc^2) \geq 0.1$, it will be necessary to use the relativistically correct formula.

And of course, you must be able to solve all parts of this problem using only units of electron volts (eV). For this purpose it is helpful to remember that $hc = 1240\,\text{eV nm}$. Also remember that for an electron, $mc^2 = 511\,\text{keV}$; for a proton, $mc^2 = 938\,\text{MeV}$; for a neutron, $mc^2 = 939\,\text{MeV}$.

Waves, Energy, and Localization

Two important consequences follow from the wave nature of particles. First, the smaller the object you study, the more energetic must be the probe. Second, particles are not sharply localized in space.

Probing Small Objects Requires Large Energies

Remember from what you learned about waves that when you probe a structure you cannot learn about details that are smaller than the wavelength of your probe. Because the de Broglie wavelength is inversely proportional to momentum, the smaller you make the probe, the greater you must make its momentum.

Kinetic energy is proportional to the square of the momentum in the nonrelativistic case and proportional to momentum in the ultrarelativistic case. This means that the smaller the thing you wish to probe, the larger must be the energy of the particles with which you do the probing. Since everything we know about tiny structures we learn by probing with something—photons, protons, electrons, or a number of other particles we have not even talked about—the need for higher and higher energies to look at smaller and smaller objects is important to keep in mind.

Let's look at one consequence of this idea. Suppose you wish to study the internal structure of an atom with a beam of electrons. How energetic must the beam be?

EXAMPLES

7. An atom is about 0.1 to 0.2 nm in size. If you want to see structure that is 1% of this size, you will want to have electrons that have a wavelength no larger than 0.001 nm. The momentum of such electrons is

$$pc = \frac{h}{\lambda} = \frac{1240}{0.001} = 1.24 \times 10^6 \text{ eV}.$$

Remember the rule for going from momentum p to kinetic energy K: Use relativistically correct equations if $pc \geq 0.2mc^2$. Clearly, that is the case here, because $1.24 \times 10^6 \geq 0.2 \times 0.511 \times 10^6$ eV. Therefore,

$$K = \sqrt{m^2c^4 + p^2c^2} - mc^2 = \sqrt{0.511^2 + 1.24^2} - 0.511 = 0.83 \text{ MeV}.$$

In other words, to probe the structure of an atom to 1% requires you to use about 1 MeV electrons.

EXERCISES

14. The Thomas Jefferson National Accelerator Facility, in Newport News, Virginia, produces a beam of high-energy electrons for probing the interior of nuclei. It is designed to look at structures as small as 3×10^{-16} m. Roughly, what is the lowest energy of electron that this will require?

Particles Are Not Sharply Localized

The second important consequence of the wave nature of particles (and the particle nature of waves) is that we are forced to conclude that until we make a measurement, a particle is in more than one place at the same time. This strange idea turns out to apply to momentum also; i.e., until its momentum is measured, a particle can have more than one momentum at the same time. This idea of simultaneously having more than one value of some definite physical quantity generalizes to become the foundation of our description of the behavior of the atomic and subatomic world. It is discussed further in the next chapter.

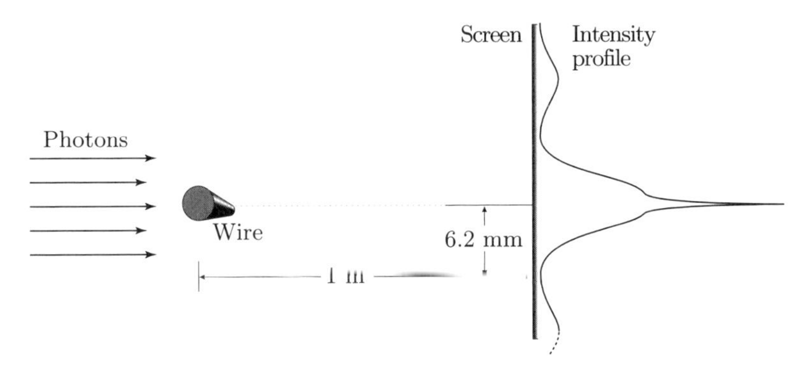

FIGURE 14.9 Light diffracting around a wire Problem 3.

PROBLEMS

1. a. If a 2 keV electron has a wavelength of 0.0274 nm, what will be the wavelength of a 32 keV electron?
   b. What will be the wavelength of a 512 keV electron?

2. Suppose you wanted to probe the structure of an atom using a beam of electrons. If you wanted to see structure on the order of the size of 0.1 of the radius of an atom, what energy of electrons would you need?

3. When you hold a wire or hair in front of a laser beam you get a diffraction pattern resembling what you get when you pass the laser beam through a slit of the same width.
   a. 2 eV photons diffracting around a wire as shown in Fig. 14.9 go on to strike a screen 1 m away. If the first minimum of the pattern occurs at 6.2 mm from the center of the pattern, what is the diameter of the wire?
   b. Suppose the wire is 50 μm in diameter and you wish to make the same diffraction pattern with electrons instead of with photons. What energy must the electrons have to produce a first minimum 6.2 mm from the center of the pattern?
   c. Suppose instead of a great, huge, thick wire, you had a wire of the diameter of a nucleus, i.e., $\approx 10^{-14}$ m. What energy would the electrons have to have in order to produce a diffraction minimum 6.2 mm away from the central maximum on a screen 1 m away?
   d. If the second maximum occurs at 9 mm, what is the angle through which the photons or electrons have been scattered to reach that point on the screen? Give your answer in radians.

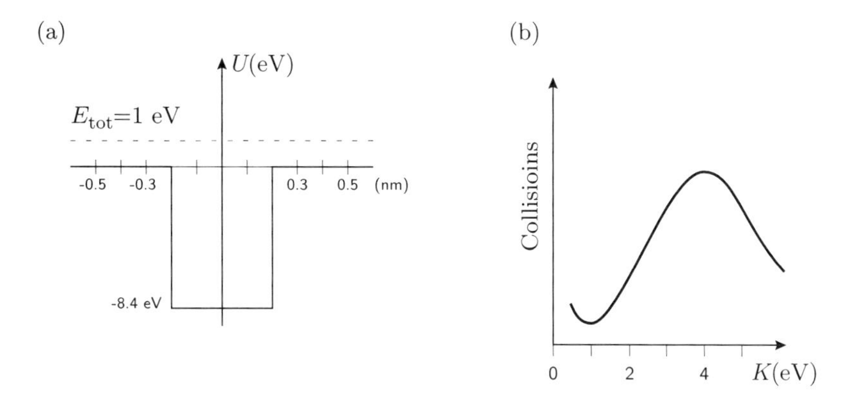

FIGURE 14.10 For Problems 4 and 5: (a) potential energy of an electron in an atom; (b) the probability of a collision between the electron and the atom as a function of the electron's kinetic energy.

4. In Fig. 14.10(a) the solid line shows a simplified version of the potential energy of an electron in an argon atom. The dashed line shows the total energy of a particular electron.

a. Is this electron bound in the argon atom? How do you know?

b. What is the electron's kinetic energy when it is at 0.05 nm?

c. What is the de Broglie wavelength of the electron when it is 0.25 nm from the origin?

d. What is the electron's de Broglie wavelength when it is at the center of the atom?

5. In the 1920s a physicist named Ramsauer sent a beam of electrons through a small amount of argon gas. He measured how many electrons collided with the argon atoms as he varied the electron kinetic energy K. His data are shown in Fig. 14.10(b).

a. Bohr showed that the collision probability should be least when the de Broglie wavelength inside the atom is equal to the diameter of the argon atom. Show that the data confirm Bohr's deduction.

b. Suppose you do the experiment with an atom like that shown in Fig. 14.10(a), except that now the P.E. inside the atom is -17.8 eV. If the minimum collision probability occurs when $K = 1.0$ eV, what is the diameter of the atom?

6. From a Bragg diffraction experiment, the wavelength of monoenergetic electrons is found to be 0.06 nm. What is the kinetic energy of these electrons?

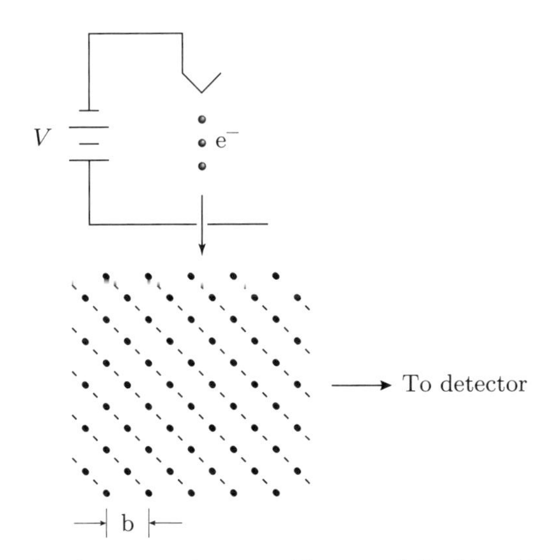

FIGURE 14.11 Arrangement for electrons to interact with a crystal (Problem 10).

7. To probe the structure of a nucleus means looking at matter 10^{-15} m in size or smaller. What wavelength of probe would you want in order to study lengths this small?

8. What energy electron has a wavelength of 10^{-15} m?

9. What energy proton has a wavelength of 10^{-15} m?

10. Electrons accelerated through a potential difference V pass through a narrow slit and strike the face of a cubic crystal, as shown in Fig. 14.11. When an electron detector (an ammeter) is placed in the position shown, a maximum current is recorded on the meter when $V = 50$ volts.

a. What is the kinetic energy of the electrons after passing through the anode slit?

b. Find the momentum of these electrons in units of eV/c. If an approximation can be used, then do so, but explain clearly why its use is justified.

c. What is the de Broglie wavelength of the electrons?

d. From your above answers and the information given in the drawing, find the interatomic spacing b in the crystal. (Hint: What atomic planes are responsible for the reflection?)

11. A beam of electrons having a well-defined energy is incident on a pair of very narrow slits whose center-to-center separation is 0.1 μm (Fig. 14.12).

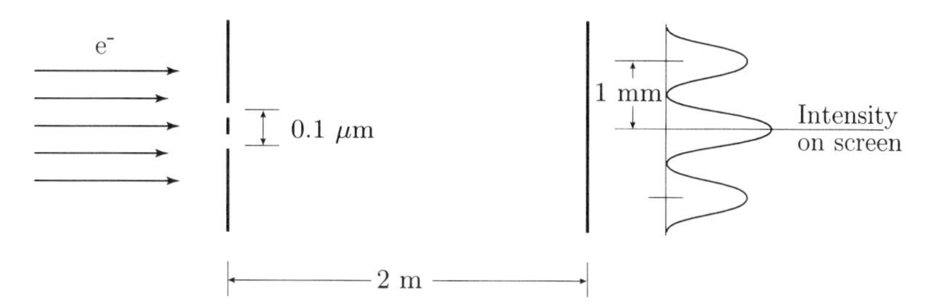

FIGURE 14.12 Electrons incident on a pair of slits (Problem 11).

On a fluorescent screen 2 m away the electrons produce an intensity pattern whose maxima are spaced by 1 mm.

a. What does the very existence of such an intensity pattern tell you about electrons?

b. What is the momentum of these electrons?

c. What is their kinetic energy?

12. A double-slit device known as an "electron biprism" is capable of producing interference patterns with electrons. With such a device, a team of experimenters recently was able to produce double-slit fringes with angular separation of 4.0×10^{-6} radians, using electrons that had been accelerated through a potential difference of 50 kV (Fig. 14.13.)

a. Show that the de Broglie wavelength of the electrons in this experiment was 5.5 pm.

b. What was the path difference between the interfering electron waves for the first minimum on either side of the central maximum?

c. What was the distance between the slits in this experiment?

d. If the accelerating voltage were increased from 50 kV to 200 kV, how would the angular separation between the fringes change? Explain briefly. A detailed calculation is not necessary.

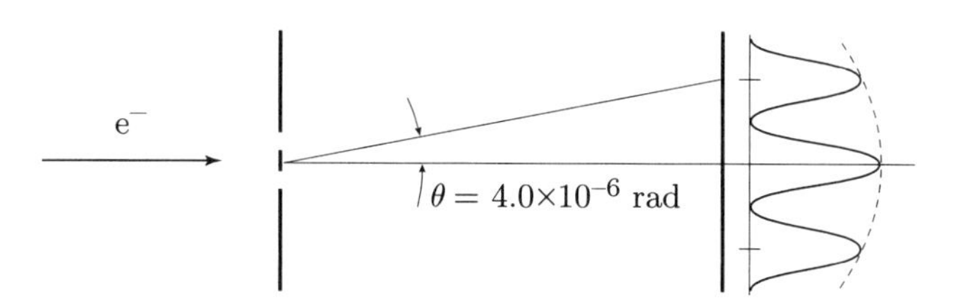

FIGURE 14.13 Experimental arrangement for producing electron-interference patterns (Problem 12).

CHAPTER 15

The Heisenberg Uncertainty Principle

15.1 INTRODUCTION

The photoelectric effect shows that waves behave like particles. A wave with a frequency f has a minimum packet, or quantum, of energy $E = hf$, where h is Planck's constant. Compton showed that when hf is comparable to the rest mass energy mc^2 of an electron, the scattering of electromagnetic radiation from electrons behaves like the scattering of one compact object from another. The particle-like behavior of light seems so prominent in these cases that we have given the quantum of light the particle-like name of "photon." Individual photons can be detected with a photomultiplier tube; such detection also suggests a degree of localization in space that is characteristic of particles rather than waves.

The fact remains, however, that all forms of electromagnetic radiation exhibit interference. Photons give rise to double-slit interference: When a beam of photons passes through two slits, the characteristic pattern of fringes appears on the screen behind the slits. Thus, depending upon the experimental setup, light can exhibit wave properties or particle properties. Niels Bohr called this combination of apparently contradictory properties "wave–particle duality."

Wave–particle duality is not just a quirk of photons. In the last chapter we saw that electrons, and all other objects that we customarily think of as particles, can exhibit wavelike interference and diffraction. Wave–particle duality is a feature of all atomic and subatomic matter. It is, therefore, a profound aspect of nature, and our theories of matter and energy must include it.

In the 1920s, physicists developed theories of microscopic matter that incorporated wave–particle duality. These theories go under the name of "quantum mechanics." Quantum mechanics revolutionized physics and changed our ideas about causality and measurement and what it means to know something about the physical world. Even now, more than two generations later, physicists argue heatedly about the meaning of quantum mechanics and the inferences one may draw from the theories. This chapter will try to give you some appreciation of why this might be so. It introduces some basic quantum ideas and their consequences for the atomic and sub–atomic worlds. In particular, it presents the Heisenberg uncertainty principle and shows you how it specifies the sizes and energies of atomic and subatomic systems.

15.2 BEING IN TWO PLACES AT ONCE

Can something be both a particle and a wave? Perhaps you don't think there is much of a problem here. If so, consider the following hypothetical experiment, which could certainly be done in principle.

A beam of electrons is allowed to strike two narrow slits close enough together to produce a nice interference pattern on a photographic plate. Imagine that the intensity of the electrons is turned down so that only 1 electron arrives at the photographic plate each hour. It would be tedious and expensive, but imagine that you exposed a plate and developed it after 1 hour; and then exposed a second plate for 2 hours and then developed it; and then exposed a third plate for 3 hours and developed; and so on. What would you see on the succession of plates?

The answer is what you would expect. On the first plate there would be one exposed grain of silver halide somewhere on the plates from the arrival of 1 electron during the 1 hour of exposure. On the second plate there would be two exposed grains of silver halide from the arrival of 2 electrons during the 2 hours of exposure; on the third there would be 3 exposed grains; on the fourth 4; and so on.

What would be the pattern of these exposed grains? At first there would be no apparent pattern. The number of exposed grains would be too few to show a pattern. But as the number of exposed grains reached into the hundreds and thousands, a distinct pattern would appear. It would be the pattern of

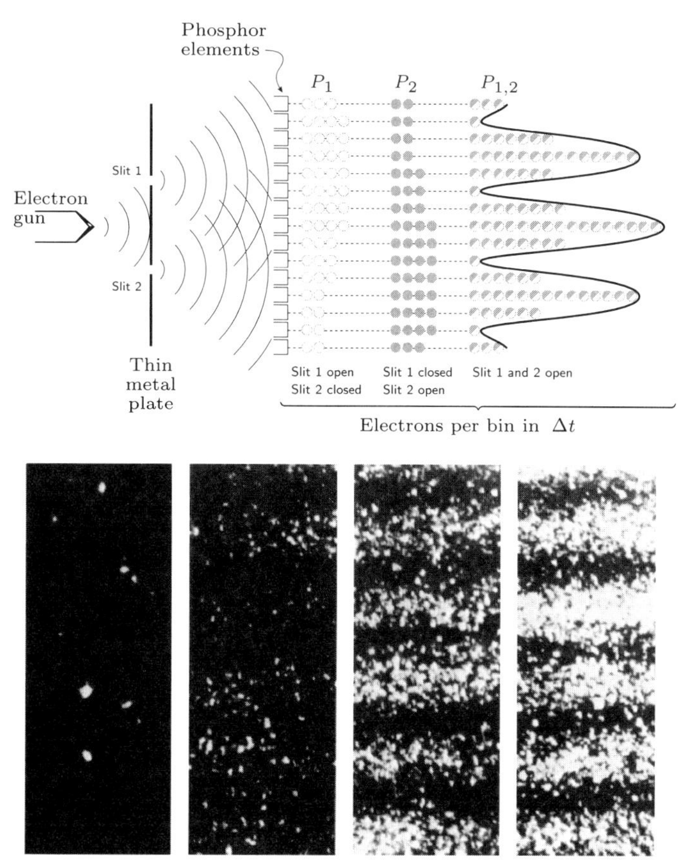

FIGURE 15.1 Hypothetical one-particle-at-a-time interference experiment. The open circles labeled P_1 are counts recorded when only slit 1 is open; the shaded circles labeled P_2 are counts recorded when only slit 2 is open; the half-shaded circles $P_{1,2}$ are the counts recorded when both slits 1 and 2 are open. The set of images in the bottom half of the figure show the gradual emergence of the double-slit interference fringes as particles are recorded over an ever longer time interval. *Taken with permission from J.G. Hey and Patrick Walters, The Quantum Universe, Cambridge University Press, 1987*

interference fringes you have come to know and love. Although the electrons arrive one at a time an hour apart, they still form an interference pattern. Electrons preferentially arrive at the places corresponding to maxima in the fringe pattern; they do not arrive at places corresponding to minima in the pattern.

The point is that interference occurs for the individual electrons; it occurs electron by electron. The same story is true for photons. Illuminate a pair of closely spaced slits with a laser beam that has been made so weak that only 1 photon goes through the slits each hour. We could detect these with the cumbersome succession of photographic plates used for the electrons, but it speeds things up if instead we imagine an array of photomultiplier tubes in place of the screen. To each of these we attach a counter so we can keep track of how many photons arrive at each photomultiplier tube. About once an hour one of these counters will go "click" and record the arrival of a photon. After many hours a pattern will emerge. The counters near the maxima of the interference pattern will have many counts; the counters near the minima will have few or none. Single photons preferentially arrive at places corresponding to maxima in the fringe pattern; they will not arrive at places corresponding to the minima. This process is illustrated schematically in Fig. 15.1.

Interference occurs for individual photons; it occurs photon by photon.

Does this bother you? If it doesn't, consider the following. Suppose we block one of the slits and do the above two experiments again. What will we see? This time we will see the single-slit diffraction pattern build up slowly one photographic grain or one click at a time. If the slits are narrow enough so that the first single-slit diffraction minimum is off the screen, then we will see only a smooth, nearly uniform distribution of exposed photographic grains or counter clicks. Electrons (or photons) *will now go to places where they did not go when there were two slits.*

Now think about this. When there are two slits, how does a single photon or electron "know" that it may not go to the places where the minima occur? When there is only one slit, how does a single electron or photon "know" that it can go to the previously forbidden places? If an electron passes through the bottom slit, how does it "know" whether or not the top slit is open?

To explain single-particle interference we are forced to assert that the particle passes through both slits. Only in this way can we make theories that are internally consistent and in agreement with experimental observation. In other words, we are led to claim that the particle can be in two different places at once. Such a claim flies in the face of all our experience and the physics of Newton, Maxwell, and Einstein. How can we maintain such a curious idea?

The idea makes us deeply uneasy because an electron—or any particle—usually has a well defined position. For example, it is detected in a well-defined, localized region of space—the photomultiplier tube or the photograph grain. That spatial localization makes apparent that the electron or the photon is a particle.

How can something so localized go through two separate slits at the same time? Quantum theory connects localization to measurement. The theory says that a particle can be prepared experimentally to be in a range of positions and that measurement has the effect of selecting out one of these. This says that an electron is in a wide range of positions as it approaches the photographic plate; then interaction of the electron with a grain of silver halide in the emulsion selects one of these positions. When there are two slits, the range of possible positions of the electron or photon does not include those positions that correspond to the minima of the interference pattern.

The urge to deny this picture is strong. "Surely," you will say, "I can measure through which slit the particle passes." You are then demanding that it pass definitely through one or the other slit as you would expect a classical particle to do. You can make such a measurement in any of several ways. For example, you might shine light on electrons as they come through one of the slits and use the resulting Compton scattering to signal the electron's passage. In other words, an electron passing through the illuminated slit will scatter light that can be detected, and the detection of a flash of light tells you that this electron passed through the illuminated slit. This possibility is consistent with our general principle: The particle is to be thought of as being in different places at the same time, but when you measure it you select out one of them. If you measure the next electron, you will find that it went through one slit or the other. The result of each measurement will be perfectly definite, but it will be unpredictable. You might find that 50% of your measurements find the electron passing through one slit and 50% through the other.

But if you can tell through which slit the particle came, then how can it interfere with itself? How can it contribute to producing the double-slit interference pattern? Well, it cannot. And here something quite interesting occurs. *If you do an experiment, any experiment, that determines through which slit the electrons pass, the double-slit interference pattern does not occur.*

This should not be so surprising, because in doing the experiment you have an effect on the electron. If no experiment is done, the electrons that come through the slits have some distribution of momenta in the y direction. But obviously, they do not have all values of p_y. For example, they do not have those values of p_y that would cause them to arrive at the minima of the interference pattern. However, in order to find through which slit the particle came, it is necessary to interact with the particle, for example, by shining light on it. The resulting Compton scattering changes the distribution of components of momentum in the y direction; it introduces components that permit the electron to arrive at places on the screen that were forbidden to it before its position was measured. These new components wash out the double-slit pattern, leaving only the broader single-slit pattern.

Here can be seen the importance of the quantum. According to classical physics we can reduce the energy of an electromagnetic wave of frequency f to an arbitrarily small amount. If we really could do this, then we could determine through which slit the electron came with light that had had its energy reduced to the point that its interaction with the electron was negligibly small. But we cannot; light of a given frequency comes with an energy of at least hf, no less, and so it has a momentum of hf/c. This is enough to add the missing components of momentum to the electron and so wipe out the interference pattern.

Of course, we could reduce the frequency f. This works. At a sufficiently low frequency the interference pattern is not affected by the interaction of the photons with the electrons. However, nothing is gained. As we reduce f, we are increasing the wavelength λ. As λ increases, the precision with which we localize the particle gets worse, because waves of wavelength λ will show you where something is to only roughly $\pm\lambda/2$. In fact, just when the photon energy is low enough so that it does not destroy the interference pattern, the wavelength has become so large that we can no longer tell through which slit the electron passed. The experiment no longer selects out a well-defined location.

15.3 HEISENBERG'S UNCERTAINTY PRINCIPLE

There seems to be a conspiracy here. Experiments that exhibit particle-like behavior of something destroy its wavelike behavior. Experiments that show the wavelike behavior of something, destroy its particle-like behavior. Although wave properties and particle properties seem mutually exclusive, nature is constructed in such a way that in principle we can never exhibit the contradictory properties at the same time. Our answer to the question "How can we tell whether something is a wave or a particle?" is that nature is constructed in such a way that the question can never be answered experimentally. As a result, the question simply makes no sense, and we reject it as physically unanswerable.

Nature *always* behaves so that particle and wave properties cannot be exhibited simultaneously. This behavior can be shown to be equivalent to the following fundamental principle. Any physical situation that forces an electron into a narrow range of positions Δx will at the same time impart to the electron a wide range of momenta in the x direction Δp_x. The converse is true. Any physical situation that forces the electron into a narrow range of momenta Δp_x gives it a broad range of positions Δx. Werner Heisenberg showed that in general, the product of Δx and Δp_x is never smaller than $h/(4\pi)$. It can be larger, but for most physical systems of interest the product is of the order of h or $h/(2\pi)$. This is the Heisenberg uncertainty principle and it is usually written

$$\Delta x \, \Delta p_x \geq \frac{h}{2\pi}. \tag{1}$$

The uncertainty principle is at the heart of quantum mechanics. It applies to all particles and things built from them. It applies to many pairs of physical quantities other than position and momentum. For example, there is an uncertainty relationship between energy and time.

You may feel there is a certain vagueness about the principle. What do Δx and Δp_x mean? Is their product greater than $h/(2\pi)$ or is it equal? When is it one or when is it the other? Should the right-hand side of Eq 1 be $h/(4\pi)$ or $h/(2\pi)$ or even, as it is often written, just h? Vagueness may be appropriate for something called the "uncertainty principle," but it is not as bad as it sounds. Δx and Δp_x have quite precise definitions, but in making order-of-magnitude arguments with the uncertainty principle, approximate values of Δx and Δp_x are often physically obvious. Also, for order-of-magnitude calculations, the particular choice of $h/(4\pi)$, $h/(2\pi)$, or h is not very important; for most purposes we will use $h/(2\pi)$ on the right side of Eq 1.[1]

EXERCISES

1. Show that $\hbar c$ equals 197 eV nm.

The precise definition of Δx is that of standard deviation of a distribution. You first met the idea of a distribution when we discussed the ideal gas law and connected pressure and temperature to molecular velocity. We argued that the molecules of a gas have a range of velocities, and we imagined dividing this range of velocities into small intervals of velocity of width dv. Such intervals are often called "bins." We labeled the velocity of the ith bin to be v_i, and then we imagined counting the number of molecules with velocities in the small range v_i to $v_i + dv$. We then called the number of molecules in the ith such bin n_i. The set of values n_i represents the distribution of velocities. Using this distribution we computed an average value of the square of the velocity $\langle v^2 \rangle$ using the relationship

$$\langle v^2 \rangle = \frac{\sum n_i v_i^2}{\sum n_i}.$$

If we had a distribution of the positions x, we could find $\langle x \rangle$ and $\langle x^2 \rangle$; if we had a distribution of the momenta p_x, we could find $\langle p_x \rangle$ and $\langle p_x^2 \rangle$. The formal definitions of Δx and Δp_x are

$$\Delta x = |\langle x \rangle^2 - \langle x^2 \rangle|^{1/2},$$
$$\Delta p_x = |\langle p_x \rangle^2 - \langle p_x^2 \rangle|^{1/2}.$$

[1] The quantity $h/(2\pi)$ is used so often that it gets its own symbol, an h with a line through it, which is written $\hbar$ and is called "h-bar." We will use it frequently. Just as it is useful to know that $hc = 1240$ eV nm, it is useful to know that $\hbar c = 197$ eV nm. For quick calculation and easy recollection many physicists take $\hbar c$ to equal 200 eV nm.

You may recognize that each of these is a quantity called "the standard deviation" of a distribution.

The mathematics of distributions and the calculation of quantities like Δx are the same for quantum mechanics as for the case of gas molecules. The meaning, however, is quite different for quantum mechanics than for the kinetic theory of gases. Kinetic theory assigns a definite value of velocity to each atom, and the distribution arises because there are many molecules with different velocities. Quantum mechanics ascribes all the velocities of the distribution to each particle and says that you will get a definite value only when you measure it.

EXAMPLES

1. Just to exercise the above definition, consider a uniform beam of electrons directed at a slit of width a. What is Δx for electrons passing through such a slit? Suppose we divide the slit into 10 intervals of width $0.1a$. To find Δx, we first need to find $\langle x \rangle$ and $\langle x^2 \rangle$.

To calculate these quantities we need to place a coordinate system on the aperture. It is convenient to place the origin at the center of the slit. Calculation of $\langle x \rangle$ is then particularly easy. The symmetry of the problem shows that $\langle x \rangle$ will then be 0, because there will be as much negative x below this origin as there is positive x above it. A similar symmetry argument is often used to calculate $\langle p_x \rangle$.

If we assume that passage through the slit is as likely at any place within the aperture as any other, calculation of $\langle x^2 \rangle$ is straightforward. Each bin is

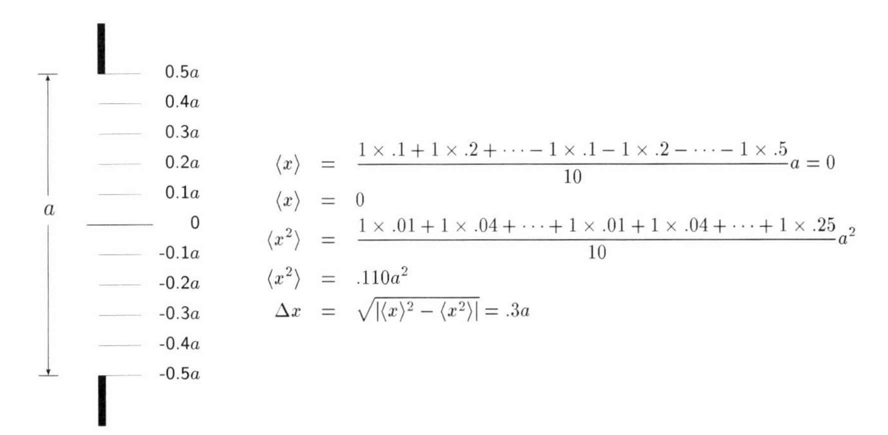

FIGURE 15.2 Calculation of $\langle x^2 \rangle$ for a slit of width a (viewed from above).

$0.1a$ wide; the values of x run from $-0.5a$ to $+0.5a$, so the values of x^2 run from $0.25a^2$ down to 0 and then back up to $0.25a^2$. Therefore, the sum runs from $0.025a^2$ down to 0 and back up to $0.025a^2$. The process is shown in Fig 15.2. We can write the sum as $0.1(0.1a)^2 2[1+4+9+16+25] = 0.11a^2$, so $\Delta x = 0.33a$.

EXERCISES

2. In the above example we chose to use the largest value of x^2 occurring in each bin. What would be $\langle x^2 \rangle$ and Δx if you chose the value of x^2 at the midpoint of each interval?

For our purposes, however, it is good enough to look at the physical setup and observe that $\Delta x \approx a$.

Average Kinetic Energy from the Uncertainty Principle

Similar arguments can be made to show how to find values for $\langle p_x \rangle$ and $\langle p_x^2 \rangle$. If the system has symmetry, $\langle p_x \rangle$ will be 0. If the particle is bound, say if it is an electron going around a proton, then $\langle p_x \rangle$ will also be 0, because on average a bound particle has no net motion in any direction. For systems with $\langle p_x \rangle = 0$, we have the very useful result that $(\Delta p_x)^2 = \langle p_x^2 \rangle = 2m\langle K \rangle$. This means that the uncertainty principle relates average kinetic energy $\langle K \rangle$ to spatial confinement, and we can use this fact to predict important features of systems built up from particles.

EXAMPLES

2. Let's consider an electron confined to a region of space about the size of an atom, so that $\Delta x \approx 0.1$ nm. Then the uncertainty principle tells us that $\Delta p_x \approx \hbar / \Delta x$. Squaring both sides, we obtain

$$(\Delta p_x)^2 = \langle p_x^2 \rangle = \frac{\hbar^2}{(\Delta x)^2}.$$

Because $\langle p_x^2 \rangle = \langle p_y^2 \rangle = \langle p_z^2 \rangle$ and $K = (p_x^2 + p_y^2 + p_z^2)/(2m)$, it follows that the average kinetic energy K of the bound electron is

$$K = 3\frac{\langle p_x^2 \rangle}{2m} = \frac{3\hbar^2}{2m(\Delta x)^2} = \frac{3(\hbar c)^2}{2mc^2(\Delta x)^2} = \frac{3 \times 197^2}{2 \times 0.511 \times 10^6 \times 0.01} = 11.4\,\text{eV}.$$

Although such calculations are approximate, the result is very informative. It says that simply by virtue of its confinement within the space of an atom, an electron will necessarily have an average kinetic energy of the order of 10 eV.

This result also tells us something about the force holding the electron in confinement. For the electron to be bound to the atom, its total energy, kinetic plus potential, must be negative. This means that in attracting the electron to the atom, the force must reduce the total energy of the electron by more than enough to offset the kinetic energy.

Does it? The hydrogen atom is formed by the electrical attraction between the positively charged proton and the negatively charged electron. These behave like point charges, so we can calculate their energy of interaction from the formula for the electrical potential energy between two point charges. For an atom ≈ 0.1 nm in diameter, the two charges are separated by a distance of 0.05 nm, and their potential energy is

$$U = \frac{-k_c e^2}{r} = \frac{-1.44}{0.05} = -28.8 \text{ eV}.$$

The total energy would then be $11.4 - 28.8 = -17.4$ eV. The total energy is negative, and the fact that the electron and the proton form a bound system about 0.1 nm in diameter is consistent with the Heisenberg uncertainty principle.

EXERCISES

3. A neutron is a neutral particle with $mc^2 = 939 \times 10^6$ eV. Neutrons occur inside atomic nuclei. The atomic nucleus has a diameter of about 10^{-14} m. Use the uncertainty principle to estimate the average kinetic energy of neutrons in the nucleus.

4. (a) Using the result of the previous problem, estimate the minimum potential energy required for a neutron to remain bound inside an atomic nucleus. (b) Speculate on what force produces this potential energy.

In the example above we assumed knowledge of the size of an atom and deduced the average kinetic energy of an electron held inside the atom. If instead of knowing the size, we know the forces between the interacting parts of the atom, we can use the uncertainty principle to estimate the atom's size. Like any physical system, the atom will configure itself to achieve the lowest possible total energy. You might think that this would occur when the distance between the two charges goes to zero, because then the electron's potential energy goes to negative infinity. However, the uncertainty principle warns us that confinement of an electron to a smaller and smaller volume of space will lead to an unbounded increase in the electron's kinetic energy. The atom takes on a size that minimizes the sum of these two effects. We find that size by expressing the total energy in terms of size and then finding the radius that gives a minimum total energy.

EXAMPLES

3. The total energy E of an electron of mass m and charge $-e$ interacting with a proton of charge e when separated by an unknown distance r is

$$E = \frac{p^2}{2m} - \frac{k_C e^2}{r}.$$

The uncertainty principle connects the kinetic energy of the electron to the size of the atom. According to the uncertainty principle, confinement to a region $\Delta x \approx r$, the radius of the atom, will cause the electron to have $\Delta p = \sqrt{\langle p^2 \rangle} = \hbar/r$. Consequently, its kinetic energy will be

$$\frac{p^2}{2m} = \frac{\hbar^2}{2mr^2}.$$

Therefore, the total energy of the atom can be written as

$$E = \frac{\hbar^2}{2mr^2} - \frac{k_C e^2}{r}.$$

To find the value of r that minimizes E, differentiate E with respect to r; set the result equal to zero; solve for r.

$$\frac{dE}{dr} = -\frac{2\hbar^2}{2mr^3} + \frac{k_C e^2}{r^2} = 0.$$

solving this for r we get,

$$r = \frac{\hbar^2}{mk_C e^2} = \frac{(\hbar c)^2}{mc^2 k_C e^2} = \frac{197^2}{0.511 \times 10^6 \times 1.44} = 0.0527\,\text{nm}.$$

This result is very satisfactory, because it is a radius of just about the size we have found atoms to have. In principle, there is no justification for stating the result to three significant figures given that it is obtained by an order-of-magnitude argument. We do it because, as you will see shortly, 0.0527 nm is precisely the radius of the hydrogen atom obtained from a model that describes many other atomic properties of hydrogen quite accurately. Obtaining the right order of magnitude for the radius of atoms using the uncertainty principle is convincing evidence that the principle is fundamental; the fact that it can be made to agree precisely with other results just makes it look good.

EXERCISES

5. Use the expression just obtained for the radius of the hydrogen atom and derive an expression for the total energy of the hydrogen atom in terms only of fundamental constants such as $e, \hbar$, c, and the mass of the electron m.

6. (a) Evaluate the expression you obtained in the previous problem to give a numerical value for the total energy. (b) Your answer should be negative. Why? What does a total negative energy mean?

15.4 A REAL EXPERIMENT

In the past few years some very clever experiments have vividly shown these strange quantum-mechanical behaviors. Let's examine an especially interesting experiment done by physicists at the University of Rochester.[2]

A schematic diagram of their apparatus is shown in Fig. 15.3. It works like a double-slit device, but instead of two slits, it has two crystals of lithium iodate, labeled NL1 and NL2 in the figure. Into each comes a beam of

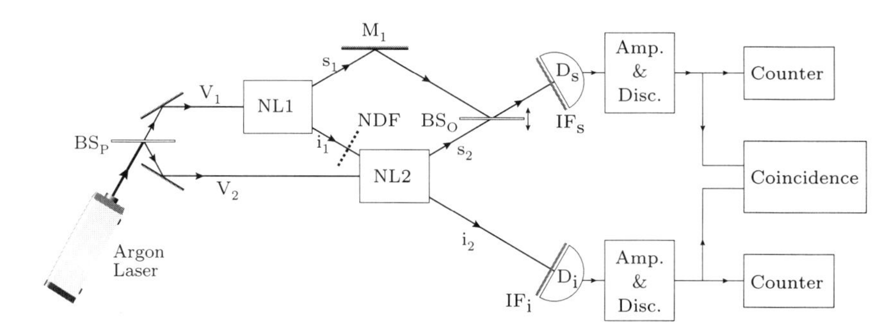

FIGURE 15.3 Schematic diagram of two-crystal analogue of double-slit interference apparatus. *Taken with permission from X.Y. Zou, L.J. Wang, and L. Mandel, Phys. Rev. Lett. 67, 318–321 (1991) ©1991 The American Physical Society*

[2]X.Y. Zou, J.L. Wang, and L. Mandel, "Induced Coherence and Indistinguishability in Optical Interference," *Phys. Rev. Lett.* **67**, 318–321 (1991).

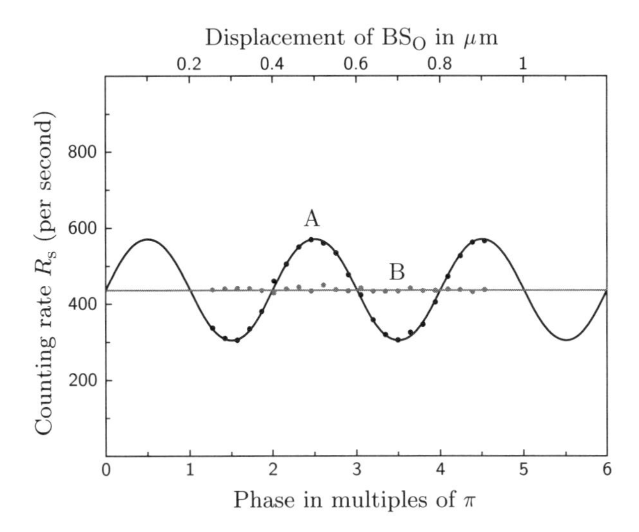

FIGURE 15.4 Measured photon counting rate as function of displacement of BS_O. The distinction between curves A and B is explained in the text. *Taken with permission from X.Y. Zou, L.J. Wang, and L. Mandel,* Phys. Rev. Lett. *67, 318–321 (1991) ©1991 The American Physical Society*

351.1 nm light brought from an argon laser by way of the beam splitter BS_P. From each crystal come two beams, an "idler" beam labeled i_1 or i_2, and a "signal" beam labeled s_1 or s_2. The signal beams are brought together by the mirror M_1 and recombined at the output beam splitter BS_O. There are two different paths, V_1 or V_2, by which signal light can reach the detector D_s, and differences in the distances along these paths will give rise to phase differences that result in interference. This device is an example of what are known as "two-beam interferometers."

After the two signal beams are combined at BS_O, the photons pass through a filter IF_s, which eliminates any idler light, and into the detector D_s. When the output beam splitter is shifted in the direction along a line from top to bottom of the diagram, the count rate in the detector varies as shown in curve A of Fig. 15.4. In other words, there is an interference pattern here like that on a screen in front of a double-slit interference apparatus.

We have already discussed that double-slit interference occurs only as long as the apparatus is set up in such a way that it is impossible to determine through which slit a photon came. In an exactly analogous way, an interference pattern occurs in the Rochester apparatus only as long as the apparatus is arranged so that there is no way to tell from which crystal a photon came before it arrived at BS_O. This is true even though the experimenters have provided an extremely clever way to tell from which crystal a photon has come.

They make use of a remarkable property of the crystals called "down conversion." When light of frequency f enters such a crystal, it absorbs it and emits two separate outgoing beams of lower frequencies f_1 and f_2 such that $f_1 + f_2 = f$.[3] The exact values of the two outgoing frequencies depend upon the geometry of the crystal and the angle of incidence of the entering beam, but by conservation of energy the two outgoing frequencies must add up to the frequency of the entering beam. In this experiment the two beams from each crystal have wavelengths of 632.8 nm and 788.7 nm.

EXERCISES

7. Show that the frequencies of the outgoing beams add up to the frequency of the incoming beam. Do this without converting the wavelengths to frequencies.

An interference pattern is formed by 632.8 nm signal photons from the two crystals. The paths available to them are labeled s_1 and s_2 (Fig 15.3). The other two beams, the idlers labeled i_1 and i_2, can be used to decide from which crystal a signal photon comes. No s_2 photon is emitted without an i_2 photon. Therefore, if a detector D_i, set up as shown in Fig. 15.3, records the presence of an i_2 photon at the same time D_s detects a photon, you know that the photon at D_s must have come from NL_2. This is just the same as knowing from which slit a photon came in a double-slit setup, although it is accomplished without any direct interaction with an s_1 or an s_2 photon. What is the result? It is shown in curve B of Fig. 15.4; there is no interference.

Well, then, how do they get interference in the first place? To get interference the apparatus must be set up so there is no possible way to distinguish which path a photon took in reaching D_s. In an apparatus in which it is *in principle* unable to distinguish which path the photon takes, interference occurs. This is like the example discussed above, where we found that in some sense an electron giving rise to an interference pattern passes through both slits at once. The experimenters arranged their apparatus to make the two paths indistinguishable by carefully aligning it so that i_1 photons passed through the crystal NL_2 and emerged exactly along the path that i_2 photons followed as shown in the diagram. Then the detector D_i cannot tell whether a photon arriving at D_s came from NL_2 or NL_1. Only under these circumstances was the interference pattern of curve A obtained.

[3] Such crystals are called "down converters" because as conservation of energy requires, the outgoing frequencies are lower than the incoming frequency.

It is important to understand that it is not the presence of D_i or the actual detection of i_2 photons that destroys the interference pattern. Just the possibility of distinguishing which crystal the interfering photons came from is enough. Nature is constructed in such a way that indistinguishable paths lead to interference. This means indistinguishable in principle, not just in practice. Perhaps the Rochester experimenters Zou, Wang, and Mandel say it better:

> Whether or not this auxiliary measurement with D_i is actually made, or whether detector D_i is even in place, appears to make no difference. It is sufficient that it could be made, and that the photon path would then be identifiable, in principle, for the interference to be wiped out.

To make this point experimentally the Rochester physicists carefully aligned their apparatus to have the i_1 and i_2 beams exactly coincide. Under these conditions they observed and recorded an interference pattern. Then they inserted an absorber to block the i_1 beam, as shown by the dashed line in Fig. 15.3. With i_1 blocked, any photon detected at D_i signals that a photon detected at the same time with D_s is from NL_2. Under these new conditions, with an absorber blocking i_1, the interference pattern disappeared, and the observers measured curve B in Fig. 15.4. Notice that D_i has nothing to do with this result; it does not have to be present at all. Blocking i_1 makes it possible in principle to distinguish the s_2 photons from the s_1 photons, and that is enough to guarantee that there is no longer any interference.

EXERCISES

8. Suppose the experimenters misaligned their apparatus just enough so that the counter at D_i could distinguish i_1 photons from i_2 photons by the slightly different directions from which they were coming. Would the interference pattern be present? Explain your answer.

9. Fig. 15.3 shows a box marked "coincidence." It represents a device that the experimenters used to tell them when photons arrived at the two counters D_s and D_i at nearly the same time, i.e., in coincidence with each other. Explain why the experimenters needed this device. Would they have observed an interference pattern without it?

15.5 SUMMARY AND CONCLUSIONS

The idea of a particle interfering with itself seems contradictory, especially when this interference must be explained by the strange idea that the particle can have different positions at the same time. For a theory of matter to be consistent and free of paradoxes, measurements must have two physically important consequences. First, whenever a measurement is made, the system must take on some particular, definite value from among the many possible. Which one of the possible values will be obtained in any given measurement is entirely unpredictable, but quantum theory quantitatively predicts the probability of each outcome, *i.e.*, it predicts how many times a given value will be obtained when a large number of measurements are made.

Second, the act of measurement of one quantity will generally cause the particle to take on a range of new values of some complementary quantity. For example, the measurement of position will introduce new values of momentum. The result will always be that the more precisely we measure position, the less precisely we know momentum, and conversely. This behavior is summarized in the Heisenberg uncertainty principle, which for the particular example of position and momentum is written

$$\Delta x \, \Delta p_x \approx \hbar,$$

where Δx is a measure of the spread of positions of a particle, Δp_x is a measure of the spread of the x-component of the momentums of the particle, and $\hbar$ is Planck's constant divided by 2π. The uncertainty principle places fundamental, general limits on what we can measure and what we can learn about nature from measurement.

In particular, the uncertainty principle guarantees that a particle can never be made to exhibit at the same time wavelike properties of interference and particle-like properties such as billiard-ball scattering. No paradox can occur. The uncertainty principle is a basic law of nature, and we were able to use it to calculate such important features of atomic and nuclear systems as size, kinetic energy of a bound particle, and binding energy. It shows that the smaller, more compact a system of particles is, the higher their kinetic energy is and the more strongly they must be bound to offset the high kinetic energy. The smaller a system of particles is, the greater the forces holding the particles together must be.

PROBLEMS

1. 650 nm light passes through two very narrow slits separated by 10 μm and then strikes a screen 1 m distant.

a. Describe what appears on the screen and where.

b. Suppose the screen were replaced with an array of photon counters, and the intensity of the incident light were reduced to 1 photon per minute. What would appear on the arrays of counters?

c. If one of the two slits were blocked, what would appear upon the screen? Assume that the width of the single slit is 1 μm.

d. What is the puzzle implicit in your answers to (b) and (c), and how does the Heisenberg uncertainty principle resolve this puzzle?

2. How does the Heisenberg uncertainty principle resolve the apparent paradox that light and electrons each exhibit wave properties in some circumstances and particle properties in other circumstances?

3. State clearly the Heisenberg uncertainty principle. Explain what the symbols mean. Explain what the uncertainty principle is saying about the behavior of physical systems.

4. Use the uncertainty principle to estimate the kinetic energy of an electron confined to the region of an atom. (You should know a reasonable dimension for an atom.)

5. Use the uncertainty principle to estimate the kinetic energy of a neutron confined to a nucleus.

6. If the average kinetic energy of an electron in the outer regions of an atom is 2 eV, what will be its average kinetic energy if it is confined to a nucleus?

7. Explain how the uncertainty principle sets the scale of energies of small, bound systems. Suppose you had reason to believe that there is a particle called the "quark," and you had reason to believe that the quark is confined to a region of about 1 fm. Estimate the momentum of the quark. What do you need to know before you can estimate its kinetic energy?

C H A P T E R 16

Radioactivity and the Atomic Nucleus

In 1896 Henri Becquerel discovered that compounds containing uranium emit radiations that can penetrate opaque paper and even thin sheets of metal and cause photographic plates to darken. Like x-rays, these emissions ionized air and caused electroscopes to discharge, but unlike x-rays, they occurred without any external source of excitation. Becquerel's student, Marie Curie, named this spontaneous emission of ionizing radiation "radioactivity."

Research soon showed that radioactivity was not rare. Within a few years, Marie and Pierre Curie, working in France, discovered two previously unknown chemical elements, polonium and radium, that were radioactive. Over the next decade or so their work and the studies they inspired identified dozens of different radioactivities. Ernest Rutherford, at first in England, later with Frederick Soddy in Canada, and then again in England characterized and identified the radiations. They and their colleagues found that radioactive atoms emit either helium ions or electrons and as a result *change into other atoms*—a stunning overthrow of the idea of immutable, eternal chemical elements.

The helium ions and electrons emitted from radioactive atoms have energies typically 10^6 times greater than the energies characteristic of chemical bonding; moreover, in some cases they continue being emitted for billions of years (Gy). This was so puzzling at first that some physicists seriously considered that energy might not be conserved.

Such a drastic hypothesis became unnecessary after it was found that a given sample does not emit particles indefinitely; the radioactivity always runs down eventually. The large magnitude of the energies became understandable later when Rutherford discovered that every atom possesses a compact, extremely dense, positively charged core—the atomic nucleus. This discovery ultimately led to the recognition that nuclei are composed of two different kinds of particles, protons and neutrons, held together by a new force, different from the electromagnetic and gravitational forces, and much stronger. As you will see, the strength of the new force and the small dimensions of the nucleus can explain the large energies released in radioactivity.

Rutherford also was one of the first to realize that each species of radioactive atom has a well-defined, characteristic probability for undergoing spontaneous disintegration and transformation. This discovery was gradually understood to mean that the moment in time at which any individual atom will decay is purely random and unpredictable *in principle*, so that radioactive decay deeply contradicts Newtonian ideas of causality. This kind of fundamental randomness required the introduction of revolutionary new ideas into physics, ideas with implications that are still surprising and mystifying physicists.

16.1 QUALITATIVE RADIOACTIVITY

Becquerel Discovers Radioactivity

In the early weeks and months after Roentgen's discovery of x-rays there was intense activity in many laboratories directed towards discovering the source and nature of the rays. When the French physicist Henri Becquerel attempted to understand the production of x-rays, he accidentally discovered an entirely new phenomenon—radioactivity.

Many substances that have been bombarded by cathode rays or illuminated by beams of light continue to emit light after the incident radiation has been turned off. This delayed emission of light is called "fluorescence." Becquerel speculated that x-rays might be associated with fluorescence. To test his idea he wrapped a photographic plate in black paper so that no light would leak in. Then he coated the outside of the wrapper with uranium sulfate salts, which were known to fluoresce strongly when illuminated by sunlight. His idea was to let sunlight strike the salts and produce fluorescence. Then any x-rays that were produced would penetrate the opaque wrapper and expose the photographic plate inside. After letting the package sit all day in the sun to give the material plenty of time to produce the supposed penetrating radiation from the fluorescence, he opened the package in the dark and developed the plate. He found that the plate had darkened in just the way that photographic materials respond to x-rays. However, another plate showed the same degree of blackening even though it had been exposed to less sunlight and therefore less fluorescence. When uranium salts were not exposed to sunlight at all and so could not fluoresce, the wrapped photographic plates still exhibited the same darkening. At this

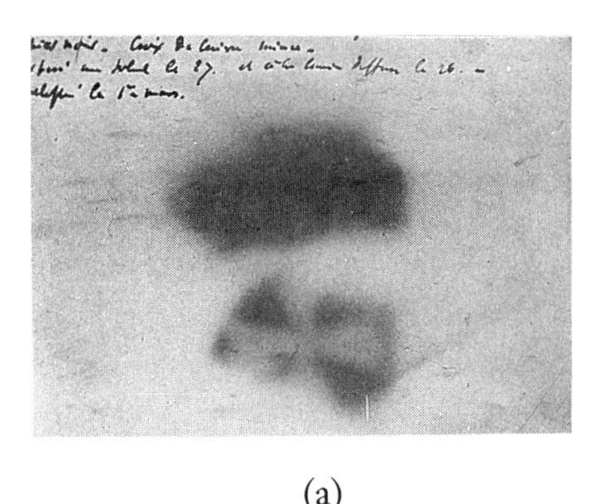

(a)

(b)

FIGURE 16.1 (a) The smudgy patches on Becquerel's photographic plate that signaled the existence of radioactivity. Notice the faint outline of a cross produced where a metal cross was placed to block radiation from the plate. (b) Some experiments were performed using uranium salts in ampoules like these. *©Bibliotheque Centrale M.N.H.N. Paris, 1998*

point, Becquerel concluded that he was dealing with something other than x-rays. Results of one of his earliest experiments are shown in Fig. 16.1. Only forty-nine years separate these smudges from the explosion of the first atomic bomb.

To show that it was the uranium in the salt that was responsible for the radiations that darkened photographic plates, Becquerel exposed the plates to a piece of uranium metal. If the radiations come from uranium, you would expect the effect on the plates to be stronger from the higher concentration of uranium of the metal than from the lower concentration in the salt. This is just what he observed.

Becquerel also established that the radiations could cause the discharge of charged electroscopes, and he began to quantify how much radiation a sample of radiating material produced. He took the amount of darkening of photographic film, or the rate at which an electroscope discharged, to be a measure of what he called the "activity" of the sample. If one sample darkened film more or discharged an electroscope faster than another, he said that the sample had a larger "activity." Quantitative measures of activity are now much more concrete and precise.

Becquerel also found that the penetrating radiation that darkened his photographic plates could be bent by magnetic fields in the same way as cathode rays. He showed that the charged particles had a velocity of about 1.6×10^8 m/s and a charge-to-mass ratio of about 10^{11} C/kg. Remember that e/m for electrons is 1.76×10^{11} C/kg. In Chapter 7 you saw how in 1900 Kaufmann also showed that one kind of radioactive ray had a charge-to-mass ratio of the order of 10^{11} C/kg. This and other evidence established that energetic electrons were being emitted.

EXERCISES

1. What is the energy of the electrons found by Becquerel?

The Curies Discover New Radioactive Elements

Becquerel's work motivated the start of two extraordinary research efforts, that of Marie and Pierre Curie, and that of Ernest Rutherford. Marie Sklodowska was a brilliant young student from Poland at the Sorbonne, in Paris, when she met and married Pierre Curie, a recent recipient of a doctoral degree. Becquerel's results became known just at the time when Marie Curie was casting about for a topic for her own doctoral research. She decided to look for other materials exhibiting radioactivity. In 1898 she found that thorium was also radioactive, but she was not alone in discovering it. She did notice, however, that the mineral ore from which uranium was extracted was several times more active than uranium itself. She deduced that there had to be other unknown substances present that were much more active than uranium. That same year she succeeded in separating a chemically distinct radioactive material that was considerably more active than uranium or thorium. As it was apparent that she had discovered a new element, she named it "polonium" in honor of her homeland.

A second radioactive substance also was observed to separate out along with barium from the ore. By heroic efforts Marie Curie extracted and purified about 0.1 g of this substance from a ton of ore. The new element was millions of times more active than an equivalent amount of uranium. She was able to determine its atomic weight, now known to be 226.0 u, and enough of its chemical properties to be sure that it was another new element. She named it "radium."

From the perspective of more than a century later you may find it hard to appreciate the great significance of the Curies' accomplishments. They showed that radioactivity could be used to identify new elements, elements that began to fill a large gap in the periodic table between bismuth ($Z = 83$) and thorium ($Z = 90$). Even more important, their discovery of new radioactive elements showed that radioactivity was more than just a peculiarity of one or two elements. It became clear that radioactivity is both widespread in nature and related in some basic way to the internal structure of individual atoms. Their work was a major contribution to the realization that radioactivity represented a new physical phenomenon profoundly different from any previously known.

The Curies also greatly advanced experimentation on radioactivity. By purifying radium, they made available a strong source of radiation that could easily be used for further research; it was no longer necessary to depend on the weakly active uranium samples that took a day to produce much darkening of a photographic plate. They also began quantitative studies of the ionizing power of radium emissions, which led to the development of a standard for expressing the activity of a radioactive substance. They used the activity of 1 g of radium as a standard. Other activities could be measured and described in equivalent grams of radium.

Alpha, Beta, and Gamma Rays

Many people had a hand in the early discoveries of radioactivity, but Ernest Rutherford was largely responsible for making sense of the phenomenon. One question that arose immediately after the discovery of radioactivity was whether the radiations were x-rays or something different. Rutherford, working in the laboratory of J.J. Thomson (of the e/m experiment) became interested in finding the answer. He soon identified two kinds of rays. The first were easily stopped by very thin foils; the others were much more penetrating. The easily stopped rays he named "α rays"; the more penetrating rays he named "β rays." A 0.02 mm thick foil of aluminum or a piece of ordinary paper would stop 95% of the radiations from uranium. These were alpha rays. Clearly, the more penetrating ionizing radiation—the other 5%—were beta rays and caused the darkening of Becquerel's photographic plates or the discharge of an electroscope.

It was natural to analyze the two kinds of rays by electric and magnetic fields (see Fig. 16.2). We have already mentioned that e/m measurements showed beta rays to be very similar to the cathode rays of J.J. Thomson. The conclusion was that beta rays are energetic electrons.

Alpha radiations bend in a magnetic field as energetic, positively charged particles with the charge-to-mass ratio of fully ionized helium (He^{++}). The fact that helium gas, relatively rare on Earth, was often found in minerals that contained significant amounts of uranium or thorium suggested to Rutherford that alpha particles are helium ions.

EXERCISES

2. What is the charge-to-mass ratio of hydrogen? Of doubly ionized helium? Would it be difficult to distinguish hydrogen from helium by their charge-to-mass ratios?

3. As part of measuring the charge-to-mass ratio of alpha rays from radium, Rutherford measured their velocity and got about 2.5×10^7 m/s. Later, more accurate measurements gave 1.53×10^7 m/s. Assume that al-

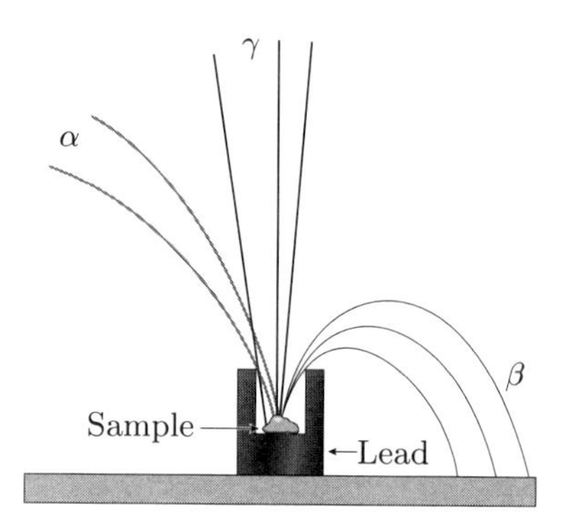

FIGURE 16.2 A schematic diagram of the behavior of different kinds of radioactive rays in a magnetic field: the α rays bend as positively charged particles with small q/m; the β rays bend as negatively charged particles with q/m of an electron; the γ rays are not affected by the magnetic field.

pha particles are helium ions, and use the later value of velocity to calculate the energy of these alphas.

At the University of Manchester, in England, Rutherford did experiments that directly showed that alpha particles are ionized helium atoms. He built an apparatus that collected the alpha particles as they stopped. Even with the much stronger sources of radioactivity that were by then available, he had to run his experiment for several months. At the end of that time he could show by optical spectroscopy that his previously empty apparatus now contained measurable amounts of helium. This was the outcome to be expected if alpha particles were helium ions and formed neutral atoms as they stopped in the apparatus.

By 1900 a third kind of ray was identified. They were named "γ rays." They are quite penetrating, but as Fig. 16.2 shows, they do not bend in a magnetic field and therefore cannot be charged particles. Later, after they were found to behave like x-rays, gamma rays were recognized to be energetic photons. They are emitted following the emission of alpha or beta rays and without radioactive transformation of the emitting atom.

EXERCISES

4. What is the direction of the magnetic field that produces the deflections shown in Fig. 16.2?

Radioactive Atoms of One Element Change into Another

It soon became evident that in the radioactive process, atoms of uranium, thorium, and so on were actually changing into other kinds of atoms, a process called "transmutation." Rutherford realized that the emitted radiation was a result of these changes. But changing atoms? The phrase is a self-contradiction because the word atom means uncuttable, and generations of chemical studies had convinced people that the elements were just that—elemental, fundamental building blocks of matter. To suppose now that atoms could transform, violated ideas built up over more than a century.

Let's consider some of the evidence for the occurrence of such changes. Uranium compounds emit both alpha and beta rays. Rutherford and Soddy[1] found that they could chemically separate the beta activity from the uranium, leaving the alpha activity almost entirely with the uranium. They called the beta-emitting fraction uranium-x and denoted it at first by UX and later by UX1. They observed that the beta activity gradually built up again in the uranium sample and could then be separated again as more UX1. Significantly, the chemical techniques to separate the UX1 were the same as those required to separate thorium from uranium, and it is now known that UX1 is the mass-234 isotope of thorium.

Second, a buildup of an alpha-emitting, radioactive gas was observed in samples of radium. Like the beta activity of UX1, this alpha-emitting substance could be drawn off a radium sample. This substance had the chemical characteristics of a noble gas. Initially called "emanation," this gas is now called "radon"—a household word in recent years.

Third, the observed radioactive elements were found to group into different families that we call "radioactive series" or "chains of radioactive decay." The members of each family have atomic masses differing by integer multiples of 4 u. Consequently, there are four possible different chains. One has atomic masses that are integer multiples of 4, i.e., $4n$—where n is any integer; another has masses that are integer multiples of 4 plus 1, i.e., $4n + 1$; another has masses that are $4n + 2$; and a fourth chain has masses that are $4n + 3$.

EXAMPLES

1. Most naturally occurring uranium atoms have an atomic weight of 238 u. To what decay chain do they belong? Divide 238 by 4; you get 59

[1]Frederick Soddy, 1877–1956, English physicist who worked with Rutherford at McGill, in Canada. They proposed the disintegration theory of radioactivity. Soddy introduced the idea of "isotope" to account for the existence of more different radioactivities than different chemical elements. He received the 1921 Nobel Prize in chemistry.

with a remainder of 2. Therefore, this uranium belongs to the $4n+2$ decay chain.

EXERCISES

5. To what radioactive series do radium atoms with atomic mass of 226 u belong?

Quite distinctive radioactive species with characteristic alpha-particle energies and alpha and beta half-lives were associated with uranium decay, but an entirely different set of species and characteristic decays were associated with thorium. Each of these different series produced products that were chemically identical to thorium or radium or other radioactive elements even though they differed greatly in their radioactive properties.

From these early studies of radioactivity, Soddy recognized that the emission of alpha or beta rays results in changes of the chemical nature of the emitting atoms. An atom that emits an alpha ray transforms into an atom two positions lower down the periodic table; beta emitters transform to elements one position up the periodic table. Thus, a uranium atom that emits an alpha particle becomes a thorium atom. And when a thorium atom emits a beta ray it becomes a protactinium atom, which can emit a beta ray and become a uranium atom. The occurrence of such radioactive "transformations" overthrew the fundamental, deeply held belief in the immutability of chemical elements. Radioactive transformation contradicted the very idea of "element."

16.2 QUANTITATIVE PROPERTIES OF RADIOACTIVITY

Measures of Activity

After it was recognized that each instance of alpha or beta radioactivity corresponds to the disintegration and transformation of a single atom, activity was defined to be the number of disintegrations per second. In honor of the Curies, a commonly used unit of activity is the "curie," abbreviated Ci. By international agreement one curie (1 Ci) of activity is 3.7×10^{10} radioactive emissions per second. This is almost the activity of one gram of radium, so it is approximately correct to say that the activity of 1 g of Ra is 1 Ci.

EXERCISES

6. Suppose a 100 μg sample of Ra causes an electroscope to discharge in 30 s, while 15 g of an unknown sample of radioactivity causes it to discharge in 100 s. What is the activity of the unknown?

The official SI unit of activity of radioactive substances is called the "becquerel" and is abbreviated Bq. A becquerel is defined to be one disintegration per second. Therefore, 1 Ci $= 3.7 \times 10^{10}$ Bq.

Radioactive Decay and Half-Life

The persistence of radioactivity posed a major mystery. Uranium and thorium samples emit very energetic particles at a steady rate. Their activities seemed to stay constant. Could a sample really keep releasing such energetic particles indefinitely?

The answer to this question came when new and different radioactive substances were separated from uranium and thorium. For example, the UX1 that Rutherford and Soddy chemically separated from uranium and that carried all the beta activity showed measurable changes in activity over time. Rutherford observed that UX1's beta activity exhibited a gradual decay after it had been separated from uranium. He measured the ionizing current produced by UX1 and observed that it dropped off over time, as shown in Table 16.1.

In about 24 days, the activity of UX1 decreased to half of its initial value and kept decreasing by half in each succeeding 24-day period. The time for the activity to fall by one half is called the "half-life," or $T_{1/2}$. Every radioactive substance has been found to have its own unique half-life, and therefore all radioactive materials do eventually lose energy; they run down. Changes in the activity of uranium, were difficult to detect because for uranium $T_{1/2} = 4.5 \times 10^9$ y, so that in any reasonable period of time the relative change of activity was not measurable. The mystery of how the materials could give off such energetic particles and so much energy was not so easily solved. We will return to that question in Section 16.4.

The data in Table 16.1 behave in a way familiar to students of growth and decay: The ionization current falls off exponentially. Using the ionizing current produced in an electroscope by a sample of UX1 as a measure of the sample's activity, Rutherford and others found empirically that relative to the initial current I_0, the ionizing current I produced by the beta radiation of UX1 diminished exponentially as a function of time,

$$I = I_0\, e^{-\lambda t}, \tag{1}$$

TABLE 16.1 Ionizing Current Caused by UX1 over Several Months

Ionizing Current (arb.units)	Elapsed Time (d)	Ionizing Current (arb. units)	Elapsed Time (d)
124.6	0	38.2	42
111.0	2	36.1	44
111.2	4	34.3	46
109.1	6	30.8	48
96.2	8	29.2	50
93.0	10	26.4	52
85.6	12	25.1	54
81.3	14	23.9	56
80.2	16	23.3	58
72.2	18	22.0	60
68.3	20	21.4	62
65.9	22	19.2	64
59.5	24	18.2	66
57.4	26	16.5	68
55.4	28	17.1	70
53.6	30	15.4	72
48.7	32	15.1	74
47.8	34	13.5	76
42.3	36	14.1	78
42.1	38	11.8	80
39.9	40		

where λ is a constant explained below.

Equation 1 is equivalent to saying that *the activity of a sample is proportional to N, the number of active atoms*. Each radioactive emission decreases the number N, and in a time Δt the radiations decrease N by ΔN. Thus, for this case it follows from its definition that the activity—the number of disintegrations per second—is the rate of decrease $\Delta N/\Delta t$. Because activity is proportional to N, you can write

$$\frac{\Delta N}{\Delta t} = -\lambda N, \tag{2}$$

where λ is the conventional symbol for the constant of proportionality in this equation. The minus sign tells you that N is decreasing. For the normal case of large N and comparatively small ΔN, you can approximate $\Delta N/\Delta t$ by a derivative. This means that you are treating the discrete decays and integer changes in numbers as smoothly varying functions. For this situation you obtain

$$\frac{dN}{dt} = -\lambda N. \tag{3}$$

Integrating this equation for N as a function of time, you get

$$N = N_0\, e^{-\lambda t}, \tag{4}$$

where N_0 is the number of radioactive atoms at time $t = 0$. Equation 4 is the law of radioactive decay: *In the absence of any source that is producing them, the number of radioactive atoms decreases exponentially over time.*

Differentiation of Eq. 4 shows why the activity of a UX1 sample was also observed to diminish exponentially. Differentiation with respect to time gives the activity A:

$$A = \left|\frac{dN}{dt}\right| = \lambda N_0\, e^{-\lambda t}, \tag{5}$$

which has the same exponential time dependence as the ionization current that Rutherford measured.

EXAMPLES

2. A good test of whether radioactivity obeys Eq. 4 and 5 is to plot the logarithm of the measured activity against time. If the result is a straight line, you know that the decay is exponential.

Figure 16.3 shows that Rutherford's UX1 data obey the relationship nicely.

The constant λ is called the "decay constant," or sometimes "the disintegration constant." Notice that if λ is large, a sample will decay quickly; if λ is small, the sample will be long-lived. Equivalently, for a given number of atoms, a large value of λ means a high activity. In other words, *the decay constant is a measure of the probability that an atom will decay.*

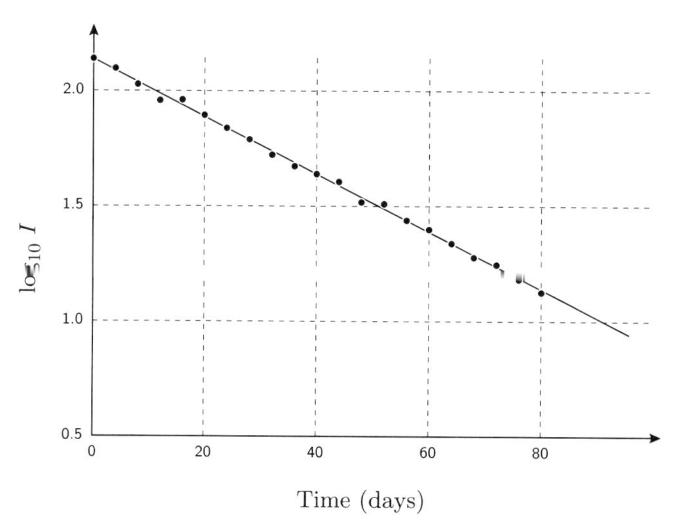

FIGURE 16.3 Plot of the logarithm (base 10) of the activity of UX1 *vs.* time. The straightness of the line shows that the decay of UX1 is exponential.

As long as the number of radioactive decays ΔN is small compared to the number of atoms present N, the probability that an atom will decay in the time interval Δt is the ratio of ΔN to N. Referring back to Eq. 2, you can see that the probability that an atom will decay in a time interval Δt is $\lambda\,\Delta t$. *As long as this probability is much less than one*, we may say that $\lambda\Delta t$ is the probability that an atom will decay within the time interval Δt.

We now have two measures of the likelihood that an atom will decay, the half-life $T_{1/2}$ and the decay constant λ. The two quantities are connected through the decay law, Eq. 4. Let's see how to find the decay constant if the half-life is known.

EXAMPLES

3. What is the decay constant of UX1?

You can find λ without knowing either dN/dt or N. All you need to know is the half life, $T_{1/2}$, of the material. For UX1, $T_{1/2}$ is 24.1 days, or 24.1×86400 seconds. During this time the activity drops to one-half of its initial value, and because the activity dN/dt is proportional to N, the number of radioactive atoms N must have dropped to half of the initial value, *i.e.*, $N/N_0 = \frac{1}{2}$. We can use this fact and Eq. 4 to write

$$\frac{1}{2} = e^{-\lambda T_{1/2}} = e^{-\lambda\, 24.1\times 86400},$$

which connects $T_{1/2}$ and λ. To solve for the constant in the exponent, take the natural logarithm of both sides of the equation. Since $\ln 2 = 0.693$, you get

$$\lambda = \frac{0.693}{2.1 \times 10^6} = 3.3 \times 10^{-7}\ \mathrm{s}^{-1}.$$

In general, the decay constant is related to the half-life by the equation

$$\lambda = \frac{\ln 2}{T_{1/2}}. \tag{6}$$

EXAMPLES

4. What is the half-life of radium?

You can use the definition of the curie to find the decay constant of radium. Remember that a curie is an activity of $3.7 \times 10^{10}\ \mathrm{s}^{-1}$ and is nearly equal to the activity of 1 g of radium. Because the atomic weight of radium is 226, one gram contains 1/226 of a mole of Ra atoms. Therefore, for a one-gram sample, N_0 is Avogadro's number divided by 226, and the activity of one gram of radium is, from Eq. 5,

$$-\frac{dN}{dt} = \lambda \frac{6.02 \times 10^{23}}{226} = 3.70 \times 10^{10}\ \mathrm{s}^{-1},$$

from which it follows that

$$\lambda = 1.39 \times 10^{-11}\ \mathrm{s}^{-1},$$

which with Eq. 6 gives

$$T_{1/2} = \frac{\ln 2}{\lambda} = 4.99 \times 10^{10}\ \mathrm{s},$$

which is 1580 years, slightly different from the currently accepted value of 1600 y for $T_{1/2}$ of Ra because modern measurements show that the activity of 1 g of Ra is 0.988 Ci, 1.2% smaller than was thought when the curie was defined.

EXERCISES

7. Derive the general relationship between $T_{1/2}$ and λ.

It can be shown that Eqs. 4 and 5 imply that radioactive decay is a purely random occurrence. Although we can measure and know the probability of decay, there is no way to predict when a particular atom will decay. It is as though the atoms are playing Russian roulette. Each atom has a revolver with a cylinder containing many empty chambers and one loaded one. In the course of a second each atom spins its cynlinder, puts the gun to its head and pulls the trigger. λ is the probability that the gun will go off. For the next second all the remaining atoms spin the cylinders of their guns and play another round. The value of λ found in Example 16.4 shows that 139 Ra atoms out of every 10^{13} lose in each round of Russian roulette. The UX1 atoms play a much tougher game. In a one-second round of Russian roulette, 33×10^5 UX1 atoms out of 10^{13} will lose.

The exponential decay law implies that in any given time interval the nucleus has the same chance of emitting a particle as in any other similar interval. This is strange. People do not age this way. Barring accidents, there is a relatively narrow range of ages in which people die. If they behaved like radioactive nuclei, half of the original population would die by a certain age, let's say forty. And half of the remaining population would die by the age of eighty, half again by one hundred and twenty and so on. Thus in a population of two million people, one-half million should be over one hundred and twenty, and nearly two thousand over four hundred years of age! We know that there are various aging processes and that people do not just die for no reason. Stars are the same. Stars like our Sun have fairly well-defined lifetimes. They do not just up and die at random. Why are nuclei so different from people and stars? How can one nucleus live for perhaps one second and another identical nucleus for ten billion years?

EXERCISES

8. The most common naturally occurring uranium atoms have a half-life of 4.47×10^9 y. What is their decay constant? What are their odds to lose at Russian roulette if they spin the cylinder once a second?

9. A certain kind of rubidium atom has a decay constant of $7 \times 10^5\ \mathrm{s}^{-1}$. What is the half-life of these atoms?

10. In the previous exercise what difficulty occurs when you try to answer the question, What is the probability that one of these special rubidium atoms will decay in a second? How can you get around the difficulty?

We have touched the edges of a very profound problem here. Today, as they have for nearly a century, physicists hotly discuss and disagree about the meaning of the apparently causeless randomness at the atomic and nuclear levels. The issue will arise repeatedly as you study more physics.

16.3 DISCOVERY OF THE ATOM'S NUCLEUS

As soon as their energies, masses, and charges were known and reliable sources were developed, alpha rays were used to probe the atom's insides. The startling result was the discovery that atoms are mostly empty space containing a tenuous cloud of electrons around a dense, compact nucleus.

Alpha Particles as Probes of the Atom

In 1906 Rutherford observed that when a well-defined beam of alpha particles passed through a sheet of mica, it made a slightly broader line on a photographic plate than when the mica was removed. He realized that the alphas were scattering from the atoms in the mica, but only through quite small-angles. In 1908 his student, Hans Geiger, examined this small-angle scattering in more detail. These first studies were interpreted using a model of an atom suggested by J.J. Thomson, who pictured the atom as an assemblage of electrons embedded in a sphere of smeared-out positive charge. Because the electrons were embedded within the positive charge like plums in a pudding, the model was called "the plum-pudding model." The Thomson idea seemed to explain some observations but not all of them.

Thomson's model implies that alpha particles passing near an atom will scatter *only* through small angles. To show this, let's estimate an upper limit for the scattering of an alpha particle from an electron. This is like a collision between a bowling ball and a ping pong ball, because the alpha particle is so much more massive than an electron. If you imagine a bowling ball just brushing the edge of a ping-pong ball, you may see that the electron is never going to scatter through an angle larger than 90°.

EXAMPLES

5. The mass of an electron is 0.511 MeV/c^2; the mass of an alpha particle is about 4×931.5 MeV/c^2. The ratio of $M_\alpha/m_e = 7300$.

Consequently, when a massive alpha particle moving with a velocity v_α and momentum $p_\alpha = M_\alpha v_\alpha$ collides head-on with an electron sitting at rest, the alpha imparts to the electron a forward velocity of 2 v_α. Thus the maximum possible change in momentum of the electron is

$$\Delta p = 2m_e v_\alpha \ll p_\alpha. \tag{7}$$

If the momentum Δp were carried away from the alpha particle at a right angle, the alpha would be deflected through through an angle θ, as shown

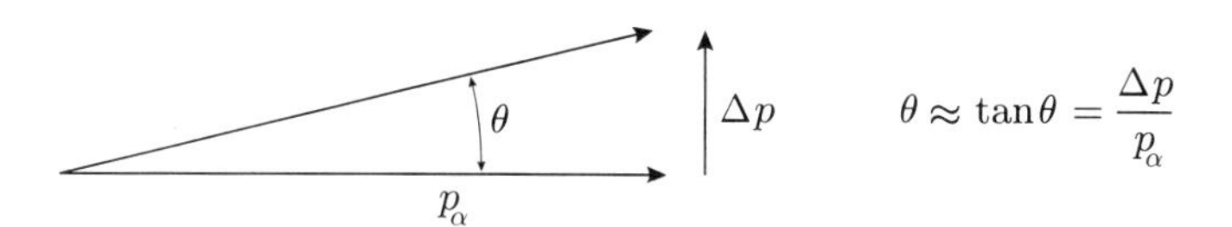

FIGURE 16.4 Upper limit of small-angle scattering produced by a small change in momentum.

in Fig. 16.4, where

$$\theta \approx \tan\theta = \frac{\Delta p}{p_\alpha}.$$

Although the alpha cannot lose this much momentum at right angles, and our calculated deflection is greater than the largest possible scattering angle, the result serves as an upper limit of what is possible. Notice that the angle is just twice the ratio of the masses of the two particles:

$$\theta = 2\frac{m_e}{M_\alpha} = \frac{1}{3650} = 0.0003 \text{ rad} = 0.015°.$$

Clearly, the electrons in an atom would be unlikely to scatter the alpha very much even if there were many such collisions.

But might the positive charge itself produce appreciable scattering? Again, a simple model is informative. Imagine that the atom is a ball of positive charge of radius R and some charge Ze, where e is the elementary charge and Z is some integer. Then the maximum force exerted on the alpha particle, which itself has a radius $r_\alpha \ll R$, by the atom's positive charge would be, from Coulomb's law,

$$F = k_c \frac{Z_\alpha Z e^2}{R^2},$$

where Z_α is the number of elementary charges on an alpha particle, i.e., 2, and k_c is the constant $9 \times 10^9 \,\text{N m}^2\,\text{C}^{-2}$ appearing in the Coulomb force law. You might think that the force would get larger if the alpha particle approached closer than the surface of the atom, but this is not so. Inside a ball of charge the force on the alpha particle is proportional to the alpha's distance from the ball's center, dropping to zero as the alpha particle moves to the center. This means that we can estimate the force by taking its value just at the surface of the sphere, where it will be largest.

To make it easy to estimate the change in momentum, we use this maximum value of the force and assume that it acts constantly over a time interval equal to the time Δt for an alpha to travel a distance equal to the diameter of the atom:

$$\Delta t = \frac{2R}{v_\alpha}.$$

This gives an upper bound on the change in momentum of

$$F\ \Delta t = k_c \frac{e^2 Z_\alpha Z}{R^2} \frac{2R}{v_\alpha},$$

which implies a maximum angle of scattering of

$$\theta = \frac{\Delta p}{p} = 2k_c \frac{e^2 Z_\alpha Z}{R M_\alpha v_\alpha^2}. \tag{8}$$

The radius R of an atom is about 0.1 nm; Z for a gold atom is 79; $\frac{1}{2}M_\alpha v_\alpha^2$ is the kinetic energy of an alpha particle, or about 5 MeV; and, of course, $k_c e^2 = 1.44$ eV nm.

EXERCISES

11. Show that the above estimate of maximum angle of scattering will be about 0.026° for alphas scattering from gold atoms.

Discovery of the Atomic Nucleus

Rutherford had another student, Marsden, look for larger-angle scattering. The method of detection was interesting. When individual alpha particles hit a screen coated with zinc sulfide they make a flash of light, or "scintillation," that can be seen by the completely dark-adapted human eye. This was a very sensitive technique for observing rare events. For the reasons given above,

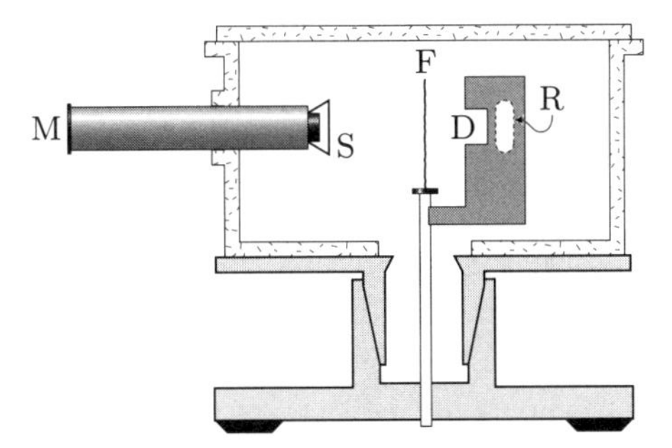

FIGURE 16.5 Apparatus used by Geiger and Marsden. Alpha particles from a source R pass through a collimator D and strike a foil F. Scintillations from scattered particles are produced on the screen S and are observed with a microscope M. The chamber is evacuated and can be rotated to different angles about the foil.

Rutherford did not expect to see any large-angle scatterings. According to the Thomson model, scatterings through large angles would have to be the result of many successive small-angle scatterings. Since each would be random, the successive scattering directions would tend to average out, and only on very rare occasions would the events add up to a significant overall deviation.

The experiment was set up using an intense source of alphas in an evacuated tube. A narrow beam was formed by a pair of thin metal plates with small holes that served to collimate the stream of alpha particles. Gold foil was used as a target in order to get massive atoms of a material that could easily be made into sheets so thin that alphas could pass through. Marsden immediately began seeing significant numbers of alphas deflected through fairly large angles. After he modified the apparatus to allow observation of alphas scattered in any direction, he saw scintillations at even larger angles, some even in the straight backwards direction. He found that 1 out of every 8000 alphas scattered through an angle of 90° or more.

This result was a complete surprise. Because a solid is a collection of atoms in contact with one another, the alphas had to pass through hundreds of atoms to get through the foil. The experiment showed that the material was very porous, that most of the alphas passed through it like bullets through a rain shower. Astonishingly, however, the same material exerted very strong forces on the few alpha particles that were bounced back. As Rutherford said, "It was quite the most incredible event that has ever happened to me in my life. It was almost as incredible as if you fired a 15-inch shell at a piece of tissue paper, and it came back and hit you."

EXERCISES

12. Show that the simple model used above to estimate the small-angle scattering will produce larger angles of scattering if you imagine that the positive charge has a much smaller radius R than assumed above.

Rutherford realized that the scattering of the alpha particles through large angles could be explained if the positive charge of the atom was concentrated very compactly in the center of the atom. Indeed, the crude approximation of Eq. 8 suggests that if R, the radius of the positive ball of charge, is 10^{-3} to 10^{-4} smaller than the radius of the atom, then θ will be large. This makes sense, because near such a compact ball of charge the electric field would be very large, and an alpha particle passing close to the compact core would experience a very large force. In the case of a head-on collision, an approaching alpha particle would slow down and come to a halt at distance r from the

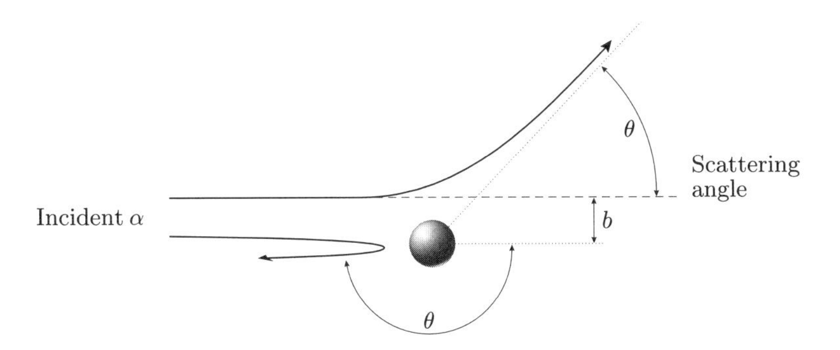

FIGURE 16.6 Trajectories of alpha particles with different impact parameters; the impact parameter is denoted by the letter "b." The dashed lines are the paths the alphas would follow if there were no force on them; the dotted line is the asymptotic trajectory of a scattered alpha particle.

charge, where the alpha's initial kinetic energy was completely converted to electrostatic potential energy; then it would be accelerated backwards away from the atom until it was far away and had regained its original speed. This would correspond to scattering through 180°, as shown in Fig. 16.6(a). Figure 16.6(b) shows how an alpha particle incident not directly head-on to the atom would also be deflected, but through smaller angles. The amount of deflection would be less the farther the alpha was from a head-on collision. This kind of scattering of one atom from another owing to the Coulomb force between their nuclei is called "Rutherford scattering" or "Coulomb scattering."

On the assumption that the scattering center, i.e., the nucleus, acted as a point charge, Rutherford calculated that the number of alpha particles scattering through an angle θ will be proportional to the *square* of the charge Ze of the scattering nuclei and inversely proportional to the *fourth power* of the sine of half of the scattering angle, $\theta/2$. He also predicted that the number scattered would be proportional to the *inverse of the square* of the energy of the alpha particles, E_α. In short he predicted

$$\Delta N \propto \frac{Z^2}{E_\alpha^2 \sin^4 \frac{\theta}{2}}, \tag{9}$$

where ΔN is the number counted.

Geiger's and Marsden's experimental data on the scattering as a function of angle could then be compared with Rutherford's theory. These data are shown in Table 16.2, which has several interesting features. For example, it shows convincingly that Rutherford's prediction that ΔN is proportional to $1/\sin^4\left(\frac{\theta}{2}\right)$ is correct. If such a proportionality holds, then the product of ΔN and $\sin^4 \theta/2$ should be a constant. Column 4 shows that the product varies

TABLE 16.2 Number of Alpha Particles Scattered through Various Angles θ from Gold

Scattering angle, θ (deg)	$\sin^4 \frac{\theta}{2}$	Number of scintillations ΔN	$\Delta N \sin^4 \frac{\theta}{2}$
150.	.8705	33.1	28.8
135.	.7286	43.0	31.3
120.	.5625	51.9	29.2
105.	.3962	69.5	27.5
75.	.1373	211.	29.0
60.	.0625	477.	
45.	.02145	1435.	
37.5	.01068	3300.	35.2
30.	.004487	7800.	
22.5	.001449	27300.	39.6
15.	.0002903	132000.	38.3
30.	.004487	3.1	
22.5	.001449	8.4	
15.	.0002903	48.2	
10.	5.77×10^{-5}	200.	
7.5	1.83×10^{-5}	607.	
5.	3.62×10^{-6}	3320.	

by no more than $\pm 20\%$, while the number of counts varies by almost four orders of magnitude.

EXERCISES

13. Fill in the blanks of column 4 of Table 16.2.

14. Compare the variation of the product with the variation of ΔN for the bottom six entries of Table 16.2.

15. Plot the data of Table 16.2 in such a way that they should all lie on the same straight line if Rutherford's theory is correct.

Another feature of Table 16.2 deserves notice. Stop and think a moment about how these data were obtained. Imagine that you are one of two people sitting in a completely darkened room holding mechanical counters and counting the flashes of light from a tiny screen only a few cm^2 in area. It took you close to an hour of sitting in total darkness before your eyes became sufficiently adapted to be able to detect the tiny flashes adequately. Even then you could reliably count no more than 90 scintillations per minute and no fewer than 5. How do you suppose someone counted the 132000 counts shown in Table 16.2? Or how did they get 33.1 counts?[2]

To deal with the limitations of these human detectors several tactics were used. The simplest was to put in front of the alpha source an aperture made small enough to limit the number of alphas until their frequency of scattering was within human capacity to count them and then correct the data to make them correspond to some standard size of aperture. Another approach was to start with a very active source, say 100 mCi, with a short half-life, e.g., ^{222}Rn, with a half-life of 3.82 d. At the beginning of the experiment observers would make measurements at the large scattering angles where the yield is small; then as the days went by and the activity dropped they would make measurements at smaller angles, where the yield was larger. Of course, they had to correct their data for the fact that the decay of the source led to a diminished number of incident alphas.

EXERCISES

16. If the counts in the two sets of data in Table 16.2 differed only because of a change in aperture size, what ratio of aperture diameters would account for the observed differences?

---

[2]In *The Theory and Practice of Scintillation Counting*, J.B. Birks, Pergamon Press, Macmillan Co., 1964, there is the following: "The story is told that at one famous laboratory during this period all intending research students were tested in the dark room for their ability to count scintillations accurately. Only those whose eyesight measured up to the standards required were accepted for nuclear research; the others were advised to take up alternative, less physically exacting, fields of study. Nowadays analogous techniques were applied to the selection of photomultipliers, from a batch of commercial tubes, for a critical experiment. Marsden, who counted hundreds of thousands of scintillations in his historic experiments on alpha particle scattering, has recalled how on train journeys his colleague Geiger would urge him not to put his head out of the window, lest a chance smoke particle should impair his efficiency as a human scintillation counter. Truly the early nuclear physicists needed to be men of vision."

TABLE 16.3 **Large-Angle Alpha Scattering Data**

θ (deg)	no target (cpm)	target (cpm)	subtract back-ground	$\Delta N \sin^4 \frac{\theta}{2}$
150.	0.2	4.95		
135.	2.6	8.3		
120.	3.8	10.3		
105.	0.6	10.6		
75.	0.0	28.6		
60.	0.3	69.2		

17. Suppose that they took 10 days to do an experiment in which ^{222}Rn was the source of alpha particles. At any given angle what would be the ratio of the number of scattered alpha particles at the beginning of the experiment to the number scattered at the end of the experiment?

Table 16.3 shows data taken by Geiger and Marsden for a study of large-angle alpha scattering.[3]

EXERCISES

18. Geiger and Marsden say that their alpha source was ^{222}Rn and that they did their experiment 51 hours after the 100 mCi source was prepared. By what factor should they multiply the initial source activity to get the activity at the time they ran their experiment?

19. Table 16.3 shows that they took data without any target in order to determine what background counts might be. Do the correction for background, and show that the corrected data obey the $1/\sin^4 \frac{\theta}{2}$ dependence expected for Rutherford scattering.

Rutherford's successful explanation of these data established the validity of the assumption that each atom contains a very small, dense core of charge.

[3] H. Geiger and E. Marsden, "The laws of deflection of α-particles through large angles," *Phil. Mag.* **25**, 604-623 (1913).

Rutherford named this compact center of the atom the "atomic nucleus," often called just the "nucleus" for short.

Nuclear Size and Charge

Rutherford scattering provides a way to determine the size of a nucleus. The scattering formula Eq. 9 is valid only when the alpha and the nucleus do not come into contact. This assumption works because it is a general result that the electric field outside a sphere of charge (the nucleus in this case) is the same as that of a point charge. Therefore, an alpha particle will experience the field of a point charge as long as it does not go inside the surface of the nucleus. If the alpha does penetrate the nucleus, however, then the force will not be that of a point charge of magnitude Ze, and Eq. 9 will no longer be correct.

The easiest case to analyze is a head-on collision, where the incident alpha particle slows down and stops when all its kinetic energy becomes electrostatic potential energy. Then the distance of closest approach is the value of r that gives a value of the potential energy equal to the alpha's incident kinetic energy

$$\frac{1}{2}M_\alpha v_\alpha^2 = k_c\frac{e^2 2Z}{r}. \tag{10}$$

Here r is the distance between the center of the alpha particle, of charge $2e$, and the center of the nucleus, of charge Ze. The other symbols are as defined for Eq. 8. Suppose the energy of the alphas is 5 MeV. Then the distance of closest approach to a gold nucleus will be

$$r = \frac{1.44 \times 2 \times 79}{5 \times 10^6} = 4.55 \times 10^{-5}\,\text{nm} = 45.5\,\text{fm}.$$

EXERCISES

20. What would be the distance of closest approach for 5 MeV alpha particles bombarding an aluminum foil?

Of course, when the distance of closest approach brings the alpha particle inside the surface of the nucleus, Eq. 10 will no longer be valid, because inside the nucleus the electric field of the charge Ze is no longer the electric field of a point charge. This means that to find the nuclear size all you need to do is keep increasing the energy of the incident alphas until the experimental results deviate from the predictions of Eq. 9. Experiments show that for gold, Rutherford's theory no longer correctly predicts the distance of closest approach when $r \leq 8.9$ fm; for aluminum the theory fails when $r \leq 5.5$ fm; for nitrogen it fails when $r \leq 4.8$ fm. We infer that the theory fails at these values of r because the outer edge of the alpha is starting to penetrate the nuclear surface, and we conclude that these values of r represent the radius of the nucleus plus the radius of the alpha particle. Many experiments have shown that nuclei have radii a few femtometers in size and, in general, the nucleus of an atom with an atomic weight A has a radius r_N that is reasonably well given by the expression

$$r_N = 1.2 \times A^{\frac{1}{3}} \text{ fm}. \tag{11}$$

EXERCISES

21. From Eq. 11 what is the radius of an alpha particle?

22. What is the radius of an aluminum nucleus? Is your answer consistent with the observation that Rutherford's theory does not hold for a head-on closest approach of 5.5 fm?

23. What is the radius of a gold nucleus? Is your answer consistent with the observation that Rutherford's theory does not hold for a head-on closest approach of 8.9 fm?

24. What is the radius of a proton?

Rutherford also realized that alpha scattering could be used to determine the charge of a nucleus. Measurements made on several different metals showed that the nucleus has a number of elementary charge units approximately equal to half of its atomic mass. In 1913 Moseley's work on atomic x-rays, described in Section 17.4, further supported the nuclear model and confirmed the Dutch amateur scientist Van der Broek's proposal that the number of elementary charges in the nucleus of a chemical element is the same as the element's atomic number Z.

Rutherford's work established the existence of the atomic nucleus. For this reason Rutherford is recognized as the discoverer of the atomic nucleus and the founder of the nuclear model of the atom. In this model a positive charge Ze is concentrated along with most of the mass of the atom in a volume about 10 fm in diameter. This is only about 10^{-4} of an atomic diameter and means that *an atom is mostly empty space*. Because atoms are electrically neutral, there must be Z electrons filling the relatively large empty volume around the nucleus. Rutherford's discoveries produced the most profound change in the concept of the atom since Dalton's ideas a little over 100 years earlier.

EXERCISES

25. Approximately what fraction of the volume of a gold atom is occupied by its nucleus?

26. Approximately what fraction of the mass of a gold atom is in its nucleus?

27. Estimate the density of nuclear matter. If all the people in your college or university were compressed into a sphere with the density of nuclear matter, how large would the sphere's diameter be?

28. What is the preferred plural of "nucleus"?

16.4 NUCLEAR ENERGIES

As noted earlier, the energies of the radiations emitted in radioactivity are astonishingly large. Individual alpha particles have energies on the order of 5 MeV. The average energy of emitted beta particles is typically a few MeV. The discovery of the nucleus made it possible to explain such large energies.

Energies of Alpha and Beta Particles

Measurements of alpha and beta particles show that they are more energetic than the highest-energy cathode rays produced. The velocities found when measuring the charge-to-mass ratios of beta rays and of alpha rays show that these particles have energies on the order of several MeV. To see how these energies can be measured you can use some of the tools you learned in Chapter 11 to calculate typical results obtained from magnetic deflection experiments.

EXAMPLES

6. For example, when the betas from UX1 enter a magnetic field of 0.04 T, many of them bend into a circular arc with a radius of about 2.8 cm. What is the energy of these beta rays?

Recall from Chapter 11 that the momentum p of particles of mass m and charge $q = Ze$ can be directly obtained from the radius of curvature R of their path in a magnetic field B in units of eV, using

$$pc = ZBRc, \tag{12}$$

where Z is 1 for an electron or a proton, 2 for an alpha particle, and so on. (As usual, Eq. 12 has been divided by a number equal to the electronic charge in order to convert joules to eV; as a result, a factor e has been removed from the equation.)

If the particle is nonrelativistic, you can calculate its kinetic energy K from

$$K = \frac{p^2c^2}{2mc^2}.$$

If the particle is relativistic, you *must* use

$$K = \sqrt{m^2c^4 + p^2c^2} - mc^2. \tag{13}$$

For our example,

$$pc = 1 \times 0.04 \times 0.028 \times 3 \times 10^8 = 3.35 \times 10^5 \text{ eV}.$$

When you compare this number to the rest energy of an electron, you find that $pc/(mc^2) = 0.7$. This is not small compared to 1, so you should use the fully relativistic Eq. 13, from which it follows that

$$K = 0.511\left(\sqrt{1 + 0.7^2} - 1\right) = 0.511 \times 0.197 = 0.101 \text{ MeV}.$$

EXERCISES

29. Calculate the kinetic energy of alpha particles that bend with a radius of $R = 0.63$ m in a magnetic field of $B = 0.5$ T. Assume that the alpha particles are doubly charged helium atoms with atomic masses of 4 u; 1 u has a mass of 931.5 MeV/c^2.

Your answer to Exercise 16.29 should be 4.8 MeV, which is the energy of an alpha particle emitted by radium. It is an extraordinary amount of energy. Because one gram of radium emits 1 Ci, or 3.7×10^{10} alphas per second, a 1 g sample of radium will give off energy at a rate that is the energy per particle times the number of particles per second $= 4.8 \times 10^6 \times 1.6 \times 10^{-19} \times 3.7 \times 10^{10}$ J/s $= 0.028$ J/s.

To appreciate the significance of such a number, compare it with more familiar quantities such as the heat from burning one mole of hydrogen gas to make water, one of the more energetic chemical reactions known (recall the *Hindenburg*). This will produce about 2.9×10^5 joules of heat, as you can verify by looking up the heat of formation of one mole of water. A mole of radium, 226 grams, gives off 6.78 J/s. Although it will take a mole of radium nearly 12 hours, or half a day, to produce the same energy as burning the equivalent number of hydrogen molecules, the radium keeps on producing energy for *thousands of years*! The energy that comes out in an alpha particle is over a *million times more* than that released when two atoms of hydrogen combine with an atom of oxygen.

EXERCISES

30. Calculate the ratio of energy released by the radioactive decay of a mole of radium to the energy released when a mole of H_2 burns:

$$2H_2 + O_2 = 2H_2O + \text{heat}.$$

31. How long will it take a mole of radium to yield 90% of its radioactive energy?

It is simpler to compare radium and H_2O using electron volts.

EXAMPLES

7. How many electron volts of energy are released in the formation of an H_2O molecule compared to the number released in the emission of an alpha particle?

We have that the formation of 6×10^{23} water molecules releases 290 kJ. Converting this amount of joules to electron volts you find that this is the same as releasing

$$\frac{2.9 \times 10^5}{1.6 \times 10^{-19}} = 1.8 \times 10^{24}\,\text{eV}.$$

The amount of eV per water molecule is this number divided by 6×10^{23}, which is 3.0 eV.

This result shows that while chemical reactions occur at energies of a few electron volts (eV) per atom, alpha decays involve millions of electron volts (MeV). Radioactivity was thus clearly revealed to be a new phenomenon profoundly different from anything previously known.

In this age of big numbers and super hype, of trillion dollar debts, nuclear bombs, and space probes (the last, of course, still depending on burning hydrogen with oxygen), it takes an effort of imagination to put oneself in the place of those early workers in radioactivity. But try to appreciate the impact of magnifying virtually overnight by a factor of a million the energies that people were used to.

Where could the energy possibly be coming from? What kind of mechanisms could convert such huge amounts of stored energy to kinetic energy? The presence of radioactive species with half-lives short enough to show detectable decreases in activity indicated that the energy had to be coming from some process drawing on a limited though large store of energy.

To see how a tiny nucleus of positively charged matter might store the high energies involved in radioactivity, consider the potential energy involved in the forces that hold the charge in place. Unlike the Thomson model, in which negative electrons were imagined to be sprinkled around, neutralizing small regions of the atom, the nuclear model has a lot of positive charge in a very small volume.

EXAMPLES

8. What is the electrostatic potential energy of an alpha particle in uranium?

For $Z = 92$, you can imagine 2 elementary charges being repelled by the other 90 at distances on the order of 7.4 fm. The potential energy will then be

$$\text{P.E.} = \frac{k_\text{c}e^2 \times 2 \times 90}{7.4} = 35\,\text{MeV},$$

where you should use the convenient fact that $k_\text{c}e^2 = 1.44\,\text{eV nm} = 1.44\,\text{MeV fm}$ so that you can use the nuclear radius directly in femtometers and get your answer directly in MeV.

This very large energy implies a strong repulsive force between the positively charged alpha and the positive charge of the rest of the nucleus. How do all these positively charged particles stay together when the forces pushing them apart are so large?

There must be very strong attractive forces in the nucleus, strong enough to overcome the large electrostatic repulsion between protons. These forces are something new. They hold the nucleus together, so we call them "nuclear forces." They produce a large negative potential energy; they bind the nucleus together; and they hold the alpha inside for the lifetime of the uranium atom before it decays.

EXERCISES

32. Repeat the above calculation to estimate the minimum negative nuclear energy required to compensate for the positive electrostatic energy in order to hold a gold nucleus together. Imagine that the two halves of the gold charge, $Z/2$, to be point charges 12 fm apart.

The scale of the energies involved makes it plausible that the nucleus is the region where most of the atom's mass is located and where most of the action occurs in the emission of radioactive particles.

Although in the first two decades of the twentieth century Rutherford and his contemporaries did not have the uncertainty principle to work with, you saw in Chapter 15 how the uncertainty principle tells us that confined particles always possess substantial kinetic energy just because they are confined. If you have not done Exercise 15.3, do it now.

EXERCISES

33. Estimate the minimum possible kinetic energy of an alpha particle confined to nucleus of diameter 7 fm.

Your answer should be on the order of the other nuclear energies we have been estimating.

16.5 THE NEUTRON

The next major advance in understanding the nucleus came in 1932, when Chadwick discovered the neutron. The identification of the neutron as a component of the nucleus on an equal footing with the proton completed the basic picture of the nucleus that we have today.

The neutron was the third basic particle to be discovered—the electron and the proton were discovered first. The neutron is electrically neutral and has a mass of 1.0086649 u, slightly larger than the mass of a proton. Although they are stable inside many nuclei, neutrons outside of a nucleus are radioactive and decay with a half-life of 10.4 min.

Every nucleus is made up of neutrons and protons. The number of protons is Z, the atomic number of the corresponding element. The number of neutrons is often designated N. The sum of these two numbers is called the "mass number" of the nucleus and is usually written as A. The mass number of an atom is the integer nearest to its atomic weight. Collectively, the neutron and the proton are referred to as "nucleons." Thus the mass number A is the number of nucleons in a nucleus,

$$Z + N = A.$$

Each combination of Z and N specifies a unique nucleus. Nuclei with the same value of Z and different values of N are called isotopes. You have met isotopes before, but now that you know about neutrons you can see how there might be a mass-one hydrogen and also a mass-two hydrogen atom. Now you can see that mass-two hydrogen has a nucleus consisting of a proton and a neutron. There is also a radioactive isotope of hydrogen with $A = 3$. The nucleus of this isotope has two neutrons and one proton.

There is a standard notation for representing different nuclei. It uses the chemical symbol to indicate the value of Z (which means that fairly often you will need to look at a periodic table of the elements to find out Z) and has the mass number as a left-hand superscript. Thus the proton is just ^{1}H. The mass-2 isotope (called "deuterium") is ^{2}H. The mass-3 isotope (called "tritium") is ^{3}H. The alpha particle is just the helium nucleus ^{4}He. (There is another, very rare, stable isotope of helium ^{3}He, which as you can see has one neutron fewer than ^{4}He.)

Sometimes the value of Z is supplied explicitly; then it is written as a left-hand subscript. For instance, there are two naturally occurring nonradioactive isotopes of carbon: $^{13}_{6}$C and $^{12}_{6}$C. There is also a radioactive isotope of carbon that occurs in nature: $^{14}_{6}$C. Occasionally, the neutron number is given also, as a right-hand subscript, e.g., $^{14}_{6}\mathrm{C}_{8}$.

The general form of this notation is $^{A}_{Z}\mathrm{X}_{N}$, where X represents any chemical element symbol and A, Z, and N are respectively the nucleon, or mass number; the atomic number; and the neutron number. The entire assemblage of neutrons, protons, and electrons is called a "nuclide." There are 272 stable nuclides plus 55 radioactive nuclides that occur naturally on Earth. About two thousand radioactive nuclides have been made artificially.

Visualizing a nucleus as a collection of Z protons and N neutrons it is easy to understand the transformations that result from radioactive decay. Alpha decay removes from the nucleus two neutrons and two protons; it reduces A by 4 units and Z by 2 units. That is why uranium-238 ($^{238}_{92}$U) turns into thorium-234 ($^{234}_{90}$Th), and in general alpha decay can be described as

$$^{A}_{Z}\mathrm{X} \rightarrow ^{A-4}_{Z-2}\mathrm{Y} + \alpha.$$

Beta decay causes a neutron in the nucleus to become a proton. As a result, Z increases by 1 unit, but A does not change. There is a different kind of beta decay we have not discussed in which a *positively charged electron*(!) (called a "positron") is emitted and a proton becomes a neutron. For this so-called positron emission A does not change, but Z decreases by one unit.

Gamma-ray emission does not change A, Z, or N. It does reduce the energy stored in a nucleus, and it changes some nuclear properties we have not talked about.

Because each nuclide is uniquely characterized by its values of Z and N, it is convenient to lay out the nuclides in a chart where Z is along one axis and N along the other. Then each nuclide can be represented as a box located at the coordinates (N,Z). A piece of such a chart is shown in Fig. 16.7; you can also find such charts on the World Wide Web.

EXERCISES

34. Get a chart of the nuclides on the World Wide Web at the URL

http : //www.dne.bnl.gov/CoN/index.html

Use the chart to find and write down the chain of decays that connects the atoms of ^{238}U to the first nonradioactive nucleus in the $4n + 2$ chain.

16.6 SUMMARY

Radioactivity led to the discovery of the atomic nucleus. This tiny core of the atom contains 99.98% of its mass in about 10^{-12} of the atom's volume. The nucleus consists of particles called nucleons of which there are two different kinds—the proton and the neutron. The number of protons in a nucleus is the atomic number Z of the atom, and it determines uniquely which chemical element an atom is. The neutron number N then determines which isotope of the element the atom is. The number of nucleons in an atom is called its mass number A, and $Z + N = A$. A nucleus with mass number A has a radius of about

$$1.2 \times A^{\frac{1}{3}} \times 10^{-15}\ \text{m}.$$

An isotope of element X is specified using the notation

$$^{A}_{Z}X_{N},$$

where the Z and N values are often omitted because they are redundant (if you *know* the atomic number of X).

Radioactivity revealed that some kinds of atoms can spontaneously transform into others. Some do this by emitting an α particle (a helium nucleus i.e., ^{4}He), which decreases Z by 2 units and A by 4 units. Others emit a beta ray (electron), which decreases Z by 1 unit but leaves A unchanged.

Nuclei of a given kind will emit alpha rays with characteristic, well-defined energies on the order of 5 MeV. Gamma radiations also are emitted with well-defined, characteristic energies just like the visible-line spectra emitted from atoms, except that gamma-ray energies can be several MeV in magnitude rather than the few eV of visible light. By contrast, beta-decay electrons do not come out with well-defined energy. Usually there is nothing like a line spectrum of electrons. In a collection of identical nuclei that undergo beta decay, some will emit low-energy electrons and some high, with energies ranging from 0 up to some maximum energy on the order of a few MeV.

Each type of radioactivity obeys the law of radioactive decay

$$N = N_0 e^{-\lambda t},$$

which shows that the emitting nuclei decay away exponentially over time. The rate of decay is specified by the disintegration constant λ, which is unique to each species of nucleus and to each type of decay. In nuclear physics it is common to use the half-life $T_{1/2}$ instead of the disintegration constant, where

$$T_{1/2} = \frac{\ln 2}{\lambda}.$$

The energies associated with nuclear properties are 5 to 6 orders of magnitude greater than those observed in the atomic processes typical of chemical interactions. These large energies and the fundamentally random nature of radioactive decay can be taken into account only by new physics: the identification of a new force in nature, the so-called "strong," or "nuclear," force, and the ideas of quantum mechanics.

PROBLEMS

1. We mentioned that Marie Curie had to process tons of ore to extract 0.1 gram of radium. The reason was that the radium isotope she discovered ($T_{1/2} = 1600$ y) was a decay product of mass-238 uranium ($T_{1/2} = 4.5 \times 10^9$ y), which over geological spans of time can exist in equilibrium only with uranium.

a. Explain why at equilibrium the activities of radium and uranium will be equal and show that under these conditions

$$\lambda_U N_U = \lambda_{Ra} N_{Ra}.$$

b. Under equilibrium conditions, what will be the ratio of the number of Ra atoms N_{Ra} to the number of U atoms N_U? Tests on old uranium deposits show that they contain this ratio of radium to uranium.

c. Marie Curie discovered radium in pitchblende, the principal uranium-bearing ore. Assume that her ore was 50% uranium. How much pitchblende (by weight) did she have to process to isolate 0.1 g of radium?

2. Estimate the kinetic energy of an alpha particle (a doubly ionized helium atom) confined to a region of space about 4 fm wide.

3. What would be the potential energy of a particle of charge $+2e$ a distance of 10 fm from a particle of charge $+90e$.

4. Two protons approach each other towards a head-on collision with kinetic energies of 1 keV each.

a. What is the total momentum of the system? Explain your reasoning.

b. Before the collision, when they are far away from each other, what are the system's total kinetic energy K_{tot} and its electrical potential energy U?

c. What is K_{tot} at the point of closest approach? Find the separation r_0 between the two protons when they are at the point of closest approach.

d. What is the force (magnitude and direction) that one proton exerts on the other when they are separated by a distance r_0?

e. What did Rutherford conclude from his analysis of Geiger and Marsden's experiment with α particles?

5. If a gold nucleus has a radius of 17 fm, what is the maximum kinetic energy that an alpha particle can have and still not penetrate the surface of the nucleus during a head-on collision?

6. What is the closest distance that a 1 MeV proton can come to a gold nucleus?

7. How much error would you make if you worked Example 16.6 nonrelativistically? Based on other values you have seen for early measurements of e/m, do you think that Becquerel would have been bothered by such a discrepancy?

8. Figure 16.7 is a small piece of a chart of the nuclides. From the chart find and write down in standard nuclide symbols:

a. Three stable nuclides that are not isotopes of one another.

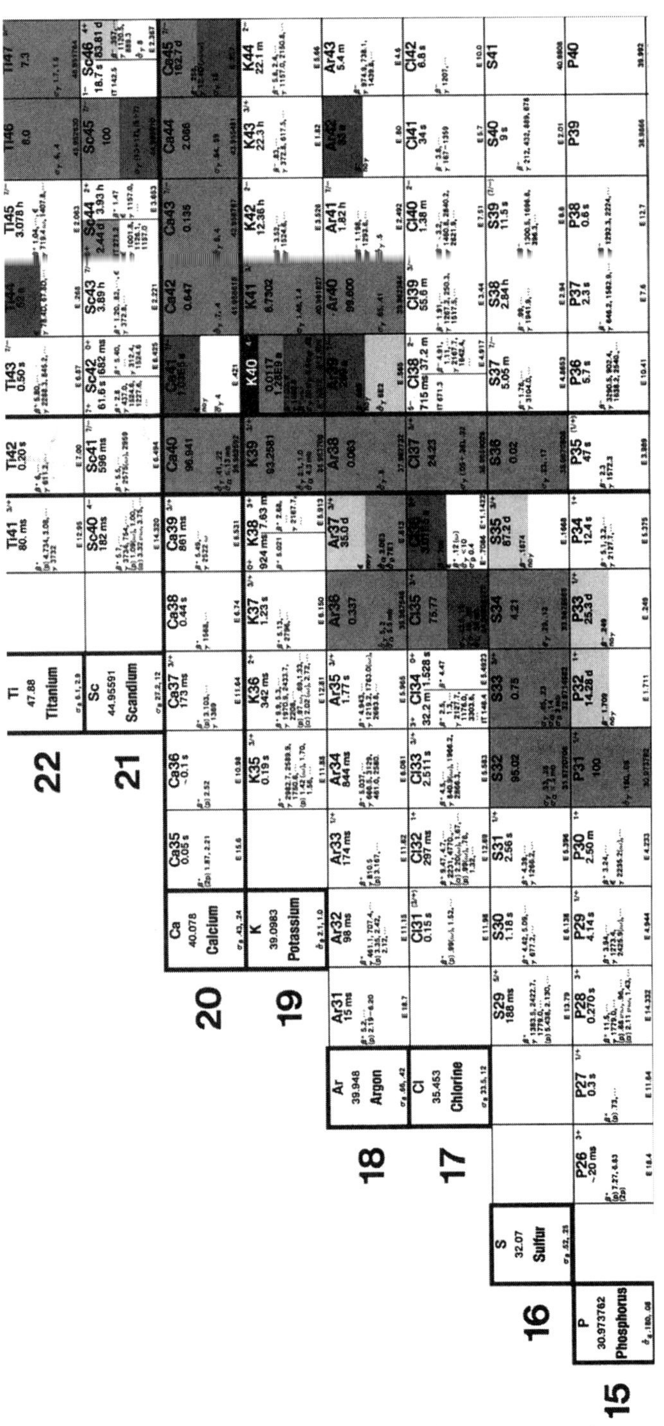

FIGURE 16.7 Each square is a distinct nuclide. The atomic number Z increases vertically from 15 to 22 here, and the neutron number N increases from left to right, going from 10 to 28 here. Problem 8.

b. Three stable isotopes of the same element; give their relative abundance in nature.

c. Three radioactive nuclides; give their half-lives and their decay products.

9. Which of the nuclides in the chart is naturally occurring and radioactive? Is it more likely to be a nuclide that was created at the time of the formation of Earth or one that is made by cosmic rays? Why?

10. Of the stable nuclides shown, how many individual nuclides are there that have both an odd number of protons and an odd number of neutrons?

11. The chart shows that for very light nuclides the number of protons and number of neutrons is roughly equal. But as the mass number increases, in any given stable nucleus there get to be more neutrons than protons. Can you suggest why this might be?

12. The mass of a deuterium atom is 2.01410177 u. How much energy is needed to break it into a hydrogen atom (^{1}H) and a neutron?

FINDING THE RADIUS OF A NUCLEUS

Introduction

With modern accelerators it is possible to impart to particles momentums high enough to correspond to a de Broglie wavelength short enough to probe the size of a nucleus. Figure 16.8 shows some particularly good data showing the diffraction pattern that arises when 800 MeV protons scatter from a nucleus of ^{208}Pb. From the theory of diffraction and these data you can determine the diameter of the ^{208}Pb nucleus.

Diffraction from a Circular Cross Section

The diffraction you have studied is from a single slit. If the slit has a width b, then there will be diffraction minima at angles θ_n such that $b \sin\theta_n = n\lambda$ where λ is the wavelength of the diffracting wave and n is any integer up to the limit determined by the sine function.

Diffraction from a circular aperture or from a circular cross section—do you remember Babinet's principle?—is slightly different from the case of a slit. The first diffraction minimum occurs when

$$b \sin\theta = 1.22\lambda.$$

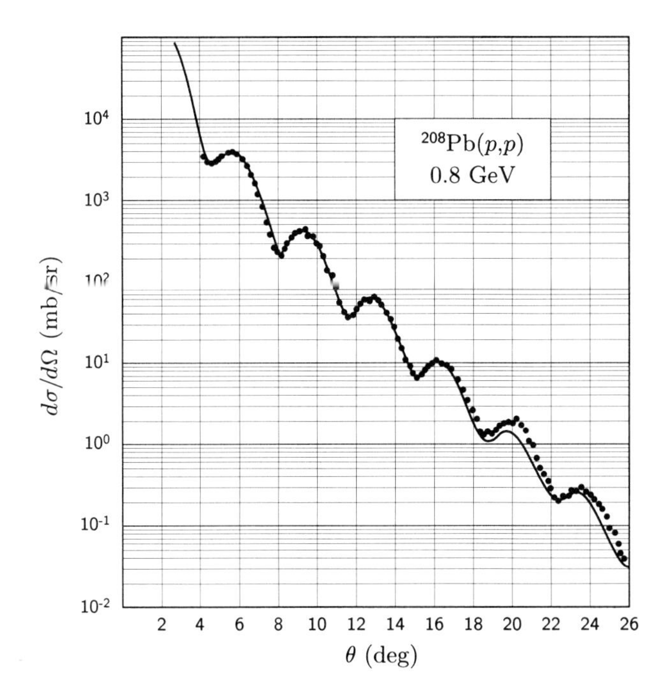

FIGURE 16.8 This is the diffraction pattern from the scattering of 800 MeV protons from ^{208}Pb. It is plotted on a semilog scale in order to show the very wide range of magnitudes of the diffraction maxima. *Taken with permission from G.F. Bertsch and E. Kashy, "Nuclear Scattering," 61, 859–859. ©1993 American Association of Physics Teachers*

The other minima occur very nearly according to the expression

$$b \sin \theta_n = 1.22\lambda + (n-1)\lambda, \tag{14}$$

where b is the diameter of the circular stop.

Consequently, for angles small enough for the small-angle approximation to hold, the angular separation of any two adjacent diffraction minima will be

$$\theta_{(n+1)} - \theta_n = \Delta\theta = \frac{\lambda}{b}. \tag{15}$$

Find the nuclear radius

a. Find the wavelength λ of 800 MeV protons.

b. Then use the data shown in Fig. 16.8 and determine the diameter of ^{208}Pb. Check to see whether the first minimum falls at 1.22 times $\Delta\theta$.

To check your answers and to see some more interesting information about nuclear radii, you may want to look at G.F. Bertsch and E. Kashy, "Nuclear scattering," *Am. J. Phys.* **61**, 858–859 (1993) from which this problem is adapted.

CHAPTER 17

Spectra and the Bohr Atom

17.1 INTRODUCTION

We come now to a new aspect of atoms: the existence of discrete energy states. Niels Bohr's idea that atoms can possess only certain well-defined amounts of energy was a major development in our understanding of atoms. In 1911 Bohr, a young Dane who had just received his Ph.D. in physics from the University in Copenhagen, came to England to visit for a year. He worked for a while in J.J. Thomson's laboratory in Cambridge, and then in early 1912 Bohr transferred to Manchester to work with Rutherford. Inspired by Rutherford's concept of the atomic nucleus, Bohr subsequently developed a nuclear model of the hydrogen atom that predicted the wavelengths emitted in the spectrum of atomic hydrogen. The agreement of his predictions with observations was startlingly good.

Bohr's nuclear model of the atom introduced two new ideas about the inner working of atoms that are basic to our present-day understanding of the atom: Atoms can exist only in special "stationary" states of well-defined energy; an atom's angular momentum comes in integer multiples of $h/(2\pi)$ (i.e., $\hbar$). Bohr's model illustrates these ideas even though the model has been replaced by quantum mechanics. Despite its fundamental defects, Bohr's model continues to be of practical use, because it often provides helpful insights into complicated problems more easily than does the full mathematical treatment of quantum mechanics. For these reasons we will examine Bohr's model—frequently called "the Bohr atom"—in some detail.

17.2 ATOMIC SPECTRA

You have already seen in Chapter 9 that each particular kind of atom emits light of well-defined characteristic wavelengths. Measurement, tabulation, and analysis of these wavelengths are the tasks of the subfield of physics called "spectroscopy." Quite soon after the discovery of the existence of these well-defined wavelengths, spectroscopists noted striking patterns and regularities among the observed wavelengths. Bohr showed that these patterns reveal much about the internal structure of the atom.

Wall Tapping and Bell Ringing

Historically, atoms have been difficult to study because they are so small. To get around this difficulty we have had to find ways to investigate objects too small to see, feel, or sense directly. One way is to collect a huge number (moles) of identical copies of the objects and look at their collective behavior. This is what you do when studying the pressure of a volume of gas or a chemical reaction. Another way is to whack the objects somehow and see whether interesting pieces break off. Cathode rays (electrons) and x-rays are some of the results of such whacking. Rutherford's alpha scattering from gold foils is another example of learning about atoms by whacking them quite hard. Although the results are interesting and informative, bashing objects lacks finesse and surely greatly modifies what you are studying.

You can also learn about an object by jiggling it gently and seeing what you can deduce from its subsequent wiggles. Have you ever tried to find a framing member of a house wall behind plaster or wallboard? A simple way is to go along the wall tapping with your finger. The sound will change when you reach the more solid area right over a support piece. The sound is a result of vibrations set up by your tapping, and the combination of vibrations changes character as the structure of the wall beneath the tapping changes. You use the quality of the sound to infer what structure lies beneath the point on which you tap.

You already use differences in the quality of sound to infer differences in structure. For example, you can easily tell the difference between a bell being struck and a piano string being hit. You can tell a banjo from a guitar, a saxophone from a tuba. The structure of each instrument determines its distinctive tone; the structure determines how much vibration of each frequency occurs whenever the object is excited in some way. In principle, it should be possible to learn something about the internal structure of the device by analyzing the combinations of frequencies present in any tone.

Something like this can be done with atoms using light rather than sound. Atoms made to vibrate will emit electromagnetic radiation, as was discovered

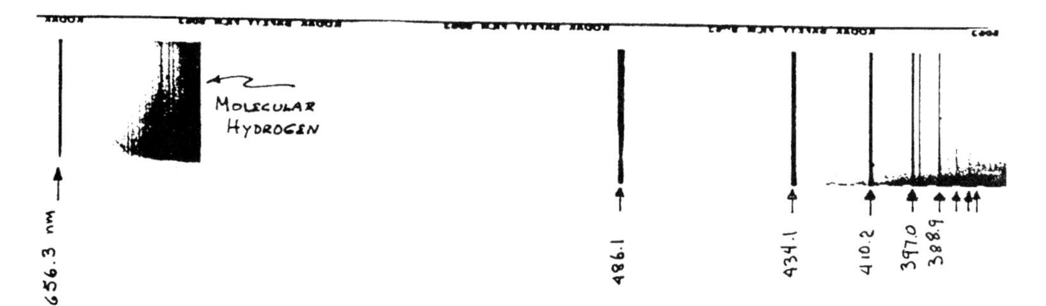

FIGURE 17.1 Photograph of the visible and near-ultraviolet spectrum of hydrogen.

in studies that began in the middle of the nineteenth century. The work of Kirchhoff, Fraunhofer, Balmer, Rydberg, and many others showed that each atomic element emits a distinctive pattern of light frequencies when properly stimulated. In Chapter 9 you learned that such a pattern is called the "spectrum" of the element.

Atomic Spectral Signatures

If each species of atom has a unique and distinctive spectrum, it is natural to think that spectra might tell us something about the internal structure of atoms. For testing this idea, hydrogen is the best choice because it has the simplest spectrum of all the atomic elements.

The simplicity is evident in Fig. 17.1, which shows a hydrogen-atom spectrum photographed by one of the authors. Notice the striking regularity with which the lines progress. They are the first three lines in a series of lines that get closer and closer together as the color goes from red (long wavelength) to violet (short wavelength) to the (invisible) ultraviolet. The series of lines approaches a well-defined limit, called the "series limit." (There are also lines in the picture that come from hydrogen molecules rather than hydrogen atoms; ignore these.)

The pattern's regularity is mathematically simple. In 1885 a Swiss schoolteacher named Johann Balmer devised a simple algebraic formula that accurately describes the sequence of observed spectral lines:

$$\lambda_n = 364.6 \, \frac{n^2}{n^2 - 4}$$

where λ_n are the wavelengths in nm, 364.6 is a constant chosen to match the formula to the observed wavelengths, and n is any integer greater than 2. A few years later Rydberg rearranged Balmer's formula into the form usually used today:

$$\frac{1}{\lambda_n} = R\left(\frac{1}{4} - \frac{1}{n^2}\right). \tag{1}$$

EXERCISES

1. Find the constant R from one of the wavelengths given in Fig. 17.1.

2. Plot a graph of $1/\lambda_n$ vs. $1/n^2$ to obtain a value for R from the slope. Check with the result of the previous problem.

3. Calculate the shortest wavelength, called the series limit, in the Balmer series.

Balmer's formula by itself was not earthshaking and did not immediately lead to any new insight into atomic structure. What it did do was to give hints and clues for discovering other series. For example, instead of writing the first term as 1/4, Rydberg generalized it to $1/n'^2$. Then by assuming other integer values of n' he predicted entirely new series with different sets of wavelengths and different series limits. One predicted series was later found by Lyman in the far ultraviolet spectrum of atomic hydrogen, and another predicted series was found in the near infrared spectrum by Paschen. The Lyman series had $n' = 1$, and the Paschen series had $n' = 3$.

EXERCISES

4. Calculate the series limits for the Lyman and Paschen series.

5. Find the first three lines (longest wavelengths) for the Lyman series.

6. Find the first three lines of the Paschen series.

7. Using the same scale of wavelengths, draw the positions of the first three lines and the series limits of all three series described above.

These regularities in the atomic spectra of atomic hydrogen provided Bohr with important clues to the inner workings of the atom. He devised a model of the hydrogen atom that exactly predicted its spectrum. Let's see how he did that.

17.3 THE BOHR ATOM

Need for a Model

J.J. Thomson's "plum pudding" model of the atom (see Section 16.3) could not predict the details of the observed spectrum of hydrogen or any other element. Initially, the nuclear model was also unpromising. In the first place, classical physics showed that electrons bound to a positively charged nucleus must revolve in orbits. But why would they have only those particular orbits that represented the special set of frequencies observed in the spectrum of hydrogen? Even more of a problem was how they could stay in orbit. Any orbiting charged particle must radiate electrical energy due to the acceleration it undergoes as it is bent into the circular, or perhaps elliptical, orbital path. Thus orbiting electrons would constantly lose energy and spiral in to the nucleus, just as Earth-orbiting satellites that are low enough to encounter some atmosphere gradually lose energy and spiral in to Earth. In the case of atomic electrons, though, the time to decay would be microseconds or less, not months or years! Classical physics had no way to explain why we have atoms at all or why they have the sizes they do.

Bohr's Ideas

Bohr was successful by being able to unstick himself from the accepted rules about how things should work. That in itself, though, is not necessarily remarkable. There is many a crank around doing the same thing. What was remarkable was Bohr's ability to invent new rules that worked and could be generalized to predict new results. We will not try to reproduce his actual steps, but will trace a similar path that is easier to follow.

Bohr decided to ignore classical physics' inability to account for stable atoms. He gave up trying to relate the hydrogen spectra directly to internal motions, and he took the nuclear model at its face value. If there had to be stable orbits, he reasoned, one should make stability a basic property of the model rather than worrying about how orbits couldn't be stable.

Turning first to the spectrum problem, you can get new insights into Rydberg's version of the Balmer formula by using the photon idea to connect wavelength to energy. Since c/λ is frequency, and the Planck constant, h, times frequency is energy, you can multiply both sides of Rydberg's equation by hc to get

$$E = \frac{hc}{\lambda} = hf = hcR\left[\frac{1}{4} - \frac{1}{n^2}\right].$$

Substituting $1.097 \times 10^{-2}\,\text{nm}^{-1}$ for R and 1240 eV nm for hc gives

$$hf = 13.6\left[\frac{1}{4} - \frac{1}{n^2}\right]\text{eV}. \tag{2}$$

In terms of photons it does not make sense to think of the hydrogen atom as "ringing" in some complicated way like a musical instrument. Any one hydrogen atom must emit a photon with essentially only one frequency for any one event. This suggests that a single photon is emitted whenever a single hydrogen atom changes its energy. This photon must have one of the wavelengths observed by spectroscopists (656,... nm). If the atom had an energy E_i before and E_f after emission of the photon, the photon energy would be:

$$hf = E_i - E_f, \tag{3}$$

where we assume $E_i > E_f$ because an atom would have to lose energy to produce a photon. Comparing Eq. 2 with Eq. 3, it is natural to think of the two terms on the right-hand side of Eq. 2 as separate energies. These different energies might be energies of different configurations of the atom, what we call "energy states" of the atom.

The important idea here is the assumption that *the atom exists only in particular definite energy states*. This is quite different from classical physics, where the energy of a system can vary continuously. It is like saying that a baseball thrown near the surface of the Earth can have 2 J of energy or 3 J of energy, but nothing in between.

Quantizing the Hydrogen Atom's Energies

Using classical mechanics Bohr derived an expression for the energy of the electron in the atom as a function of the distance r between the electron and the nucleus (proton). Then he took a giant step beyond classical physics: He invented a rule that permitted only certain values for the atoms's energy E. In this way he was able to explain Eq. 2. Today we say that Bohr "quantized the energy" of the atom.

Bohr assumed that the electron is kept in a circular path around the nucleus by the attractive Coulomb force between the electron and the proton that is the nucleus of the hydrogen atom. Why its energy did not leak away as classical physics predicted he did not know. It obviously didn't, so he set the question aside to be dealt with at some future time. The proton is so much more massive than the electron that it remains nearly stationary while the electron moves around it. Then since the Coulomb force, Eq. 1, is supplying the centripetal force, it follows that

$$\frac{mv^2}{r} = \frac{k_c e^2}{r^2}. \tag{4}$$

(Because the proton charge is $+e$ and the electron charge is $-e$, the electrical force is attractive.) Rewriting Eq. 4 in terms of momentum,

$$\frac{p^2}{mr} = \frac{k_c e^2}{r^2},$$

you get two expressions, one for momentum p and one for orbital kinetic energy K:

$$p = \sqrt{\frac{mk_c e^2}{r}} \tag{5}$$

and

$$K = \frac{p^2}{2m} = \frac{k_c e^2}{2r}. \tag{6}$$

Because electrostatic potential energy U is

$$U = \frac{-k_c e^2}{r},$$

the total energy, i.e., the sum of the kinetic and potential energies, is

$$E_{\text{tot}} = K + U = -\frac{k_c e^2}{2r}. \tag{7}$$

Notice that the total energy depends only on r. Therefore, for the total energy to take on only certain values, there must be some rule that correspondingly constrains r. Here Bohr took a bold step. In effect he said that the rule is that the product of the radial position r and the momentum p must always be some integer multiple n of the Planck constant divided by 2π, i.e.,

$$rp = n\frac{h}{2\pi}. \tag{8}$$

There was no precedent for this rule, although there surely had to be some connection to the Planck constant. After all, something had to take on specific values, had to be quantized, and h was already associated with the quantum of light, the photon. There are several equivalent ways to rationalize Eq. 8. One is to recognize that rp is the so-called angular momentum of the system. Equation 8 then states that the allowed values of the angular momentum are integer multiples of $\hbar$, i.e., *the angular momentum is quantized in units of* $\hbar$. This assertion is validated by modern quantum theory. Because we have not discussed angular momentum in this book, let's consider an argument that makes use of the wave nature of matter introduced in Chapter 14. Although not logically sound, it is a convenient mnemonic.

Consider the implications of the electron having wave properties. In the presumed "orbit" such a wave must come back on itself. If after each circling of the nucleus the wave's phase has changed by an exact multiple of 2π, the succession of waves will reinforce constructively. If the phase difference is anything else, there will be destructive inteference over time. In other words, only those circular orbits will exist for which the circumference is an integer number of wavelengths $2\pi r_n = n\lambda$. (It is stretching things to use the idea of a wave along the circumference of a circle and yet assume an exact radius.) Using the de Broglie relation between wavelength and momentum,

$$p = \frac{h}{\lambda} = \frac{hn}{2\pi r_n},$$

and substituting for p from Eq. 5 yields a special set of values of r, which we denote as r_n:

$$r_n = \frac{n^2 h^2}{4\pi^2 m k_c e^2}. \tag{9}$$

We can use these values of r_n to find the allowed energies of the hydrogen atom. Replacing r_n in the total energy equation, Eq. 7, by Eq. 9 gives a set of discrete energies E_n one for each integer value of $n = 1, 2, 3 \ldots$:

$$E_n = -\frac{2\pi^2 m k_c^2 e^4}{n^2 h^2}. \tag{10}$$

These special values of E that can occur are called "energy states." The lowest energy E_1, i.e., $n = 1$, is often referred to as the "ground state" of the atom. Because of this association of the energy states of the hydrogen atom with the integers n, it is customary to label the energy states with the index n. Because the restriction of n to integer values forces there to be a finite difference, a *quantum* of difference, between energies, the index is called a "quantum number." For atoms more complicated than hydrogen, n continues to be very important, but other such indices are needed as well, and they are all called quantum numbers. To distinguish the n quantum number from the others, it is called the "principal quantum number," or, sometimes, the "radial quantum number."

The difference in energy between a lower-energy state of quantum number n' and a higher-energy state of quantum number n, is just

$$E_n - E_{n'} = \frac{2\pi^2 k_c^2 e^4 m}{h^2}\left[\frac{1}{n'^2} - \frac{1}{n^2}\right].$$

Therefore, if, as Eq. 3 proposes, a photon of energy hf is emitted when the atom changes from state E_n to state $E_{n'}$, the photon's energy would be

$$hf = E_n - E_{n'} = \frac{k_c^2 e^4 m 2\pi^2}{h^2}\left[\frac{1}{n'^2} - \frac{1}{n^2}\right],$$

which has exactly the same form as Eq. 2 when $n' = 2$. Putting in values for the constants to find the factor multiplying the bracketed terms gives

$$\frac{2\pi^2 k_c^2 e^4 mc^2}{(hc)^2} = \frac{2\pi^2 1.44^2\ 511 \times 10^3}{1240^2} = 13.6\,\text{eV},$$

exactly as observed experimentally.

This is a spectacular result. Until Bohr Eq. 1 was only an empirical guess, and the value of R was an experimentally determined number. Bohr's model of the internal structure of an atom yielded an expression for R in terms of fundamental constants from several areas of physics that when evaluated numerically is in exceptionally good agreement with the experimental value. Clearly, Bohr's result cannot be just a fluke. Therefore, although the model with its ad hoc assumptions has serious flaws, some of its elements must be correct and must play a role in a full theory of the atom.

Although the wave nature of the electron makes the concept of an orbit with a well-defined value of r questionable, the calculated values of r_n do correspond roughly to the atom's size. The smallest one, r_1, is often given a special symbol, a_0,

$$a_0 = \frac{h^2}{4\pi^2 mk_c e^2} = 0.0528\,\text{nm},$$

and is called the "first Bohr-orbit radius." All other possible Bohr-model radii are thus

$$r_n = n^2\, a_0 = 0.0528\, n^2\ \text{nm},$$

where the quantum number n can take on any integer value. Notice that the value of a_0 is what we obtained in Chapter 15 when we used Heisenberg's uncertainty principle to estimate the minimum energy of the hydrogen atom. This means that the Bohr model is consistent with this fundamental principle.

Do not attribute too much significance to these values of r_n. Do not think that the electrons move in well-defined orbits in the atom. The uncertainty principle emphatically denies the possibility of such orbits. The electron in an atom is much more wavelike than particle-like. It does not and cannot have a well-defined location. Thus, although the value of r_n gives an indication of the spatial extent of a hydrogen atom that has an amount of energy E_n, r_n does not label the path of an electron inside the atom the way a planetary radius labels the planet's orbit around the Sun.

Energy-Level Diagrams

It is possible to describe *any* atom in terms of its energy states. An important graphical aid for representing the energies of an atom is the "energy-level

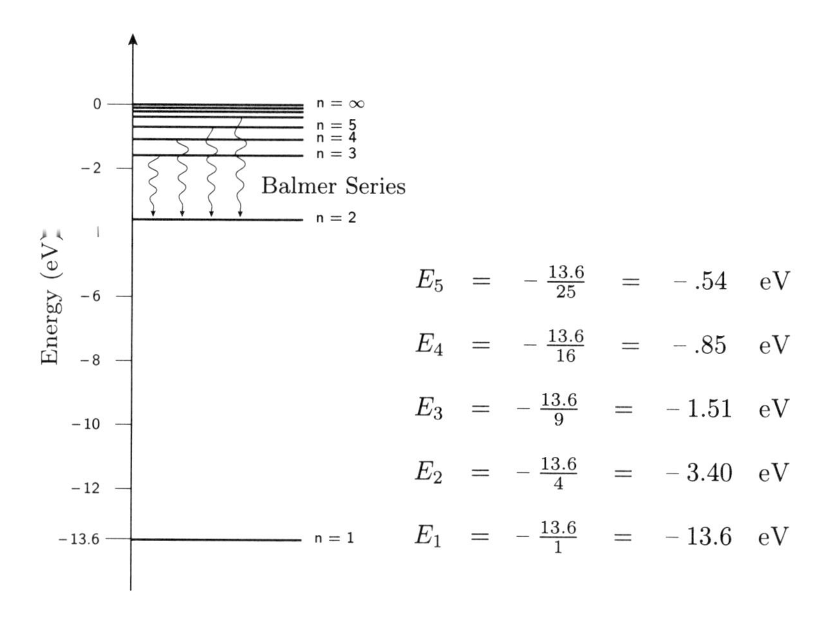

FIGURE 17.2 Energy-level diagram of the hydrogen atom.

diagram." As an example, consider the hydrogen atom. You can calculate the energies E_n of its quantum states from the relation

$$E_n = -\frac{13.6}{n^2}$$

and arrange them vertically on a scale, as shown in Fig. 17.2. The lines represent the "energy levels" of the atom. In the diagram shown here, the zero of the scale is the energy for n equal to infinity. Note that from Eq. 9, this condition corresponds to an infinite separation between the electron and the nucleus, i.e., the hydrogen atom would be ionized. This means that 13.6 eV of energy is required to free an electron from the hydrogen atom in its ground state.

The energy-level diagram looks like a ladder with "rungs" getting closer together as they approach the top. *A photon is emitted by an atom when it drops from one energy level to a lower one*. The energy of the emitted photon is just equal to the difference in energy between the two energy levels of the atom. We say that the atom has undergone a "transition" from one energy state to another.

You can use this picture to understand the various spectral series that are emitted by gaseous atomic hydrogen. A hydrogen atom can undergo a transition from any higher energy state to any lower one. Different transitions have different probabilities of happening. If a hydrogen atom is ionized, an unbound electron, i.e., an electron outside of the atom, can make a transition to any energy level of the atom.

In the laboratory when an electric current is passed through hydrogen gas in a thin glass tube, the resulting electric discharge ionizes some hydrogen atoms and excites the electrons in others to higher energy states. After excitation an atom loses its energy by a cascade of successive jumps. Which jumps occur differ from one atom to the next, and there is no guarantee that any given atom will go from any particular energy state to another. The process is quite random, and the atom can de-excite by skipping steps. (Modern quantum theory can predict the probabilities of these transitions, but the Bohr model cannot.) The result is that when a collection of atoms de-excites, photons of many wavelengths will be emitted by different atoms. For any given atom, the whole sequence of photon emissions down to the ground state usually takes a very short time, from microseconds to nanoseconds. An observed spectrum is made up of photons emitted from a variety of different transitions occurring in many different hydrogen atoms.

The transitions that produce the Balmer series of spectral lines are shown in Fig. 17.2 by the vertical arrows connecting higher energy states with the $n = 2$ energy level. Note that the final state of the Balmer line transitions is not the lowest possible energy state of the atom. After a Balmer transition, the hydrogen atom will change into its lowest, most stable, energy state—its ground state. In other words, it will make a transition from $n = 2$ to $n = 1$, and it will emit a photon that has a wavelength too short (energy too high) to be visible to the eye.

The Balmer series is apparent in the laboratory because it is the only series that consists of wavelengths of visible light. Even so, only three or four of the possible transitions to the $n = 2$ level are visible; the rest are in the ultraviolet and beyond the sensitivity of the eye.

The diagram suggests other possible series. For example, there should be a series that has a final state $n' = 1$. This is the Lyman series, consisting entirely of spectral lines with photon energies of 10.2 eV or more.

EXERCISES

8. Redraw the diagram in Fig. 17.2, and show on it the possible Lyman series transitions.

9. In the hydrogen spectrum the lines of the Paschen series are transitions to $n' = 3$. Use Fig. 17.2 to determine the energies of the first three lines of this series.

EXAMPLES

1. What photon energies would be produced from atoms excited to $n = 3$?

The electron could go to the ground state by two routes. One would be to jump directly to $n = 1$, emitting the second line of the Lyman series, or it could go from 3 to 2 and then to 1. These transitions would be the first lines of the Balmer and Lyman series, respectively. Thus there are three possible photon energies:

$$E_3 - E_1 = -1.51 + 13.6 = 12.1\,\text{eV},$$
$$E_3 - E_2 = -1.51 + 3.4 = 1.9\,\text{eV},$$
$$E_2 - E_1 = -3.4 + 13.6 = 10.2\,\text{eV}.$$

EXERCISES

10. Eventually, two more hydrogen series in the infrared were identified: the Brackett series with $n' = 4$, and the Pfund series with $n' = 5$. Find the photon energies of the first two lines in each series.

11. What are the wavelengths of the lines calculated in the previous problem?

12. Suppose a group of hydrogen atoms are all excited to $n = 5$. How many different photon energies are emitted as the assemblage decays?

13. Find the energies of all the photons that could be emitted from hydrogen atoms excited to $n = 4$.

17.4 CONFIRMATIONS AND APPLICATIONS

Bohr's model produced some remarkably accurate results. But in important ways it was *ad hoc*, introducing some strange new concepts just to explain a limited set of data. For the new ideas to become credible, there had to be further predictions that could be confirmed by experiment.

Energy Levels

Existing only in one well-defined state of energy or another, an atom can lose energy only in discrete amounts. But if energy is emitted only in certain well-defined amounts, energy also can be absorbed only in discrete amounts. If a particular atomic transition emits a photon of a certain energy, the same energy must be absorbed in order to reverse the transition and raise the atom from a lower to a higher energy level.

You can add energy to atoms just by shining light on them. If a photon in a beam of light meets an atom and if the photon's energy hf is *exactly equal* to the difference between the atom's present energy state and some higher energy state, the photon can be absorbed by the atom, and the atom will be excited to the higher energy level. The absorbed photon is then lost from the light beam. If nothing is in the path of a beam of light traveling from a hot body such as a lamp filament and passing through a slit and a diffraction grating, the grating spreads the beam out into a continuous spectrum of colored light ranging from red to violet. If there is a gas between the source and the spectrometer slit and if that gas absorbs light at certain well defined wavelengths, the intensity of light will be reduced in the spectrum at the positions corresponding to the absorbing wavelengths. As a result there appear in the spectrum thin strips of reduced intensity that are dark compared to the brighter parts of the spectrum. These dark strips are called "absorption lines."

It is also possible to make a beam of light of a single wavelength that can be varied. If as the wavelength is changed it takes on a value corresponding to the photon energy that matches the energy of a transition, the beam will be sharply absorbed and its passage through the gas will be much diminished.

Suppose you were to send a beam of visible light through ordinary hydrogen gas in a transparent container. Would photons at the wavelengths of the Balmer series be absorbed? No, not if the Bohr idea is correct. Aside from the complications of the hydrogen molecule, absorptions corresponding to the Balmer series must all start from the $n = 2$ level, but at ordinary temperatures there would be no H atoms excited to this state and capable of absorbing the light that is present.

A gas of hydrogen atoms could absorb only the much higher energy photons of the Lyman series. These are the only transitions that start from the lowest energy state of the atom.

Rydberg Atoms

You can imagine using a mixture of ordinary visible light and ultraviolet. The ultraviolet might excite an atom to its $n = 2$ energy level, and then a visible photon could excite a Balmer transition from this state to some higher state. This is not practical using ordinary light sources because they are too weak: The atom will return to its ground state long before a visible photon arrives to induce a Balmer transition upwards. With lasers, however, it is possible to induce double absorption. Lasers can produce enormous numbers of photons with very well defined frequency. The frequency of one laser can be adjusted to exactly the right energy to excite atoms from their ground states to some particular excited state. Then if these excited atoms are illuminated with a large number of photons from a second laser set to a wavelength to induce a transition upward from the excited state, it becomes likely for such a transition to occur before the atoms de-excite to the ground state.

By such two-step excitation it is possible to make atoms in states with very high values of n. Any atom, not just hydrogen, in a high-n energy state obeys the Rydberg formula, Eq. 2. Consequently, such highly excited atoms are called "Rydberg" atoms, and the energy states with high values of n are called "Rydberg states." It is as though the excited electron is so far from the nucleus that down below it the negative charge of the other electrons neutralizes all but one of the positive charges of the nucleus. As a result, the atom's nucleus and innermost electrons look to the excited electron like a hydrogen nucleus, and the Bohr model describes it quite well. Atoms have been prepared with n as large as 400!

EXERCISES

14. Find the Bohr radius of a hydrogen atom with $n = 400$.

If you have done the calculation of the previous problem correctly, you found the dimensions of the $n = 400$ atom to be as large as objects that can be seen with a high-powered microscope. However, you could not "see" such a Rydberg atom. An atom in a high-n state is very fragile; any photon of visible light would ionize it.

EXERCISES

15. Confirm the previous statement by finding the energy of the $n = 400$ level and comparing the ionization energy to that of the lowest-energy photon that is visible (wavelength around 700 nm).

16. Suppose you irradiated hydrogen atoms with light consisting of photons with a continuous distribution of energies from 0 to 14 eV. What spectral series would be observed in absorption, assuming that you have the right equipment and that double-photon absorption is negligible.

Note that we have discussed only hydrogen *atoms* even though the gas comes as diatomic molecules. Unfortunately, molecules are beyond the scope of the Bohr theory. However, the usual way of exciting hydrogen is to pass current at high voltages through a sample of gas. The ions created have enough energy to dissociate the molecules, and we get enough single atoms to produce the atomic spectrum. Usually, radiation from excited molecules is also present, and we just ignore it when studying the atomic spectral series.

The Franck–Hertz Experiment

You can also add energy to atoms by bombarding them with energetic particles. Electrons accelerated through an electric potential difference of a few volts will excite atoms when they collide with them. Just as photons do, colliding electrons lose energy in discrete amounts. That is, only the exact energy difference between two levels is absorbed, because there is no way for an atom to exist at any energy in between. There is an important difference in the way photons and electrons lose energy to atoms. *For photons it is all or nothing*. They either lose all their energy and disappear, or they lose nothing and continue on. Only when the photon energy exactly matches the transition energy is the photon absorbed. *Electrons, on the other hand, can lose part of their energy to the atom and retain the rest.* They can lose an amount equal to the energy of the transition that is induced in the atom, and continue on with whatever energy is left over. Electrons with any energy over the minimum necessary to induce a transition can and will induce transitions.

A year after the Bohr model was published, J. Franck and G. Hertz bombarded mercury atoms with energetic electrons to exhibit directly the existence of discrete atomic energy levels. They accelerated electrons through a vapor of mercury and showed that the electrons lost energy to the mercury atoms in discrete amounts. Figure 17.3 shows how this was done. Electrons were accelerated from the filament to the accelerating electrode. The accelerating electrode was a mesh, so that the accelerated electrons passed through it. They were then decelerated by the retarding voltage V_{ret} of 2 to 3 V between the accelerating electrode and the collecting electrode.

As the accelerating voltage V_{acc} was increased from 0 to about 6 V, the number of electrons reaching the collecting electrode, measured by the current meter, increased, but instead of rising steadily as V_{acc} was increased further, the number dropped to a minimum at $V_{\text{acc}} \approx 8\,\text{V}$. As Fig. 17.4 shows, the number of electrons reaching the collecting electrode varied regularly as the accelerating voltage was increased towards 40 V. The collector

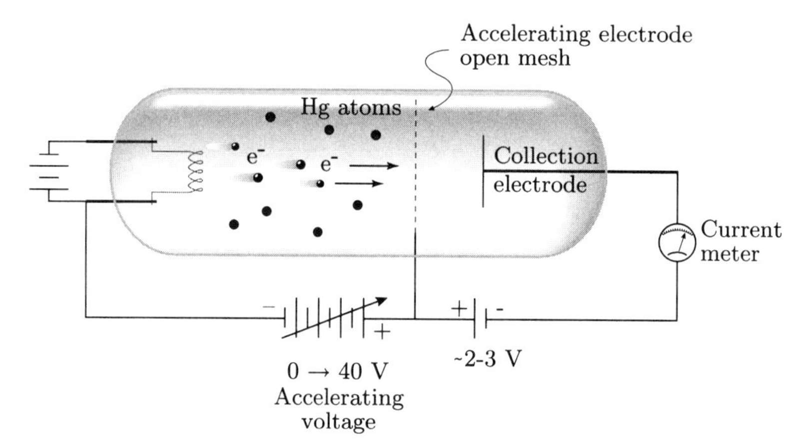

FIGURE 17.3 Schematic of a version of the Franck–Hertz experiment. The arrangement of applied voltages with typical values is shown.

current fell to a minimum every time the accelerating voltage changed by 4.9 V.

The regular fluctuations in the collector current are just what is expected if the electrons give up their energy to mercury atoms in quanta of 4.9 eV. An electron accelerated through 6 V would have a kinetic energy of 6 eV. For example, this would be enough to permit it to pass through a retarding potential of 3 V and reach the collecting electrode.

EXERCISES

17. What would be its kinetic energy when it reaches the collecting electrode?

But if a 6 eV electron loses 4.9 eV of energy in a collision with a mercury atom somewhere between the filament and the accelerating electrode, it will reach the accelerating electrode with only 1.1 eV of kinetic energy. Then it would not have enough kinetic energy left to overcome the retarding potential and reach the collecting electrode, and it will not contribute to the current measured by the meter in Fig. 17.3. This absorption of energy by the Hg atoms will cause the electron current reaching the collecting electrode to decrease.

The most striking feature of the graph in Fig. 17.4 is the sharp drop in the current at periodic intervals of the accelerating voltage. The analysis above suggests that every time an electron gains another 4.9 eV or so in the space between the electrodes, it will have enough energy to excite another mercury atom. It is then highly likely that it will lose the energy it has gained and then

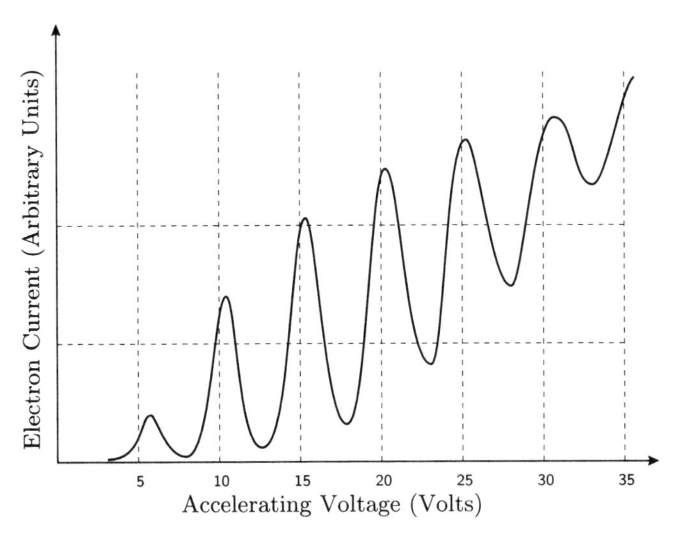

FIGURE 17.4 Typical current vs. accelerating voltage data in a Franck–Hertz experiment on mercury vapor.

not have enough to pass through the retarding potential. This means that the mercury atom must have an excited state that is 4.9 eV above the ground state. Some complications having to do with the work function produce an offset of ≈ 2 V in the voltage of the first dip, but after that, as the accelerating voltage is increased, the electron will acquire enough energy to lose 4.9 eV in one collision and then gain enough more to lose another 4.9 eV in another collision. The succession of dips in Fig. 17.4 are the result of losses of energy from such multiple collisions. The diagram in Fig. 17.5 shows schematically a possible version of successive gains and losses of energy by an electron as it passes from the filament to the accelerating electrode.

The Franck–Hertz experiment was an important confirmation of the existence of energy levels in atoms. In a later experiment Hertz observed the light emitted by the mercury atoms as they decayed back to the ground state. As you would expect, these photons had 4.9 eV of energy.

EXERCISES

18. For defining the voltage differences it is more reliable to pick the voltages of the highest currents just before the start of the dips. Find the voltages of peak currents in the graph above and plot a graph of V_p versus n, the number of collisions. Take the slope to obtain the best value for the voltage difference.

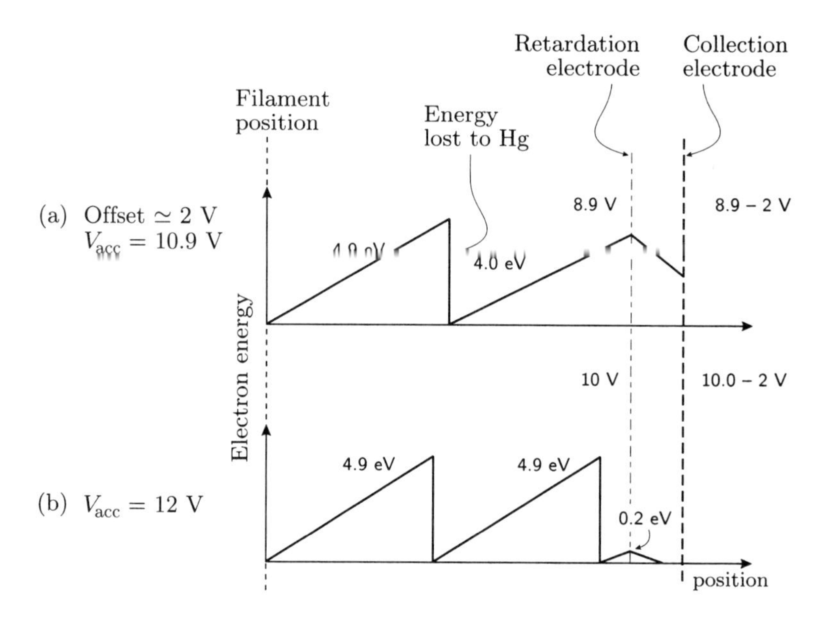

FIGURE 17.5 Diagram showing the mechanism of multiple collisions in a Franck–Hertz experiment. Because the collection and screening electrodes have work functions different from that of the filament, the effective acceleration voltage is 2 V less than the applied voltage.

Hydrogen-like Ions

The Bohr model only works well for simple two-body systems like hydrogen atoms or Rydberg atoms. When the atom has more electrons than the hydrogen atom, the potential energy of each electron becomes very complicated, arising now from the constantly changing distribution of the other electrons as well as the nuclear charge. However, there are some other simple two-body combinations of charged particles besides the hydrogen atom. For example, if the helium atom, which has a doubly charged nucleus, is singly ionized, it will look like a hydrogen atom with two units of nuclear charge. It is easy to include the effect of a different nuclear charge in Eqs. 4 to 10.

Recall that in these equations the product $q_1 q_2$ was written as $-e^2$ because the charge of the hydrogen nucleus is just e. If there were a larger charge on the nucleus, it would have to be written as Ze, where Z is the atomic number of the atom. When the factor of e^2 is replaced with Ze^2, Eqs. 4, 7, 9, and 10 become, respectively,

$$\frac{mv^2}{r} = \frac{Zk_{\mathrm{c}}e^2}{r^2}, \tag{11}$$

$$K = \frac{Zk_{\mathrm{c}}e^2}{2r}, \tag{12}$$

$$r = \frac{n^2\hbar^2}{mk_cZe^2}, \tag{13}$$

$$hf = \frac{mZ^2k_c^2e^4}{2\hbar^2}\left[\frac{1}{n'^2} - \frac{1}{n^2}\right] = 13.6Z^2\left[\frac{1}{n'^2} - \frac{1}{n^2}\right] \text{ eV.} \tag{14}$$

Equation 14 applies to any low-mass atom that has had all but one of its electrons removed. For example, it describes quite accurately the lines of the spectrum of He^+ and of Li^{++}.

EXERCISES

19. For singly ionized helium, calculate the energies of the first three lines of the series equivalent to the Lyman and Balmer series.

20. Show that some of the "Balmer-series" lines of singly ionized helium are the same as lines of the hydrogen Lyman series. Which ones are they, in general? This correspondence was a source of confusion when spectral lines of He ions were first observed in the solar spectrum; some people speculated that they were seeing a new form of hydrogen not found on Earth.

21. Find the energies of the first lines of the series in doubly ionized lithium that are equivalent to the Lyman and Balmer series.

X-Ray Line Spectra and the Bohr Model

The discussion of x-rays in Chapter 13 concentrated on the continuum x-ray spectrum, and the data shown there were chosen because of their simplicity. Most x-ray spectra also have prominent sharp spikes on top of the continuum, as illustrated by Fig. 17.6. The Bohr model offered an explanation of these so-called x-ray lines.

X-ray line spectra appear for every element, and unlike the continuum, the line x-rays have the same energy regardless of the energy of the electrons bombarding the anode of the x-ray tube. If x-ray lines are present, their wavelengths depend only on what the anode was made of. They are characteristic of that element, and for this reason they are called "characteristic x-rays." Physicists, chemists, geologists, biologists, and many other scientists often induce x-ray emission from substances of unknown composition, and then from measurements of the wavelengths of the characteristic x-rays they identify the elements present in the sample. The x-ray lines labeled K_α and K_β in Fig. 17.6 are characteristic x-rays of zirconium.

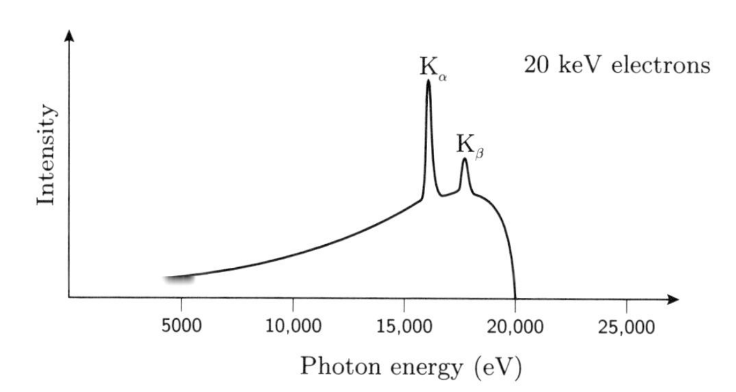

FIGURE 17.6 Typical x-ray spectrum from element of atomic number 40.

The existence of characteristic x-rays suggested a further elaboration of the internal structure of the atom—the idea of energy shells. The characteristic x-rays could be explained by assuming that the energy states of each atom occur in groups. In each group the states are close together in energy, but between the groups there is a considerable separation in energy. These well defined groups of energy states are called "shells" of energy. There is one shell for each value of the principal quantum number $n = 1, 2, \ldots$. Furthermore, a shell with principal quantum number n cannot contain more than $2n^2$ electrons. This rule is a version of the "Pauli exclusion principle." Together, shell structure and the exclusion principle can describe the regular progression of elements in the periodic table. For example, H consists of a single electron in the $n = 1$ shell, and He consists of 2 electrons in the $n = 1$ shell. But following the rule that there cannot be more than $2n^2$ electrons in a shell, Li with $Z = 3$ has to consist of 2 electrons in the $n = 1$ shell plus one electron in the $n = 2$ shell.

Shell structure and the exclusion principle also explain the generation of characteristic x-rays. Because of the exclusion principle, electrons in higher-energy shells cannot make transitions to lower-energy shells when these shells are filled with electrons. But when electrons accelerated in an x-ray tube slam into the atoms of the anode, some of the collisions will blast electrons out of the lowest-energy shell of some of these atoms. These collisions make vacancies in the $n = 1$ shell, which can then be filled by the transition of an electron from a higher shell to the lower one. When such a transition occurs, a photon is emitted with an energy equal to the difference between the energies of the two shells, as illustrated by Fig. 17.7. These photons are the characteristic x-rays, and they form a unique "fingerprint" for each chemical element.

The Bohr model can predict rather accurately the energy of a characteristic x-ray photon. For example, consider the energy of an electron in the

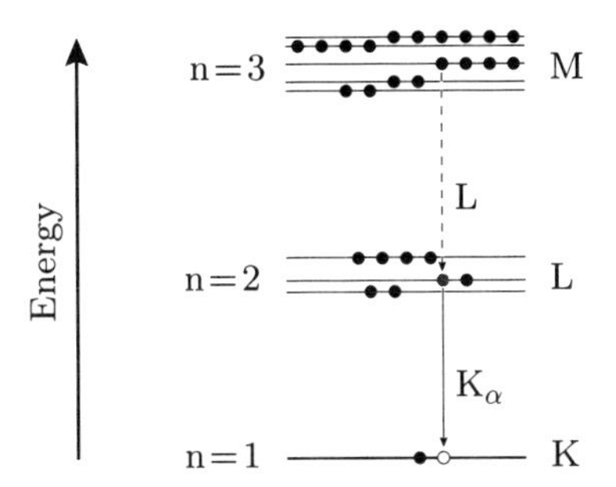

FIGURE 17.7 Schematic representation of energy shells of an atom, illustrating how x-rays can be produced by transitions from a higher-energy shell to a hole in the lower-energy shell.

$n = 1$ shell of an element with nuclear charge $Z = 40$ (zirconium). This energy will be about $-40^2 \times 13.6 = -22000$ eV, where we are guessing that an $n = 1$ electron will see the full nuclear charge. (Remember, as Eq. 14 shows, that the Bohr model scales as Z^2.) The electron that drops from a higher-energy shell to the empty space in the lower one will most probably come from the next highest shell, where its energy is -5500 eV.

EXERCISES

22. Show that the energy of an electron in the $n = 2$ shell of a $Z = 40$ nucleus would be on the order of -5500 eV.

The result of a transition of an electron from an $n = 2$ shell to an $n = 1$ shell would be the emission of a photon with energy around 15 keV—a rough answer that we will refine shortly. Photons of this magnitude of energy are x-rays, and as you learned in Chapter 13, you can use the Braggs' technique of crystal diffraction to measure the x-rays' wavelengths and thus their energies.

Most elements emit several groups of characteristic x-ray lines. The two closely spaced lines with the shortest wavelengths are called the K_α and K_β x-ray lines. In the Bohr model they arise from the $n = 2$ to $n = 1$ and $n = 3$ to $n = 1$ transitions. There are also characteristic x-ray lines with longer wavelengths (lower energy) than the K x-rays; these are known as L x-rays. In terms of the Bohr model they arise from transitions of electrons from higher-energy shells into a hole in the $n = 2$ shell. In particular, the L_α x-rays correspond to transitions from $n = 3$ to $n = 2$.

Moseley, the Atomic Number, and the Periodic Table

Just as Bohr published his model, H.G.J. Moseley,[1] working first with Rutherford, at Cambridge, and then later on his own, at Oxford, began a systematic study of the characteristic x-rays of as many elements as he could get samples of. He was powerfully motivated by his discovery that the energy of the characteristic x-rays increased in a very regular fashion as the samples progressed through the periodic table of the elements from lighter to heavier elements. In little over a year Moseley measured the K or L x-rays of nearly forty elements in the periodic table. His results explained the ordering of the periodic table, revealed that the number of elements is limited, and showed that only a few remained to be discovered.

Some of his results are shown in Table 17.1. Moseley's results showed that if he plotted the square root of the energy of the K_α x-rays vs. the emitting substance's atomic number Z, the result was a remarkably straight line, as shown in Fig. 17.8.

Because the energy of the K_α x-rays increased as he worked his way up the periodic table, their production soon required greater voltages than were

TABLE 17.1 X-ray Data for Ten Elements and Moseley's Law

Z	Element	λ (nm)	hf (eV)	$\sqrt{\frac{hf}{10.2}}$	$\sqrt{\frac{hf}{10.2}}+1$
20	Ca	0.3357	3694	19.01	20.0
22	Ti	0.2766	4483	20.97	22.0
23	V	0.2521	4919	21.97	23.0
24	Cr	0.2295	5403	23.02	24.0
25	Mn	0.2117	5854	23.96	25.0
26	Fe	0.1945	6375	25.00	26.0
27	Co	0.1796	6904	26.02	27.0
28	Ni	0.1664	7451	27.03	28.0
29	Cu	0.1548	8010	28.02	29.0
30	Zn	0.1446	8575	28.99	30.0

[1] The story of Moseley's extraordinary achievement is well told in Bernard Jaffe's *Moseley and the Numbering of the Elements*, Doubleday & Co., Garden City, NJ, 1971. The 26-year-old Moseley did his work between April 1913 and May 1914, working alone with remarkable ingenuity and fierce intensity. His work surely deserved a Nobel Prize, but Nobel Prizes are not awarded posthumously. In August 1915 Moseley was killed in battle in Gallipoli, one of the 55,000 young British, Australian, and New Zealand men who died in the failed attempt to establish a military front in the Balkans during World War I.

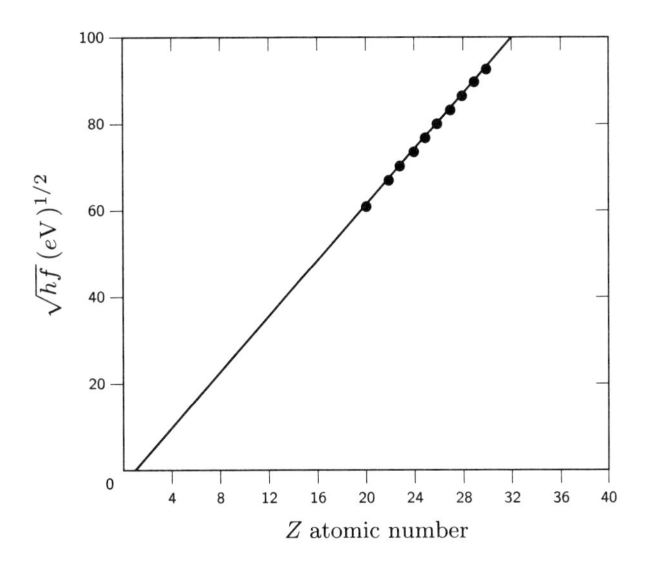

FIGURE 17.8 A plot of the square root of measured x-ray energies *vs.* atomic number Z.

available for his x-ray tube. However, the L x-rays, which from small-Z elements were too low in energy to be easily detected, also increased in energy as Z increased and could be produced with available tube voltages when the K x-rays could not. Moseley's data show that a plot of the square root of the energy of the L x-rays against the emitting element's atomic number Z also gives an extremely straight line. The slope and the intercept of the straight line are different from those of the K x-rays, but the straightness is still remarkable.

In algebraic terms, Moseley found that for K_α x-rays,

$$\sqrt{f} = m(Z - s),$$

where m is the slope of the line and s is its x-intercept. The form of this equation can be understood from the Bohr model. If the K_α lines come from the change in energy between the $n = 2$ and the $n = 1$ states of an atom with a nuclear charge Ze, then from the Bohr model,

$$hf = 13.6Z^2\left[\frac{1}{1^2} - \frac{1}{2^2}\right].$$

This might not be quite right, because there may be other electrons in the $n = 1$ shell, which would have the effect of reducing by an amount s the nuclear charge seen by the electron undergoing a transition from the $n = 2$ shell to the $n = 1$ shell. This so-called screening correction would give

$$hf = 13.6(Z - s)^2\left[\frac{1}{1^2} - \frac{1}{2^2}\right]. \qquad (15)$$

Putting in numerical values for the constants and taking the square root of both sides gives

$$\sqrt{hf} = \sqrt{10.2}(Z - s),$$

which predicts that $\sqrt{f}$ will depend linearly on the atomic number Z, as Moseley observed. The slope also agrees well with Moseley's data, and the intercept corresponds closely to $s = 1$.

The last two columns of Table 17.1 are obtained by solving Eq. 15 for Z and using $1/\sqrt{10.2}$ for the slope and $s = 1$ for the intercept. The fact that choosing $s = 1$ produces such good agreement between the model and the data suggests that there is one other electron in the $n = 1$ energy state. The generalization of Bohr's model to the shell model of the atom is thus borne out by experiment. The data of Moseley and many other experimenters show that the number of electrons in a shell does not depend on the element, and the energies of the innermost shells are determined mainly by the nuclear charge.[2]

Until Moseley's work, the atomic number was defined by the order of the chemical elements in the periodic table, an ordering loosely based on the elements' atomic weights and chemical and physical properties. Moseley's measurements showed that K and L x-ray energies correlate precisely with increases in the quantity Z in the Bohr model. From Rutherford's work this quantity was known to be the nuclear charge, which led Moseley to summarize in the paper in which he published his results:

> We have here proof that there is in the atom a fundamental quantity, which increases by regular steps as we pass from one element to the next. This quantity can only be the charge on the central positive nucleus, of the existence of which we have already definite proof.[3]

By his brilliant work Moseley established the concept of atomic number.

Another triumph of Moseley's work was the identification and ordering of the rare earth elements. Attempts to identify rare earth elements (La, Ce, ..., Lu) by chemical means had produced spurious identifications and much confusion because of their nearly identical chemical properties. He also cleared up some smaller uncertainties such as the order of iron, cobalt, and nickel in the periodic table. In the larger scheme of things, he showed

---

[2]Modern calculations show that the energies of interactions of the electrons among themselves are quite important. They shift the energies of the shells away from the values predicted by the Bohr model. Fortuitously though, the energy difference between the shells remains very close to the value predicted by the Bohr model.

[3]H.G.J. Moseley, "The high-frequency spectra of the elements," *Phil. Mag.* **26**, 1024–1034 (1913).

that the number of elements was limited and that up to $Z = 92$ all of them had been found except for $Z = 43, 61, 72$, and 75. Elements 72 (hafnium) and 75 (rhenium) were found shortly after the end of World War I; all isotopes of elements 43 (technetium) and 61 (promethium) are radioactive and too short-lived to occur in nature; their discovery had to wait until they could be made by nuclear physicists and nuclear chemists.

EXERCISES

23. Figure 17.8 shows a plot of the square root of the photon energies from Table 17.1 as a function of the atomic number of the elements. Repeat the plot in the neighborhood of Fe, Co, Ni, reversing the atomic numbers of Ni and Co to show how evident a misordering would be on Moseley's graph.

17.5 SUMMARY

The Bohr Model

Bohr explained the spectral series of the hydrogen atom. He hypothesized that the energy of the photons emitted by hydrogen was equal to the difference in energies of specific, well-defined energy states. The allowed states were those for which the angular momentum of the electron was an integer multiple of $\hbar$. When in these states, the electron interacted with the nucleus by means of the Coulomb force and satisfied the law of conservation of mechanical energy. These three assumptions justify the following four equations:

$$hf = E_{n'} - E_n,$$
$$rp = n\hbar,$$
$$\frac{mv^2}{r} = \frac{k_c Z e^2}{r^2},$$
$$E = \frac{p^2}{2m} - \frac{k_c Z e^2}{r},$$

where the symbols are as defined earlier in the chapter. The integer quantity n is the principal quantum number.

From these assumptions Bohr's model predicts that

$$E_n = -\frac{2m\pi^2 k_c^2 Z^2 e^4}{n^2 h^2},$$
$$hf = \frac{k_c^2 Z^2 e^4 m 2\pi^2}{h^2}\left[\frac{1}{n'^2} - \frac{1}{n^2}\right]$$

$$= -13.6Z^2 \left[\frac{1}{n'^2} - \frac{1}{n^2}\right] \text{eV},$$

$$r_n = \frac{n^2 h^2}{4\pi^2 m k_c Z e^2} = n^2 a_0,$$

$$a_0 = 0.0528\,\text{nm}.$$

Limitations of the Bohr Model

Bohr's major innovation was the idea of well-defined, discrete energy levels. His model worked well for single-electron atoms, but he as well as others recognized that it was an unsatisfactory theory. It met with very little success in atoms with more than one electron. It was completely unable to predict how rapidly transitions between states occur nor could it explain the relative intensities of the various spectral lines. Moreover, it was inconsistent in its approach. It used classical derivations for a very nonclassical situation. The idea of having orbits to which classical mechanics was applied, but then forbidding electrons to radiate, was definitely an uncomfortable way of producing a theory. Using the wave nature of electrons to help suggest why the orbiting electrons do not radiate is no more satisfactory. To talk of an electron with a specific *orbit* but then consider it to be a wave mixes two different kinds of descriptions, and the good result does not justify the bad logic.

A complete, consistent theory of the atom came with the development of quantum mechanics. This theory, which used several of Bohr's fundamental ideas, completely superseded the Bohr model.

X-Ray Line Spectra

The existence of x-ray line spectra, sharp peaks ("lines") in the x-ray spectrum, was explained by an extension of Bohr's model. Characteristic x-ray lines suggested the existence of shells of energy in the atom. The exclusion principle limits the number of electrons in shell n to $2n^2$. This means that there cannot be transitions into a full shell. However, when holes are created in a lower-energy shell, for example by bombardment with energetic particles, transitions become possible, and they result in the emission of characteristic x-rays. Transitions to the $n = 1$ shell give rise to K x-rays; transitions to the $n = 2$ shell yield L x-rays.

Moseley's Law, the Atomic Number, and the Periodic Table

Moseley's measurement of K and L x-ray line spectra from many different elements showed that the square root of the energy of these x-ray photons increases linearly with atomic number Z. His work established that the atomic number is the parameter that orders the elements in the periodic table. Moseley's work also showed how to use x-rays as a practical, reliable way to identify the chemical composition of a substance.

PROBLEMS

1. **a.** Bohr showed that the possible energy states of a hydrogen atom are accurately given by the expression $E_n = -\frac{13.6}{n^2}$, where n is any integer $1 \leq n < \infty$.

b. Draw to scale a level diagram of hydrogen showing the first 5 or 6 energy states. Label the energies of the lowest 4 states.

c. Show on your diagram the transitions that give rise to the first three lines of the Balmer series.

d. What is the energy of the photon emitted when a hydrogen atom goes from its first excited state to its ground state?

e. What is the wavelength of that photon?

2. Atoms of the never-discovered element fictitium (Fi) have energy states as shown in Fig. 17.9.

a. What would be the energies of photons emitted after a vapor of Fi is bombarded with 3.7 eV electrons?

b. What would be the energies of photons emitted after a vapor of Fi is bombarded with 3.7 eV photons?

c. Assume that Fi is in its ground state. What is the longest-wavelength photon that this atom can absorb?

d. Fi has a nucleus with a radius of 4 fm. Use the Heisenberg uncertainty principle to estimate the kinetic energy of a proton confined within this region.

3. Bohr derived the energy levels and the corresponding radius r of the hydrogen atom from the following three relations:

$$pr = n\hbar,$$

$$E_{\text{tot}} = \frac{-k_{\text{C}}e^2}{2r},$$

$$\frac{p^2}{2m} = \frac{k_{\text{C}}e^2}{2r},$$

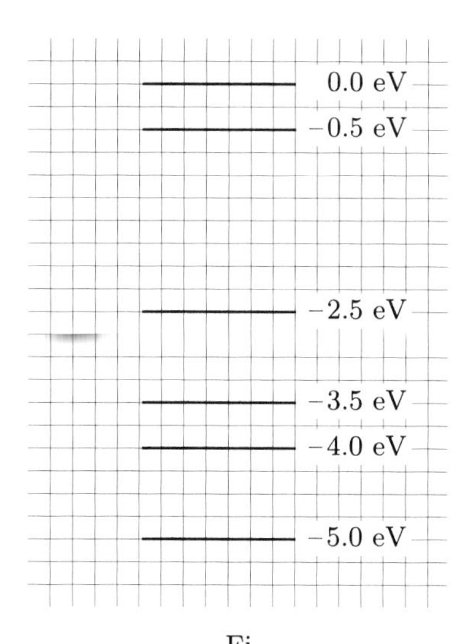

FIGURE 17.9 Energy levels of a fictitious element (Problem 2).

where p is the momentum of an electron of mass m and charge e orbiting a distance r from the center of the positively charged nucleus, n is any integer $1, 2, \ldots$, k_c is the Coulomb force constant, and $\hbar$ is Planck's constant divided by 2π.

a. Obtain an expression for the radius of the lowest Bohr energy state of an atom in terms of fundamental constants. Evaluate your results.

b. Use the above relations to show that the possible energy states of an H atom are

$$E_n = -\frac{mk_c^2 e^4}{2\hbar^2 n^2}.$$

c. Show that the value of this expression = $\frac{-13.6}{n^2}$ eV.

d. Draw to scale a level diagram of the H atom showing the four lowest of these energy levels. Show on your diagram the transitions that produce spectral lines in the Balmer series, and calculate the wavelength of the lowest-energy photon emitted in the Balmer series.

4. What is meant by the "stationary states" that Bohr postulated?

5. Give the names of five series of spectral lines that appear in the spectra of hydrogen atoms.

6. What are two fundamental assumptions necessary for Bohr's model of the hydrogen atom but that are in conflict with the ideas of classical mechanics?

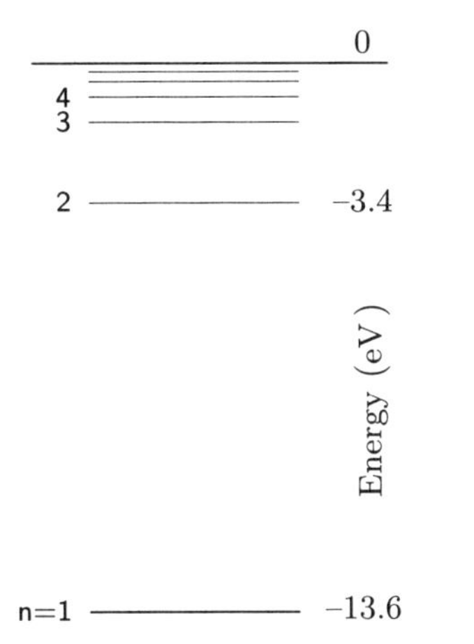

FIGURE 17.10 Energy levels of the hydrogen atom (Problem 9).

7. Who was Henry Gwyn Jeffreys Moseley and when, where, and how did he die?

8. How did Moseley's work relate to Bohr's model of the atom?

9. Figure 17.10 is a diagram representing the possible energy states of a hydrogen atom. The lowest energy state ($n = 1$) has an energy of -13.6 eV. The next lowest state ($n = 2$) has an energy equal to $-13.6/2^2 = -3.40$ eV, etc.

- **a.** What are the energy and wavelength of the photon emitted when a hydrogen atom changes from its $n = 2$ to its $n = 1$ state?
- **b.** Identify the transitions that produce the H_α, H_β, and H_γ lines of the Balmer spectrum. These are three visible lines easily observed in laboratory. What is the color of each of the three lines?

10. What is meant by the binding energy of a system? Give a numerical example of the binding energy of an atom. Of a nucleus.

11. Consider an x-ray tube with an anode made of a Ni–Cr alloy and to which 20 kV has been applied.

- **a.** If $Z_{Cr} = 24$ and $Z_{Ni} = 28$, what are the energies of the K_α lines ($n_i = 2$ to $n_f = 1$) emitted by each element?
- **b.** Make a sketch of the x-ray spectrum (I vs. λ) you would expect to see emitted by the Cr–Ni anode. Your wavelength scale should be accurate. Explain the main features of the spectrum.

TABLE 17.2 L_α **X-ray Lines Measured by Moseley**

Element	Z	λ (nm)	Element	Z	λ (nm)
Zr	40	0.6091	Sm	62	0.2208
Nb	41	0.5749	Eu	63	0.2130
Mo	42	0.5423	Gd	64	0.2057
Ru	44	0.4861	Dy	66	0.1914
Rh	45	0.4022	Er	68	0.1790
Ag	47	0.4170	Ta	73	0.1525
Sn	50	0.3619	W	74	0.1486
Sb	51	0.3458	Os	76	0.1397
La	57	0.2676	Ir	77	0.1354
Ce	58	0.2567	Pt	78	0.1316
Pr	59	(0.2471)	Au	79	0.1287

c. If you analyze the spectrum by sending the x-rays to a crystal with $d_{100} = 0.3$ nm, at what angle (2θ) will 0.28 nm x-rays be diffracted by the (100) plane?

d. Find the momentum in eV/c of electrons with a de Broglie wavelength of 0.28 nm.

e. Sketch an apparatus that could be used to generate electrons with the momentum of part (d). If there are hydrogen atoms in the apparatus, could they be ionized by those electrons? Explain.

12. Moseley measured the L_α x-ray lines of 24 elements from zirconium to gold. Some of his data are given in Table 17.2.

a. Show that plotted as Moseley would have plotted them, these data lie on a straight line. From your graph determine which elements would produce L_α x-rays with wavelengths of 0.2382 nm and 0.4385 nm.

b. Determine the slope and the screening correction that satisfy

$$hf = A(Z - s)^2.$$

c. Compare your value of A with what you would expect from the Bohr-model explanation of L x-rays.

d. Discuss the significance of the value of s that you obtain.

CHAPTER 18

Epilogue

This book has introduced you to the physicist's special way of looking at and trying to understand nature. Out of the many ways to make such an introduction, we chose to present and develop important evidence, ideas, and reasoning that have led to our present-day conception of the atom. We chose this approach partly because we think it is interesting physics and partly because the idea of the atom is so important. It is the basis of all our modern technologies, from computers to gene manipulation, from pharmacology to agriculture, mining, manufacturing, transportation, communications, and management of the environment. The atom as it has been elaborated in this century is fundamental to physics, chemistry, biology, geology—to all the natural sciences. It is arguably one of the most important ideas in human history.

Despite all the details, calculations, exercises, and explanations offered to familiarize you with the atom, we have left an immense amount unsaid. But if all has gone as we hope, you now know enough to be able to explore the richness of the atomic idea further on your own.

For example, there is within the atom more and deeper structure than we have begun to describe. There are other fields than the familiar electric and magnetic fields. There are particles within particles. Physicists have studied the protons and neutrons that lie within the atomic nucleus and found that they, too, have parts, which have been given the unlikely names of quarks and gluons. There is a successful theory of these entities called quantum chromodynamics. Some aspects of this realm of matter are nicely described in the closing chapters of Sheldon Glashow's *From Alchemy to Quarks*, Brooks/Cole Publishing Co., 1993. Steven Weinberg connects this deeper structure to cosmology and the structure of the universe in *The First Three Minutes*, Basic Books, 1988. Some of what physicists are thinking about the very large as well as the very small can be found in *The New Physics*, edited by Paul Davies, Cambridge University Press, 1989.

But even without going deeper into the atom, there is much more to be said. For example, atoms are tiny magnets. Their magnetic properties have remarkable consequences. Most of the knowledge of the world is now stored on magnetic tapes and disks. These would not exist without our understanding of the magnetic behavior of atoms. Great advances in medical imaging have followed from our understanding of the magnetism of atoms and their nuclei. It is from our understanding of the magnetism of atoms that we understand such wonderful objects as pulsars. To learn more, read James D. Livingston's *Driving Force: The Natural Magic of Magnets*, Harvard University Press, Cambridge, MA, 1996.

Our understanding of atoms is being advanced by a revolution in experimental control and manipulation. We can now hold a single atom in a trap made of electromagnetic fields and light waves and then use lasers to prod and probe it with extreme precision. We can map the interaction of individual atoms from moment to moment as they combine chemically. We can watch quantum jumps in a single atom. Using the technology of trapping we can create a large-scale, directly observable quantum state called Bose–Einstein condensation. Quantum interference between beams of atoms has been observed, and there is the prospect of building an atom laser. Hans von Baeyer's *Taming the Atom: The Emergence of the Visible Microworld*, Random House, New York, 1992, gives you the flavor of these developments.

One of the most striking advances in single-atom manipulation is a device called the "scanning tunneling microscope," or STM. It can be used to image and manipulate single atoms. When a sharp tungsten point, as sharp as a single atom at its tip, is brought near atoms sitting on a surface, an electric current flows. The quantum properties of the flow of electrons restrict the current to such a small region of space that as the tip is moved across the surface, the variation in the current can outline the presence of single atoms. Figure 18.1 shows a pair of STM scans. The first shows two conical mounds; each is a molecule of O_2 sitting on a flat surface of platinum atoms. Before the second scan was made, the tip of the STM was brought down to within 0.6 nm of one of the mounds and a small voltage pulse was applied. The second scan shows that the effect of the voltage pulse was to divide the mound into two smaller mounds—single oxygen atoms. The STM has revealed the dissociation of a single molecule into its constituent atoms. Other remarkable examples of this kind of manipulation are shown at http://www-i.almaden.ibm.com/vis/stm/catalogue.html, IBM's gallery of STM images.

The strange mysteries and ambiguities of the quantum nature of the atom promise further remarkable changes in our technology and society. You have seen the evolution of our picture of the atom from the tiny, hard, featureless

FIGURE 18.1 *Upper left*: Two O_2 molecules are revealed by the scan of an STM; *lower right*: Scan of the same two molecules after a voltage pulse has been delivered to one. The scan shows that the O_2 molecule has been separated into two O atoms. *Picture courtesy of Wilson Ho, Laboratory for Atomic and Solid State Physics, Cornell University*

ball that explains the gas laws to a complicated assembly of electrons and a nucleus made up of protons and neutrons. You have seen that as we learned more about the atom, its inner parts got fuzzy. The particle-like behavior of light and the wavelike behavior of particles blurred the insides of the atom. Bohr's model was correct in its idea of well defined internal states of energy that can be represented by a level diagram, but it was wrong in its simple planetary images. There are no well defined orbits. In Tom Stoppard's play *Hapgood* the physicist Kerner says

> So now make a fist, and if your fist is as big as the nucleus of one atom then the atom is as big as St. Paul's, and if it happens to be a hydrogen atom then it has a single electron flitting about like a moth in the empty cathedral, now by the dome, now by the altar.... Every atom is a cathedral. I cannot stand the pictures of atoms they put in schoolbooks, like a little solar system: Bohr's atom. Forget it. . . . an electron does not go round like a planet, it is like a moth which was there a moment ago, it gains or loses a quantum of energy and it jumps, and at the moment of quantum jump it is like *two* moths, one

> to be here and one to stop being there; an electron is like twins, each one unique, a unique twin.

The atom, which began as a hard, featureless ball, is now a moth-filled cathedral that can be depicted as a level diagram.

The fuzziness, the flittering uncertainty, the property of being in more than one state at the same time, are all integral parts of our contemporary understanding of the atom, and they may be the bases of some surprising practical uses. Objects that can be in several different states at the same time may make possible quantum cryptography with unbreakable codes that will warn their users when someone tries to listen in. There is a prospect of designing computers that use these multiple-state systems to achieve massively parallel computation with speeds and capabilities that are impossible in principle with the kinds of computers we now use. Such possibilities are described in *Schrödinger's Machines: The Quantum Technology Reshaping Everyday Life* by Gerard J. Milburn, W. H. Freeman and Co., New York, 1997.

With traps and cooling and scanning probe microscopes, with ever more refined lasers, with a deeper appreciation that quantum mechanics means what it says, our understanding of the atom improves day by day. This understanding has already had extraordinary consequences for human society; the future promises unimaginably more. We hope that our book has helped to prepare you to understand that future.

Of course, there is more to physics than atoms. One of the powerful attractions of physics for physicists is its universal applicability. Whether you are studying the collision of quarks or galaxies, the flow of electrons or the sliding of sand piles, a plasma in a star or a vortex in superfluid helium, the bending of beams in a building or the folding of proteins, the laws of physics apply. We invite you to participate further in the exciting enterprise of revealing and savoring this universality. We invite you to study more physics.

APPENDIX

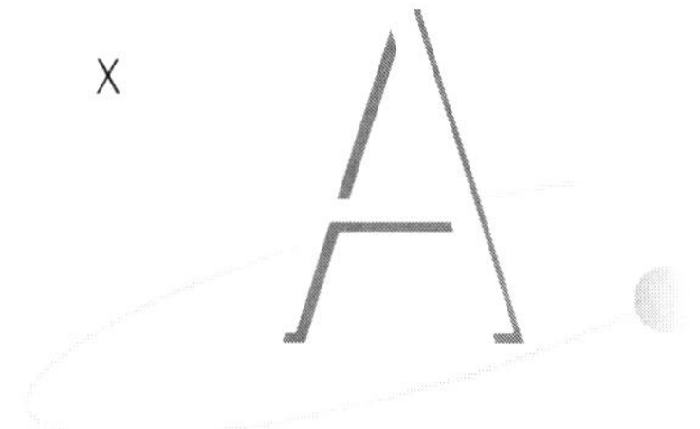

Useful Information

Just as you need to know your name and address and telephone number and e-mail address to locate yourself in the world, so must you know some basic information to locate yourself in physics. Like competent professionals in any field, a practicing physicist carries a large amount of factual baggage. Starting out in physics you will need only the small backpack of facts presented in Tables A.1 and A.3.

Then there is information that you need occasionally. Some of that is collected here for your convenience. If you don't find what you want here or in the text, try the library. Ask a reference librarian to help you find what you want to know, or look in the *Handbook of Chemistry and Physics*. You can also use the World Wide Web to find constants:

http://physics.nist.gov/PhysRefData/codata86/codata86.html

will supply you the very latest, most precise values from NIST (National Institute of Technology and Standards).

A.1 SI PREFIXES

You need to know the SI prefixes. They tell you the order of magnitude of the units of whatever physical quantity they are attached to. It is absolutely essential that you know them. They are widely used, and when you are wrong about them, you make mistakes of factors of thousands! Maybe you can absorb them by osmosis as you use them; maybe you need to get them by heart by the purest of rote learning; maybe you can come up with a clever mnemonic; maybe (and this would be best) you can learn them attached to particular physical situations and quantities, as suggested in Chapter 2. However you do it, learn them! They are listed in Table A.1.

TABLE A.1 SI Prefixes

Factor	Prefix	Symbol	Factor	Prefix	Symbol
10^{18}	exa	E	10^{-1}	deci	d
10^{15}	peta	P	10^{-2}	centi	c
10^{12}	tera	T	10^{-3}	milli	m
10^{9}	giga	G	10^{-6}	micro	μ
10^{6}	mega	M	10^{-9}	nano	n
10^{3}	kilo	k	10^{-12}	pico	p
10^{2}	hecto	h	10^{-15}	femto	f
10^{1}	deka	da	10^{-18}	atto	a

A.2 BASIC PHYSICAL CONSTANTS

You need to know some basic physical constants. These set the scale of the phenomena of the physical world. Which ones are most important depends on the physical situation under consideration. In this book, with its emphasis on atoms and their parts, the elementary charge; the masses of the electron, the proton, and the neutron; and the values of the Planck and Boltzmann constants are very important. When you deal with macroscopic quantities of atoms in the laboratory, Avogadro's number and Earth's gravity are important. For convenient reference Table A.2 lists the official values of these constants in SI units. Table A.3 lists the ones you need to know in the units in which you need to know them.

These constants are not just special, precise numbers given in SI units. One goal of this book is to show how these constants interrelate and how they specify the scale of observed effects and phenomena. They are essential elements out of which we build a consistent and informative picture of the microphysical world and its connection to the macrophysical world where we live and do physics.

A.3 CONSTANTS THAT YOU MUST KNOW

You need to be able to calculate quickly and easily with these constants. For this purpose, you need the constants expressed as much as possible in terms of units chosen to match the natural scale of atoms. Electron volts (eV) and nanometers (nm) are convenient for atoms, while megaelectron volts (MeV)

TABLE A.2 Basic Physical Constants

Name of constant	Symbol	Value
Atomic mass unit	m_u or u	1.661×10^{-27} kg
Avogadro constant	N_A	6.022×10^{23} mol^{-1}
Bohr radius	a_0	5.292×10^{-11} m
Boltzmann constant	k_B	1.381×10^{-23} J·K^{-1}
		8.617×10^{-5} eV·K^{-1}
Charge-to-mass ratio of electron	e/m	-1.759×10^{11} C·kg^{-1}
Coulomb constant	k_c or $\frac{1}{4\pi\epsilon_0}$	8.988×10^{9} N·m^2·C^{-2}
Electron mass	m_e	9.109×10^{-31} kg
Elementary charge	e	1.602×10^{-19} C
Faraday constant	F	96485 C·mol^{-1}
Intensity of Earth's gravitational field	g	9.82 N·kg^{-1} (m·s^{-2})
Molar gas constant	R	8.315 J·mol^{-1}·K^{-1}
Neutron mass	m_n	1.675×10^{-27} kg
Planck constant	h	6.626×10^{-34} J·s
		4.136×10^{-15} eV·s
	$\hbar = \frac{h}{2\pi}$	1.055×10^{-34} J·s
		6.582×10^{-16} eV·s
Proton mass	m_p	1.673×10^{-27} kg
Rydberg constant	R_∞	1.09737×10^{7} m^{-1}
Speed of light	c	2.99792458×10^{8} m·s^{-1}

and femtometers (fm) are a good choice for nuclei. It is also often simpler to work with masses in units of eV/c^2.

Table A.3 gives constants, combinations of constants, and masses in terms of these more convenient units. The combinations simplify calculations of energies, wavelengths, and frequencies that are frequently made in this course. The Remark column tells you when the constant is one that you absolutely need to know. No kidding!

TABLE A.3 Constants in Convenient Energy Units

Name	Symbol	Value	Remark
Planck constant	h	4.14×10^{-15} eV·s	
	hc	1240 eV·nm	know this one!
Reduced Planck constant: $\frac{h}{2\pi}$	$\hbar$	6.58×10^{-16} eV·s	
	$\hbar c$	197 eV·nm	know as $\approx$ 200 eV nm
	$\hbar c$	197 MeV·fm	know as $\approx$ 200 MeV fm
Coulomb force numerator	$k_c e^2$	1.44 eV·nm	know this
Thermal energy at $T = 300\,\mathrm{K}$	$k_B T$	0.0259 eV	remember as $\approx 1/40$ eV
Bohr radius	$a_0 = \frac{\hbar^2}{ke^2 m_e}$	0.0529 nm	
Fine structure constant	$\alpha = \frac{ke^2}{\hbar c}$	1/137.036	no units
Rydberg energy	hcR_∞	13.61 eV	know this one
Electron mass	$m_e c^2$	511 keV	know this
Proton mass	$m_p c^2$	938.3 MeV	know $\approx$ 938 MeV
Neutron mass	$m_n c^2$	939.6 MeV	know m_n is 1.29 MeV $> m_p$
Atomic mass unit	u	931.50 MeV/c^2	remember 1 u $\approx m_p$
Speed of light	c	3×10^8 m·s^{-1}	know this
Elementary charge	e	1.6×10^{-19} C	know this

A.4 SOME UNITS AND THEIR ABBREVIATIONS

There are seven units that form the basis of the SI. In this book we use six of them. Table A.4, which gives their names, symbols, and definitions, is provided here just for your general information. You will find it more useful and informative to remember the looser definitions that are given in the chapters where they are introduced.

For quick reference Table A.5 lists units and abbreviations. The entries in the table are in alphabetical order according to their abbreviations.

TABLE A.4 SI Base Units

Name	Symbol	Definition
meter	m	The meter is the length of path traveled by light in vacuum during a time interval of 1/299 792 458 of a second.
mass	kg	The kilogram is the unit of mass. It is equal to the mass of the international prototype of the kilogram. (The international prototype is a platinum–iridium cylinder kept at the BIPM in Sèvres (Paris) France.)
second	s	The second is the duration of 9 192 631 770 periods of the radiation corresponding to the transition between the two hyperfine levels of the ground state of the cesium-133 atom.
ampere	A	The ampere is that constant current that if maintained in two straight parallel conductors of infinite length, of negligible circular cross section, and placed 1 meter apart in vacuum, would produce between these conductors a force equal to 2×10^{-7} newton per meter of length.
kelvin	K	The kelvin is the unit of thermodynamic temperature. It is the fraction 1/273.16 of the thermodynamic temperature of the triple point of water. (The Celsius temperature scale is defined by the equation $t = T - T_0$, where T is the thermodynamic temperature in kelvins and $T_0 = 273.15$ K.)
mole	mol	The mole is the amount of substance of a system that contains as many elementary entities as there are atoms in 0.012 kg of carbon-12.
candela	cd	The candela is the luminous intensity, in a given direction, of a source that emits monochromatic radiation of frequency 540×10^{12} hertz and that has a radiant intensity in that direction of 1/683 watt per steradian.

A.5 ATOMIC MASSES

Table A.6 lists some useful chemical atomic masses and densities of elements that are solids at room temperature. If you need to know the density of any gaseous element, you can calculate it.

A.6 MASSES OF NUCLIDES

For determining how nuclei will behave, the difference between masses of atoms may be important. When this is the case, you need to know the individual atomic masses quite precisely. Table A.7 lists some of the more important elements and their nuclides and their masses.

TABLE A.5 Commonly Used Units and Abbreviations

Quantity	Name	Abbrev.	SI units
current	ampere	A	A
length	angstrom	Å	10^{-10} m
pressure	atmosphere	atm	101.3 kPa
area	barn	b	10^{-24} m^2
pressure	bar	bar	100 kPa
energy	calorie	cal	4.1858 J
electric charge	coulomb	C	A·s
viscosity	centipoise	cp	10^{-3} Pa·s
energy	electron volt	eV	1.602×10^{-19} J
magnetic field	gauss	G	10^{-4} T
frequency	hertz	Hz	s^{-1}
energy	joule	J	kg·m^2·s^{-2} =N·m
temperature	kelvin	K	K
mass	kilogram	kg	kg
volume	liter	L	10^{-3} m^3
length	meter	m	m
pressure	millimeters of mercury	mm Hg	133.32 Pa
volume	cubic meter	m^3	m^3
amount	mole	mol	mol
force	newton	N	kg·m·s^{-2}
electric field	newton per coulomb	N·C^{-1}	N·C^{-1}
pressure	pascal	Pa	N·m^{-2}
viscosity	pascal seconds	Pa·s	
angle	radian	rad	rad
time	second	s	s
magnetic field	tesla	T	kg·s^{-1}·C^{-1}
pressure	torr	torr	133.32 Pa
mass	atomic mass unit	u	1.6605×10^{-27} kg
electric potential	volt	V	J·C^{-1}
electric field	volts per meter	V·m^{-1}	N·C^{-1}
power	watt	W	J·s^{-1}
angle	degree	°	1.7453×10^{-2} rad

TABLE A.6 Some Chemical Atomic Masses

Element	Symbol	Z	Mass (u)	Phase	Density ($g\ cm^{-3}$)
hydrogen	H	1	1.00797	gas H_2	
helium	He	2	4.0026	gas He	
lithium	Li	3	6.939	solid	0.534
beryllium	Be	4	9.0122	solid Be	1.848
boron	B	5	10.811	crystalline B	2.34
carbon	C	6	12.01115	amorphous C	≈ 2.0
nitrogen	N	7	14.0067	gas N_2	
oxygen	O	8	15.9994	gas O_2	
fluorine	F	9	18.9984	gas F_2	
aluminum	Al	13	26.981538	solid	2.6989
silicon	Si	14	28.0855	solid	2.33
iron	Fe	26	55.844	solid	7.874
cobalt	Co	27	58.93320	solid	8.9
nickel	Ni	28	58.69	solid	8.902
copper	Cu	29	63.546	solid	8.96
zinc	Zn	30	65.40	solid	7.133
tantalum	Ta	73	180.9479	solid	16.6
silver	Ag	47	107.8681	solid	10.5
gold	Au	79	196.96654	solid	19.32
lead	Pb	82	207.2	solid	11.35
uranium	U	92	232.0289	solid	18.95

A.7 MISCELLANEOUS

Table A.8 contains some constants used occasionally in this course, including constants having to do with Earth, Moon, and Sun. Table A.9 contains conversion factors between some especially common English units and their metric equivalents.

TABLE A.7 Masses of Some Nuclides

Name of Nuclide	Symbol	Z	Nuclide Mass (u)	Natural Abundance (%)	Half-life
hydrogen	^{1}H	1	1.007825	99.985	
deuterium	^{2}H or D	1	2.01410	0.015	
tritium	^{3}H or T	1	3.016050		12.26 y
helium-3	^{3}He	2	3.016030	0.00013	
helium-4	^{4}He	2	4.002603	100.0	
lithium-6	^{6}Li	3	6.015125	7.42	
lithium-7	^{7}Li	3	7.016004	92.58	
beryllium-9	^{9}Be	4	9.012186	100.	
boron-10	^{10}B	5	10.012939	19.78	
boron-11	^{11}B	5	11.009305	80.22	
carbon-12	^{12}C	6	12.000000	98.89	
carbon-13	^{13}C	6	13.003354	1.11	
carbon-14	^{14}C	6	14.003242		5730 y
nitrogen-14	^{14}N	7	14.003074	99.63	
nitrogen-15	^{15}N	7	15.010599	0.37	
oxygen-16	^{16}O	8	15.994915	99.759	
oxygen-17	^{17}O	8	16.999133	0.037	
oxygen-18	^{18}O	8	17.999160	0.204	
fluorine-19	^{19}F	9	18.998405	100.0	

TABLE A.8 Miscellaneous Occasionally Used Constants

Name	Symbol	Value	Units	Remarks
Earth's mass	$M_\oplus$	6×10^{24}	kg	10 moles of kilograms
Earth-Sun distance	R_{ES}	1.5×10^{11}	m	1 A.U.
Earth radius	$R_\oplus$	6.366×10^{6}	m	$2\pi R_\oplus = 40\,\mathrm{Mm}$
Earth–Moon distance	R_{EM}	3.82×10^{8}	m	$60\, R_\oplus$
Moon's mass	$M_)$	$0.01234\, M_\oplus$		$M_\oplus/81$
Sun's mass	$M_\odot$	2×10^{30}	kg	$333{,}000\, M_\oplus$
Viscosity of air	η	18.2	μPa·s	at 20°C
Speed of sound in air	v_s	343	$\mathrm{m\,s^{-1}}$	at 20°C

TABLE A.9 Some Conversion Factors Between English and Metric Units

English		English		Metric	
1	in			2.54	cm
1	ft	12	in	30.48	cm
1	mile	5280	ft	1609.3	m
3.28	ft			1	m
0.396	in			1	cm
1	mph	1.467	ft/s	0.447	m/s
0.621	mph	0.911	ft/s	1	km/hr
2.24	mph	3.28	ft/s	1	m/s
1	lb	16	oz	453.5	g
1	oz			28.3	g
2.205	lb			1	kg

Index